Creditor Reporting System

AID ACTIVITIES IN SUPPORT OF WATER SUPPLY AND SANITATION 2001-2006

World Water Council
World Water Forum

ORGANISATION FOR ECONOMIC CO-OPERATION AND DEVELOPMENT

The OECD is a unique forum where the governments of 30 democracies work together to address the economic, social and environmental challenges of globalisation. The OECD is also at the forefront of efforts to understand and to help governments respond to new developments and concerns, such as corporate governance, the information economy and the challenges of an ageing population. The Organisation provides a setting where governments can compare policy experiences, seek answers to common problems, identify good practice and work to co-ordinate domestic and international policies.

The OECD member countries are: Australia, Austria, Belgium, Canada, the Czech Republic, Denmark, Finland, France, Germany, Greece, Hungary, Iceland, Ireland, Italy, Japan, Korea, Luxembourg, Mexico, the Netherlands, New Zealand, Norway, Poland, Portugal, the Slovak Republic, Spain, Sweden, Switzerland, Turkey, the United Kingdom and the United States. The Commission of the European Communities takes part in the work of the OECD.

OECD Publishing disseminates widely the results of the Organisation's statistics gathering and research on economic, social and environmental issues, as well as the conventions, guidelines and standards agreed by its members.

THE WORLD WATER COUNCIL

The World Water Council is an independent, international organization that promotes sustainable water management throughout the world. It has more than 300 member organisations including Governments, non-governmental organisations, businesses, professional networks and research institutions, based in over 50 countries.

The World Water Council brings sound information and knowledge on water-related issues into the public sphere. It strives to raise awareness and create political will in order to improve access to water and sanitation and to promote wise management of water resources.

Every three years, the World Water Council organizes the World Water Forum. It is the world's premier event on water and gathers over 15 000 participants from around the world to generate concrete solutions to the world's water challenges.

The World Water Council's headquarters is in Marseille, France.

This work is published on the responsibility of the Secretary-General of the OECD. The opinions expressed and arguments employed herein do not necessarily reflect the official views of the Organisation or of the governments of its member countries or those of the World Water Council.

Also available in French under the title:
Système de notification des pays créanciers
ACTIVITÉS D'AIDE DANS LE SECTEUR DE L'APPROVISIONNEMENT EN EAU ET L'ASSAINISSEMENT
2001-2006

Corrigenda to OECD publications may be found on line at: *www.oecd.org/publishing/corrigenda*.

Foreword

This report examines data on aid in support of water supply and sanitation over 2001-2006. It is based on donors' reporting on Official Development Assistance (ODA) commitments and disbursements to the Creditor Reporting System (CRS) Aid Activity Database.

The report has been prepared by the Secretariat of the Development Assistance Committee (DAC) in collaboration with the DAC Working Party on Statistics and the World Water Council (WWC). The work has been carried out in the context of the OECD Horizontal Water Programme. Main findings will be incorporated in the OECD synthesis report *Sustainable financing to ensure affordable access to water supply and sanitation* that will be made available at the 5th World Water Forum in Istanbul (March 2009).

The report contains i) an **analysis of aid to water supply and sanitation** over 2001-2006 including trends in donors' aid, the degree of targeting of countries most in need, and a methodological note; ii) **individual donor profiles** with summary statistics in the form of charts and tables, and descriptions of donors' development co-operation policies in the water sector; and iii) **the record of aid activities** for the water supply and sanitation sector, with amounts and descriptions, reported by bilateral and multilateral donors in 2006.

Data for 2007 will only become available at the end of 2008, and updated statistics on aid to the water sector will then be posted on the DAC website dedicated to water at **www.oecd.org/dac/stats/crs/water**.

The complete CRS Aid Activity database contains records from 1973 onwards. It is available on the yearly *International Development Statistics* CD-ROM, and at ***www.oecd.org/dac/stats/idsonline***.

Table of contents

INTRODUCTION: Basic aspects of the CRS Aid Activity database

(Creditor Reporting System)

The CRS Aid Activity database includes data on Official Development Assistance (ODA), and other official flows to developing countries.

The CRS was established in 1967, jointly by the OECD and the World Bank, with the aim of "supplying the participants with a regular flow of data on indebtedness and capital flows". Calculating capital flows and debt stock remain key functions of the system, but others have evolved in the course of years. In particular, the CRS aid activity database has become the internationally recognised source of data on the geographical and sectoral breakdown of aid and is widely used by governments, organisations and researchers active in the field of development. For DAC members the CRS serves as a tool for monitoring specific policy issues, supplementing the information collected at the aggregate level in the annual DAC Statistics.

> The CRS Aid Activity database comprises commitment and disbursement data on Official Development Assistance (ODA) activities in developing countries submitted by members of the DAC and multilateral institutions (50 000 – 65 000 transactions in recent years). Data in this publication refer to commitments, unless otherwise stated.

PART 1

Analysis of aid to the water supply and sanitation sector

2001-2006

www.oecd.org/dac/stats/crs/water

Key findings

Aid for water supply and sanitation has risen since 2001 after a temporary decline in the second part of the 1990s. In 2005-2006, DAC members' bilateral annual aid commitments to the water sector and sanitation rose to USD 5 billion, double the 2001-2002 figure in real terms. Taking into account multilateral agencies' outflows, the total was USD 6.2 billion.

In 2005-2006, DAC members dedicated 9% of their total sector allocable aid to water supply and sanitation activities. This illustrates renewed emphasis on the water sector in members' aid programmes, after a drop to 6-7% in years 2001-2004.

Among DAC members, the largest donors in 2005-2006 were Japan (on average USD 1.6 billion per year), the United States (USD 903 million) and the European Commission (USD 730 million). The bulk of Japanese aid related to ODA loans for infrastructure projects in China, Costa Rica, India, Indonesia and Malaysia. On their own, these projects represented one fourth of total DAC members' aid for water. Reconstruction projects in Iraq by the United States also made up a significant proportion (15%) of the total.

The main recipient regions over the whole period 2001-2006 were Asia (55%) and Africa (32%). The region most in need of improved access to water supply and sanitation, Sub-Saharan Africa, received 24% although this declined to 18% in 2005-2006. The next most needy region, South and Central Asia, received 19% of total aid for water over the whole period.

Twelve members refer to specific policy documents to guide their interventions in the water sector, while a number of other donors do not have a specific co-operation policy for water. Some consider that the implementation of the principles of harmonisation and alignment of the Paris Declaration is not compatible with specific donor strategies or investment targets in the water sector.

www.oecd.org/dac/stats/crs/water

Introduction

Context

This report examines data on aid flows in support of the water supply and sanitation sector over the years 2001-2006. It has been prepared by the Secretariat of the Development Assistance Committee (DAC) in collaboration with the DAC Working Party on Statistics and the World Water Council (WWC). The work is part of the OECD Horizontal Water Programme and will contribute to the OECD Synthesis Report on Water ("*Sustainable financing to ensure affordable access to water and sanitation*") to be presented at the 5th World Water Forum (WWF) in March 2009 in Istanbul.

Improving water supply and sanitation is one of the world's major development challenges, with over one billion people still lacking access to safe drinking water and 2.5 billion lacking access to basic sanitation. *Target 7.C. of the Millennium Development Goals is to halve, by 2015, the proportion of people without sustainable access to safe drinking water and basic sanitation.*

Structure of the report

This report updates the 2003 study *"CRS Aid Activities in the Water Sector, 1997-2002"*[1]. It is structured in three parts. Part 1 is an **analysis of aid to water supply and sanitation** over 2001-2006 including trends in donors' aid, and the degree of targeting of countries most in need. It includes a methodological note on the scope for improving the data. Part 2 contains **individual donor profiles** with summary statistics on aid to water supply and sanitation in the form of charts and tables, and descriptions of donors' development co-operation policies in the water sector. Part 3 is the **list of aid activities** for the water supply and sanitation sector, with amounts and descriptions, reported by bilateral donors and multilateral agencies in 2006.

Statistics shown include data up to 2006. Data on 2007 flows will become available at the end of 2008, and will be summarised in a separate factsheet in time for the WWF.

Source of the information

Statistics are based on donors' reporting on Official Development Assistance (ODA) commitments and disbursements to the Creditor Reporting System (CRS) Aid Activity Database. As regards separate figures for water supply and sanitation, complementary data and methodological comments were obtained through a special survey sent to members in Spring 2008. Textual information on donors' development co-operation policies in the water sector were prepared by the WWC in collaboration with DAC members.

Data that support this report are available from the DAC website

www.oecd.org/dac/stats/water
www.oecd.org/dac/stats/idsonline

---

[1] This original study was further updated through two successive factsheets in 2005 and 2006 that respectively covered periods 1999-2004 and 2000-2005 (see www.oecd.org/dac/stats/crs/water).

I. Trends in aid to the water supply and sanitation sector

I.1 Definitions

Coverage of "water supply and sanitation"

The DAC defines aid to **water supply and sanitation** as including water resources policy, planning and programmes, water legislation and management, water resources development, water resources protection, water supply and use, sanitation (including solid waste management) and education and training in water supply and sanitation.

The water supply and sanitation sector is divided into the sub-sectors shown in Box 1. This classification does not distinguish between aid flows for water supply and aid flows for sanitation. Possibilities for introducing such a differentiation in future data collection are discussed in section III.

The definition of aid for water supply and sanitation therefore excludes dams and reservoirs primarily for irrigation and hydropower and activities related to river transport which are coded elsewhere in the classification (aid to agriculture, energy and transport respectively). Statistics shown in the present section are all based on the DAC "narrow" definition of water supply and sanitation. However, the *Donor profiles on aid to the water supply and sanitation sector* in Part 2 of this report include data based on a "wider" definition (see table on "aid to all water-related sectors").

DAC statistics classify humanitarian aid as a separate category (the main purpose being to save lives in an emergency context), and do not record the ultimate sector of destination of humanitarian interventions (water, health, education, etc.). Statistics shown in this section therefore do not take into account donors' expenditures on water supply and sanitation that occurred in the context of humanitarian aid.

Box 1. Aid to the water supply and sanitation sector: definition and sub-sectors

Water resources policy and administrative management (CRS purpose code 14010)
Water sector policy, planning and programmes; water legislation and management; institution capacity building and advice; water supply assessments and studies; groundwater, water quality and watershed studies; hydrogeology; excluding agricultural water resources (31140).

Water resources protection (CRS purpose code 14015)
Inland surface waters (rivers, lakes, etc.); conservation and rehabilitation of ground water; prevention of water contamination from agro-chemicals, industrial effluents

Water supply and sanitation - large systems (CRS purpose code 14020)
Water desalination plants; intakes, storage, treatment, pumping stations, conveyance and distribution systems; sewerage; domestic and industrial waste water treatment plants.

Basic drinking water supply and basic sanitation (CRS purpose code 14030)
Water supply and sanitation through low-cost technologies such as handpumps, spring catchment, gravity-fed systems, rain water collection, storage tanks, small distribution systems; latrines, small-bore sewers, on-site disposal (septic tanks).

River development (CRS purpose code 14040)
Integrated river basin projects; river flow control; dams and reservoirs [excluding dams primarily for irrigation (31140) and hydropower (23065) and activities related to river transport (21040)].

Waste management/disposal (CRS purpose code 14050)
Municipal and industrial solid waste management, including hazardous and toxic waste; collection, disposal and treatment; landfill areas; composting and reuse.

Education and training in water supply and sanitation (CRS purpose code 14081)

Note: To assist in distinguishing between "basic drinking water supply and basic sanitation" on the one hand and "water supply and sanitation – large systems" on the other, consider the number of people to be served and the per capita cost of provision of services.
• Large systems provide water and sanitation to a community through a network to which individual households are connected. Basic systems are generally shared between several households.
• Water supply and sanitation in urban areas usually necessitates a network installation. To classify such projects consider the per capita cost of services. The per capita cost of water supply and sanitation through large systems is several times higher than that of basic services.

Terminology

Sector-allocable aid

In order to better reflect the sectoral focus of donors' programmes, when calculating the share of aid for water in total bilateral aid, contributions not susceptible to allocation by sector are excluded from the denominator (general budget support, actions relating to debt, humanitarian aid, administrative costs and other internal transactions in the donor country).

DAC members' imputed multilateral contributions

DAC members, in addition to undertaking bilateral aid activities in the water sector also contribute to multilateral agencies active in the field of water. In order to provide the most complete picture possible of the total ODA effort the donor makes in respect of aid to the water sector, data for DAC members' imputed multilateral aid for water can be compiled.

The calculation of imputed multilateral contributions is done in two steps[2]: 1/ the share of each multilateral agency's aid flows for water supply and sanitation in their total sector-allocable aid is calculated (core resources only); 2/ the share obtained in step 1 is multiplied by a member's contribution to the core resources of the agency concerned; the resulting amount is the imputed flow from that donor to the water supply and sanitation sector through the multilateral agency concerned. It can only ever be an approximation.

I.2 Bilateral and multilateral aid to water supply and sanitation

This section is dedicated to the presentation of statistics on aid to water supply and sanitation by donors, bilateral DAC countries and multilateral institutions.

Chart 1 below illustrates the long-term trend in aid to water supply and sanitation. Annual figures highlight sharp increases in DAC countries' bilateral ODA commitments to the water sector during three successive years – 2003, 2004 and 2005. Aid however slightly dropped the following year, 2006, to **USD 3.9 billion**. Over the period 2002-2006, bilateral aid to water increased at an average annual rate of 24% (real terms), and the volume of aid more than doubled over the period.

Multilateral aid also rose swiftly over the period 2002-2006 (21% annually), and reached **USD 2.5 billion** in 2006.

Figures based on 5-year moving averages smooth the fluctuations in commitments and identify long-term trends. Bilateral aid to the water sector regularly increased over the period 1973-2006, except for a temporary decline in the second part of the 1990s. It is now at an all-time high.

---

[2]   See also "The OECD methodology for calculating imputed multilateral ODA" under www.oecd.org/dac/stats.

Chart 1. Trends in ODA to water supply and sanitation

1973-2006, 5-year moving averages and annual figures, constant 2006 prices

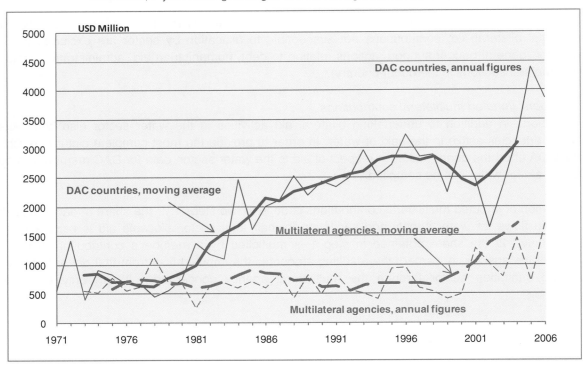

Table 1 below provides the breakdown of aid to water supply and sanitation by individual donors and multilateral institutions.

Out of the total DAC members' aid for water of USD 5 billion over 2005-2006, Japan is the predominant donor accounting for one third, followed by the United States (18%), and the European Commission (EC) (15%). The other major bilateral donors are Germany, Netherlands and France which together account for 20% of total DAC members' aid.

The EC, ranking third among DAC members, also became in 2005-2006 the largest multilateral donor to the water supply and sanitation sector, accounting for 38% of multilateral aid. In previous years, among the multilateral agencies that report to the CRS, IDA was the largest multilateral agency, and it still represented close to 38% of multilateral aid in 2005-2006. The African Development Fund (AfDF) and the Asian Development Fund (AsDF) together account for a further one fourth of multilateral aid.

The largest relative increases in aid commitments to the water sector over the period 2001-2006 came from Japan, the United States, the EC and Denmark which all at least tripled their aid in real terms over the period; Finland doubled its aid commitments; the AfDF and the Netherlands also showed sizeable increases.

The share of aid to water in total bilateral sector-allocable aid is an indication of the extent to which donors' aid programmes focus on water issues, and of the priority these issues are being given. For DAC countries, the share has regularly increased over the period 2001-2006, rising from 6% in 2001-2002 to 9% in 2005-2006. However, similar levels were already observed in the mid 1990's, so there was no increased prioritisation of the water sector over the whole decade 1996-2006.

A number of individual donors and multilateral agencies dedicate relatively high proportions of their aid to projects in the field of water supply and sanitation, and are above the 9% average: the AfDF and Japan dedicated approximately 20% to water in 2005-2006, the AsDF 16%, IDA and Denmark 11%, Finland and the Netherlands 10%.

Table 1. Aid to water supply and sanitation by bilateral and multilateral donors

2001-2006, annual average commitments and shares in total sector-allocable aid,

constant 2006 prices

	Commitments, USD million			% of Donor Total			% All donors		
	2001-2002	2003-2004	2005-2006	2001-2002	2003-2004	2005-2006	2001-2002	2003-2004	2005-2006
Australia	27	33	6	3	3	1	1	1	0
Austria	15	22	19	6	10	7	0	1	0
Belgium	44	23	63	5	3	8	1	1	1
Canada	31	108	32	3	6	2	1	2	1
Denmark	37	166	123	5	17	11	1	4	2
Finland	22	11	44	7	3	10	1	0	1
France	208	192	186	6	5	4	6	4	3
Germany	392	416	453	10	10	9	11	10	7
Greece	1	2	1	1	1	1	0	0	0
Ireland	16	22	17	7	7	5	0	0	0
Italy	38	35	63	7	6	9	1	1	1
Japan	497	823	1626	6	11	20	14	19	26
Luxembourg	15	14	12	13	12	9	0	0	0
Netherlands	179	142	334	8	8	10	5	3	5
New Zealand	2	2	4	2	2	2	0	0	0
Norway	60	25	39	4	2	2	2	1	1
Portugal	1	1	2	1	1	1	0	0	0
Spain	73	79	58	6	7	4	2	2	1
Sweden	58	54	97	6	5	5	2	1	2
Switzerland	31	35	50	5	5	7	1	1	1
United Kingdom	122	63	112	4	2	3	3	1	2
United States	303	566	903	4	4	5	9	13	15
Total DAC countries	**2173**	**2835**	**4244**	**6**	**7**	**9**	**61**	**65**	**69**
AfDF	142	168	264	12	12	21	4	4	4
AsDF	203	156	197	13	9	16	6	4	3
EC	227	412	730	4	6	8	6	9	12
IDA	772	780	709	12	9	11	22	18	11
IDB Sp.Fund	0	0	23	0	0	5	0	0	0.4
UNICEF	32	19	23	7	4	4	1	0.4	0.4
Total Multilateral	**1376**	**1536**	**1945**	**8**	**8**	**9**	**39**	**35**	**31**
Total	**3548**	**4370**	**6189**	**6**	**7**	**9**	**100**	**100**	**100**

DAC members' imputed multilateral contributions to the water sector

As explained in section I.1, DAC members' imputed multilateral contributions can be estimated (Table 2) and added to their bilateral contributions (Table 3).

The shares of aid to water in total outflows from multilateral agencies that were taken into account for the compilation of figures shown in Tables 2 and 3 are as follows: 5.6% for EC, 8.4% for IDA, 15.1% for the AfDF, 12.8% for the AsDF, 2.7% for UNICEF and 15% for the Global Environment Fund (GEF). Shares were calculated based on CRS data for the multilateral agencies concerned, except for GEF where data were derived from its online project database. The analysis was thus limited to those multilateral institutions reporting to the CRS (plus GEF), and could be further refined in future to take into account other UN agencies (e.g. contributions to UNDP, UN-Habitat and WHO are quoted by a number of members in their policy briefs shown in Part 2).

Table 2 indicates that, out of DAC countries' total core contributions to multilateral agencies, **USD 1.3 billion** can be imputed to water over 2005-2006. The EC is the largest multilateral channel representing 41% of this amount, followed by IDA with 37%.

The total for imputed multilateral contributions, USD 1.3 billion, does not exactly correspond to the total multilateral outflows shown in Table 1 (USD 1.9 billion) because, in general, multilateral flows in a given year are not exactly imputable to donors' contributions in that year. Among the reasons for this are the delays between receipt of funds and their disbursement. Also, in the case of international financial institutions (IFIs), lending in any one year is much greater than donors' contributions (grants), since it also draws on reflows of principal, interest receipts, and transfers of funds within the IFIs.

Table 2. DAC members' imputed multilateral contributions for water supply and sanitation

annual average 2005-2006, USD million, constant 2006 prices

	through EC	through IDA	through Afr. DF	through AsDF	through UNICEF	through GEF	Total imputed multiltateral contributions
Australia		12.2			0.1	0.5	12.8
Austria	13.0	6.1	4.1	0.9		0.4	24.5
Belgium	21.6	12.3	3.6	0.7	0.1	1.5	39.8
Canada		20.5			0.8	4.1	25.4
Denmark	11.8	6.3	2.6	1.0	1.0	1.4	24.0
Finland	8.3	3.6	1.9	0.3	0.5	1.2	15.7
France	106.6	32.0	22.2	4.9	0.4	6.0	172.1
Germany	122.9	24.8	17.5	6.3	0.2	4.5	176.2
Greece	9.2	2.0				0.7	11.9
Ireland	6.7	6.2			0.4	0.2	13.4
Italy	73.1	30.6	13.3		0.3	9.6	126.9
Japan		129.7	15.6	17.5	3.5	6.0	172.3
Luxembourg	1.4	0.8		1.1		0.2	3.5
Netherlands	24.5	11.3	11.8		1.0	5.8	54.3
New Zealand		0.7			0.1	0.2	1.0
Norway		10.1			3.4		13.6
Portugal	7.2	1.1	1.5	0.7		0.1	10.5
Spain	46.9	15.0	4.0	3.4	0.4	3.2	72.8
Sweden	12.6	13.8	10.4	1.5	1.6	6.7	46.5
Switzerland		12.9	6.4	1.4	0.4	2.0	23.0
United Kingdom	76.6	68.5	18.3	6.7	1.2	6.3	177.6
United States		71.3	18.3	12.9	3.4	10.9	116.8
Total DAC countries	**542.4**	**491.7**	**151.3**	**59.3**	**18.7**	**71.6**	**1334.9**
EC		*4.2*					*4.2*

Table 3 gives credit to donor countries' support for water through multilateral agencies by showing total commitments to the water sector as the sum of bilateral commitments and imputed multilateral contributions. For a number of countries, taking imputed contributions into account makes a significant difference in the total: for Australia, Austria, Greece, Italy, Portugal, Spain and the United Kingdom, imputed multilateral contributions represent more than half of their total support for water.

Table 3. DAC members' total aid to the water supply and sanitation sector

2001-2006, bilateral commitments, imputed multilateral contributions, bilateral disbursements, annual averages, USD million, constant 2006 prices

	Bilateral commitments			Imputed multilateral contributions			Total commitments to the water sector			Bilateral Disbursements	
	2001-2002	2003-2004	2005-2006	2001-2002	2003-2004	2005-2006	2001-2002	2003-2004	2005-2006	2003-2004	2005-2006
Australia	27	33	6	15	10	13	42	43	19	29	20
Austria	15	22	19	9	18	25	24	40	43	12	14
Belgium	44	23	63	21	24	40	65	47	103	18	43
Canada	31	108	32	12	22	25	43	129	57	40	55
Denmark	37	166	123	19	21	24	55	187	147	51	101
Finland	22	11	44	9	11	16	31	21	60	7	15
France	208	192	186	104	135	172	311	327	358	141	194
Germany	392	416	453	105	187	176	497	603	629	347	363
Greece	1	2	1	6	7	12	7	9	13	2	1
Ireland	16	22	17	5	6	13	21	28	31	22	17
Italy	38	35	63	60	65	127	98	101	190	12	24
Japan	497	823	1626	98	99	172	595	923	1798	823	926
Luxembourg	15	14	12	1	2	3	16	17	16	15	13
Netherlands	179	142	334	30	52	54	210	195	388	88	126
New Zealand	2	2	4	1	1	1	3	3	5	2	2
Norway	60	25	39	20	19	14	80	44	52	34	37
Portugal	1	1	2	8	9	11	9	10	12	1	2
Spain	73	79	58	31	45	73	104	124	131	100	58
Sweden	58	54	97	29	25	47	87	79	144	42	67
Switzerland	31	35	50	11	20	23	43	55	73	38	35
United Kingdom	122	63	112	89	129	178	211	193	290	73	76
United States	303	566	903	115	117	117	418	684	1019	241	979
Total DAC countries	**2173**	**2835**	**4244**	**797**	**1026**	**1335**	**2970**	**3861**	**5579**	**2138**	**3170**
EC	227	412	730	25	9	4	252	422	734	65	321
Total DAC members	**2400**	**3247**	**4974**							**2203**	**3490**

The United Kingdom commissioned a study to further assess DFID spending in the water sector which suggested a rather higher level of spending, £ 128.8 million in financial year 2005-2006 (approximately USD 238 million). The study tried to identify all activities related to some extent to water. The search was based on water-related sector codes including when aid for water was extended as part of humanitarian aid, water-related objectives (activities reported under non-water-related sectors but marked as contributing to water-related objectives), and keywords. For each identified activity, the proportion of amount to take into account as "aid for water" was estimated.

The use of the multilateral channel by DAC members is not limited to core contributions to multilateral agencies' regular budgets. DAC members also make earmarked contributions to multilateral agencies' funds dedicated to water, or in view of financing specific aid activities. These earmarked aid activities are reported as bilateral aid. In the CRS/DAC classification of international organisations, the *Global Water Partnership* (GWP) is classified as a "public-private partnership" (and not as a multilateral organisation); contributions to the GWP also fall under bilateral aid. In 2005-2006, DAC members reported only 6% of their total bilateral aid for water as channeled through multilaterals and GWP. This figure takes into account only those members that reported on the "channel of delivery", which represented 58% of aid for water. Based on contributions in the form of grants, the share of aid channeled through multilaterals rises to 8% (5% for aid channeled through NGOs). The largest bilateral earmarked contributions benefited Africa and were reported by the Netherlands (*UNICEF WASH programme, AfDB trust fund, UN-Habitat for drinking water supply and sanitation*).

DAC members' disbursements

DAC members' disbursements to water supply and sanitation regularly increased over the period 2002-2006, and amounted to USD 3.5 billion in 2005-2006. The ranking of donors is slightly modified when based on disbursements instead of commitments: Japan is still the largest donor followed by the United States, but Germany ranks third instead of the EC (4th). France (5th) and the Netherlands (6th) also change places.

The EC ranked third on a commitment basis in 2005-2006 following a considerable increase in its aid commitments to the water supply and sanitation sector. The reason why the EC ranked only fourth on a disbursement basis is that only a share of the new large commitments undertaken in 2005-2006 were disbursed during the same period; the remainder will be paid over several subsequent years.

Chart 2 illustrates the relation between commitments and disbursements at the DAC total level. The fact that commitment figures are higher than disbursements over the period indicates increased focus on aid to water in donor programmes in recent years. Commitments are multi-year and subsequent disbursements spread over several years. An increase in aid allocations (commitments) to the water supply and sanitation sector will thus be visible in disbursement data with a few years' time lag.

Commitment data reflect donors' programming and changes in their policies; they thus give an indication about future flows. Disbursements show actual payments in each year. A closer analysis of 2006 commitments and disbursements[3] shows that only 10% of 2006 commitments were disbursed the same year; the remaining 90% will only be disbursed in subsequent years. 2006 disbursements related mostly to commitments undertaken before 2006 (85%), and a significant proportion related to projects committed before 2000 (20%). The disbursement pattern illustrates that infrastructure projects in the water supply and sanitation sector take several years to implement (on average at least 8 years according to an OECD study conducted in 2004).

Chart 2. DAC countries' bilateral and imputed multilateral aid for water supply and sanitation,

2002-2006, constant 2006 prices

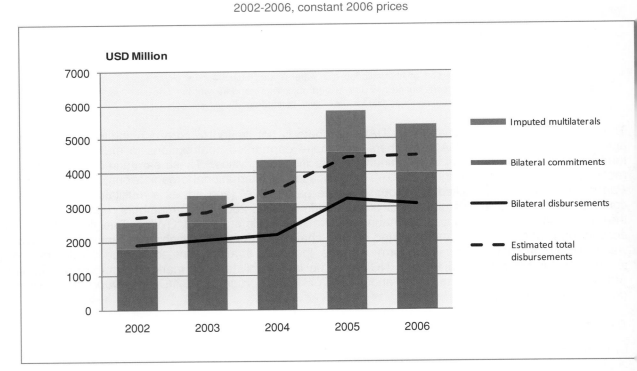

CRS AID ACTIVITIES IN SUPPORT OF WATER SUPPLY AND SANITATION – ISBN 978-92-64-05172-0 – © OECD/WWC 2008

I.3 Main characteristics of DAC members' bilateral aid to the water sector

Geographical focus and concentration of aid

Table 4 below summarizes the geographical focus of DAC members' aid for water. Aid amounts are concentrated in a relatively small number of very large commitments: the 50 largest activities (above USD 40 million) reported by members for 2005-2006 accounted for half the total aid committed during these two years.

Activities of more than USD 50 million included[4]:

- loans to finance investments in infrastructure by Japan in India (Bangalore – **USD 601 million**, Varanasi – **95**, Hyderabad – **66**), China (Changsha - **90**, Guiyang - **103**, Harbin - **64**, Kunming - **109**, Shaanxi - **133**), Malaysia (Pahang/Selangor - **698**), Indonesia (Semarang - **140**, East Java - **80**) and Costa Rica (San Jose - **129**). The total of Japanese loans to these five recipient countries represented one fourth of total DAC members' aid for water (80% of Japanese aid).

- reconstruction projects by the United States in Iraq amounted to **1.5 billion USD**, and represented 15% of total DAC members' aid (82% of the United States' total aid for water);

- other commitments of more than USD 50 million were undertaken in Egypt (Water sector programme of **USD 102 million**) and India (State partnership in Rajasthan of **USD 94 million**) by the EC; Bangladesh (*Karnaphuli water supply project in Chittagong* by Japan for **USD 105 million**; *Water supply, sanitation and hygiene (WASH) component of the Bangladesh Rural Advancement Committee (BRAC)* by the Netherlands for **USD 68 million**; *Sanitation, hygiene education and water supply (SHEWA-B)* by the United Kingdom for **USD 66 million**); Sub-Saharan Africa (Dutch contribution of **USD 59 million** to the UNICEF WASH international programme); Morocco (German loan for a water sector programme - **USD 51 million**); and Viet Nam (Danish sector budget support for **USD 50 million**).

Aid activities in the water sector are spread over all recipient countries. While a number of donors made disbursements for the water sector to over 70[5] recipient countries during the last five years 2002-2006 (Japan, Germany, Ireland, Spain, France and the Netherlands), others were active in less than 30 countries (Australia, Greece, New Zealand, Finland, Denmark, Luxembourg and Portugal). For example, over 2005-2006, Japan extended new loans to 9 different recipient countries, and made disbursements (loans and technical co-operation grants) to more than 100 countries; France's disbursements over the same period were concentrated on fewer recipient countries (50).

---

[4]   Figures indicated in parenthesis are cumulative amounts committed during 2005-2006.

[5]   This includes donors' funding of NGO activities.

Table 4. Main recipients and donors of DAC members' bilateral aid to the water supply and sanitation sector
2005-2006 average commitments in million USD, constant 2006 prices

USD million, average 2005/2006	Japan	United States	EC	Germany	Netherlands	Other DAC members	Total DAC members	% of water aid to all recipients
Iraq	2	744	0	0	0	2	748	15%
India	401	16	47	3	0	22	490	10%
China	369	0	0	6	0	11	387	8%
Malaysia	350	0	0	0	0	0	350	7%
Bangladesh	53	0	0	0	52	66	172	3%
Indonesia	115	0	0	2	30	10	157	3%
Morocco	19	0	27	49	0	49	144	3%
Viet Nam	27	0	0	8	22	52	109	2%
Jordan	12	53	0	44	0	0	109	2%
Palestinian Adm. Areas	3	60	0	5	0	19	87	2%
Other recipients*	276	29	656	335	229	698	2224	45%
Total amount	1626	903	730	453	334	929	4974	100%
% of water aid from all DAC members	33%	18%	15%	9%	7%	19%		

Financial instruments

At DAC total level, aid activities in the water supply and sanitation sector are financed primarily in the form of **grants to developing countries which represent 63% of total aid**. However, the bulk of a number of members' support to this sector is through concessional lending: Italy (90% of water aid is in the form of loans); Japan (86%); France (71%). The other members that make use of loans are Germany (39%), Belgium (28%), Spain (27%) and the EC (4% through the European Investment Bank).

The average grant element of ODA loans in the water sector was 74% in 2005-2006, reflecting favorable financial terms and long repayment periods (32 years on average)[6]. However, loans as opposed to grants cannot be considered as "ultimate sources of revenue" for the water sector, as they ultimately need to be paid back by water users (or public budgets). Concessional loans therefore are useful in helping water utilities "bridge" the financing gap that is created by the need for large up-front infrastructure investment (i.e. to expand infrastructure in order to achieve the MDG target), but they will contribute to increasing infrastructure costs in subsequent periods due to repayment of those loans and increased operations and maintenance costs due to expanded infrastructure[7].

DAC members also carry out financial operations that associate ODA with export credits for projects in the water sector. The objective is to reduce the cost of the export credit operation for the recipient country through either an ODA contribution complementary to the export credit, or an ODA subsidy to soften the terms of the export credit. Total ODA reported by Austria, Belgium, Denmark, Finland, Japan and Spain for these "associated financing packages" in the water sector during 2005-2006 was **USD 720 million**. On the basis of a minority of operations for which the export credit counterpart amounts were also reported,

---

[6] It is the full face value of an ODA loan that is recorded as an ODA commitment (however, the amortisation of the loan - repayments received from the recipient - will be deducted from the donor's net ODA). The grant element measures the concessionality of a loan, expressed as the percentage by which the present value of the expected stream of repayments falls short of the repayments that would have been generated at a given reference rate of interest. The reference rate is 10% in DAC statistics. Thus the grant element is nil for a loan carrying an interest rate of 10%; it is 100% for a grant; and it lies between these two limits for a loan at less than 10% interest.

[7] Even though the conditions of concessional loans are more affordable than market conditions and will therefore increase infrastructure costs by less than a loan from commercial financial markets would have.

the leverage effect of ODA can be estimated at 1.5 (the export credit amount equalled one and a half times the ODA amount).

For the time being, CRS/DAC statistics do not allow identification of other forms of leveraging. For example, the extension of guarantees by donors in order to attract private investors in the water supply and sanitation sector does not qualify as ODA as there is no flow of resources (if the guarantee is called upon, the resulting payment may be reportable as ODA). In their policy briefs (see Part 2), members also mention other mechanisms put in place to facilitate private sector participation: investment guarantees (e.g. Japanese ODA loans combined with USAID investment guarantees), and public-private partnerships between public authorities and local, national or international private firms.

As regards aid modalities, CRS reporting indicates that aid in the water sector is mainly used to finance investment projects. In their policy briefs, a number of donors mention the growing use of budget support, programme based approaches, sector wide approaches, and sector programmes. These aid modalities cannot currently be properly identified in CRS reporting, but donors will be requested to report on them starting with 2010 flows.

Sub-sectors

The analysis of allocation of aid by sub-sectors reveals the predominance of projects for "large systems", which accounted for almost half of total DAC members' contributions to the water supply and sanitation sector in 2005-2006. It is this particular sub-sector which suffered the most from the temporary decline in aid to the water sector in the second part of the 1990s while aid for "water resources policy" and for "river development" received an increasing share of total aid for water (respectively 18% and 12% in 2005-2006). The EC in particular reported large amounts of aid under "water resources policy" related to the *ACP-EU Water facility*.

"River development" activities include contributions for dam repair, basin management (San Julian river in Venezuela, Kabul basin in Afghanistan) and transboundary management (e.g. Mekong River Commission, Nile Basin Initiative, and other river basins such as the Niger and Senegal in Western Africa and the Kura in Central Asia). The order of magnitude for these contributions of a regional nature was on average USD 25 million per year over 2005-2006 for Africa (of which USD 15 million for Sub-Saharan Africa) and USD 12 million for Asia.

Box 2. Examples of transboundary management inititiatives

The Nile Basin Initiative (NBI) is a partnership among the Nile riparian states. It was formally launched in 1999 by the water ministers of 9 countries that share the river - Egypt, Sudan, Ethiopia, Uganda, Kenya, Tanzania, Burundi, Rwanda, the Democratic Republic of Congo, as well as Eritrea. The NBI aims to "achieve sustainable socioeconomic development through the equitable utilisation of, and benefit from, the common Nile Basin water resources."

The Mekong River Commission (MRC) was formed in 1995 by an agreement between the governments of Cambodia, Lao PDR, Thailand and Viet Nam. The MRC is an international, country-driven river basin organisation that provides the institutional framework to promote regional co-operation in order to implement the Agreement on Co-operation for the Sustainable Development of the Mekong River Basin. Its members agreed on joint management of their shared water resources and development of the economic potential of the river.

Chart 3. DAC members' aid to water supply and sanitation by sub-sector,

average 2005-2006, constant 2006 prices

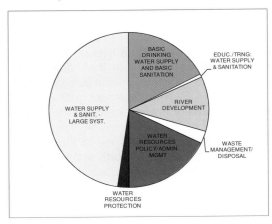

Gender equality focus

Information on the extent to which members have taken particular measures in their aid activities for water supply and sanitation to advance gender equality and women's empowerment is captured through the "gender equality policy marker". For members that report on this aspect of their aid programme[8], one third of aid for water supply and sanitation targeted the objective of gender equality (among activities that members actually screened against the objective); if the largest loans for infrastructure are excluded (for which the gender-equality focus may be more difficult to assess at the commitment stage), the share rises to 45%. Chart 4 shows the individual members with the highest shares of gender-equality focused aid, along with the DAC total.

Chart 4. Gender focus of aid activities by DAC members in the water supply and sanitation sector,

average 2005-2006, constant 2006 prices

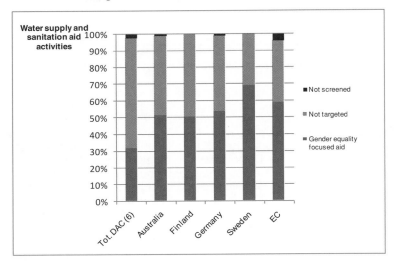

---

[8]   Countries which do not report on the gender equality policy marker, or for which the marker coverage over 2005-2006 is too low to draw conclusions are: France, Ireland, Italy, Luxembourg, Spain, Switzerland, United States.

Policies for development co-operation in the water supply and sanitation sector

Individual members' policies are presented in Part 2. Twelve members refer to a specific policy on water: Australia (ODAW); Austria; EC (EUWI); France (Long-term Framework Strategy for Water Supply and Sanitation and four Strategic papers); Germany (Water sector strategy 2006); Japan (Water and sanitation broad partnership initiative: WASABI); Netherlands; Spain (Environment and Sustainable Development strategy paper); Sweden (Pure Water – Strategy for water supply and sanitation); Switzerland (Orientation Paper Water 2015); United Kingdom (Water Action Plan); United States (Water for the Poor Act).

Other members do not have a distinctive co-operation policy for water (Italy; Portugal). In the context of the Paris Declaration process, Canada stated that the alignment of country programmes to priorities identified within the PRSPs made it inappropriate to predetermine investment targets for specific sectors; and Denmark would no longer develop its own global strategy papers like the Danish Sector Policy Guidelines for Water Supply, Sanitation and IWRM, but would continue to produce best practice papers.

Water is quoted as a thematic priority by Austria (one out of six thematic priorities); Belgium (one of the strategic pillars); Denmark (key sector of Denmark's development assistance); France (one of the seven priority intervention sectors of French ODA); Germany and Japan (one of the top priority sectors); Luxembourg (traditional priority focus); the Netherlands (important component of development co-operation since the 1970s) and Norway (thematic priority).

II.   Degree of targeting to countries most in need

Water supply and sanitation is still one major development issues with 1.1 billion individuals lacking access to safe drinking water and 2.5 billion lacking access to basic sanitation.

MDG 7.C addresses this issue: *Halve, by 2015, the proportion of people without sustainable access to safe drinking water and basic sanitation.* The last UN report on the MDGs assessed that the water supply goal would be met overall, though not in Sub-Saharan Africa and Oceania, but that the sanitation goal would be missed, with South Asia and Sub-Saharan Africa both lagging behind. In Sub-Saharan Africa 300 million people lack access to improved water sources and 450 million lack adequate sanitation. South Asia is on track regarding the provision of water but progress has been slower in providing sanitation.

Tables 5a. and 5b. show the top ten recipients of aid for water supply and sanitation in volume and per capita terms. In this section, the focus is on a recipient country perspective, and figures represent total ODA flows including both bilateral donors and multilateral agencies.

In 2005-2006, Iraq was the top recipient both in terms of volume and per capita as a result of the reconstruction programme financed by the United States. The high rankings of China, Costa Rica, India, Indonesia, Malaysia and Tunisia are attributable to the large Japanese infrastructure loans described in section I.3, while India and Indonesia also benefited from IDA projects, and Tunisia from Germany's projects. For a number of other recipient countries, the high ranking is also due mainly to one or two major donors (Jordan to the United States and Germany; Lesotho to the EC; Nigeria to IDA), while for others (Albania, Bangladesh, Benin, Lebanon, Viet Nam) contributions are made by a larger number of donors.

Tables 5a. and 5b. Top ten recipients of aid for water

2005-2006 average

Table 5.a Aid in volume

		Total USD amount USD million	Per capita USD
1	Iraq	748	26.5
2	India	624	0.6
3	China	389	0.3
4	Malaysia	350	13.8
5	Viet Nam	339	4.1
6	Indonesia	229	1.0
7	Bangladesh	206	1.4
8	Nigeria	168	1.2
9	Morocco	144	4.7
10	Jordan	109	19.9
	Other	*2883*	
	Total	6189	

Table 5.b Aid per capita[9]

		Per capita USD	Total USD amount USD million
1	Iraq	26.5	748
2	Palestinian Adm. Areas	25.0	92
3	Jordan	19.9	109
4	Costa Rica	15.6	68
5	Lesotho	14.5	26
6	Lebanon	13.9	56
7	Malaysia	13.8	350
8	Albania	10.3	32
9	Benin	9.7	83
10	Tunisia	6.2	63
	Other		*4563*
	Total		6189

Charts 5a. and 5b. illustrate the distribution of aid commitments by region and income group over the period 2001-2006. Aid flows for water supply and sanitation primarily targeted Asia (55%) and Africa (32%). The region most in need both in terms of access to water supply and sanitation, Sub-Saharan Africa, received a significant share of total aid (24%) although this decreased in recent years (18% over 2005-2006). The other region suffering the most from a lack of sanitation services, South Asia, was also a relatively large recipient of aid for water (19% of the total[10]).

Least Developed Countries (LDCs) received one quarter of total aid for water, and Other Low Income Countries (OLICs) a further quarter. Loans represented 40% of total water aid to LDCs, and 75% of the total to OLICs.

Charts 5a. and 5b. Aid for water, distribution by region and income group

Chart 5a. Distribution by region
average 2001-2006

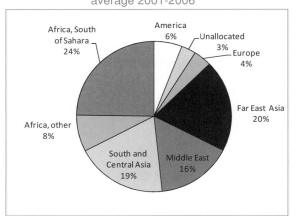

Chart 5a. Distribution by income group
average 2001-2006

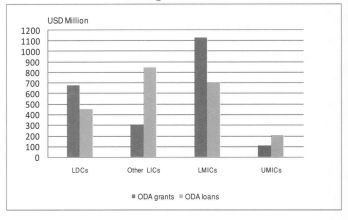

---

[9] Amongst countries of more than 1 million inhabitants.

[10] The figure refers to South and Central Asia.

CRS AID ACTIVITIES IN SUPPORT OF WATER SUPPLY AND SANITATION – ISBN 978-92-64-05172-0 – © OECD/WWC 2008

The *Joint Monitoring Programme* compiles indicators on the proportion of population using an improved drinking water source and on the proportion using an improved sanitation facility. These indicators help identify the countries in most need, and can be compared with the level of donors' support for water supply and sanitation in these countries.

Maps 1 and 2 present the indicators and illustrate the difficult situation of Sub-Saharan Africa (for both water supply and sanitation) and South Asia (sanitation). Map 3 shows aid flows for water on a per capita basis and spotlights the higher per capita levels in Sub-Saharan Africa (though only Benin, Lesotho and Nigeria appear amongst the top ten).

Map 1. Proportion of population using an improved drinking water source
2004, source: Joint Monitoring Programme

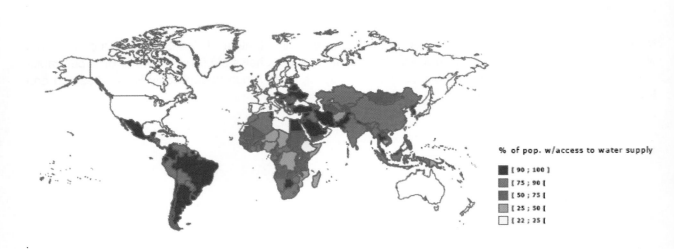

% of pop. w/access to water supply

- ■ [90 ; 100]
- [75 ; 90 [
- [50 ; 75 [
- [25 ; 50 [
- □ [22 ; 25 [

Map 2. Proportion of population using an improved sanitation facility
2004, source: Joint Monitoring Programme

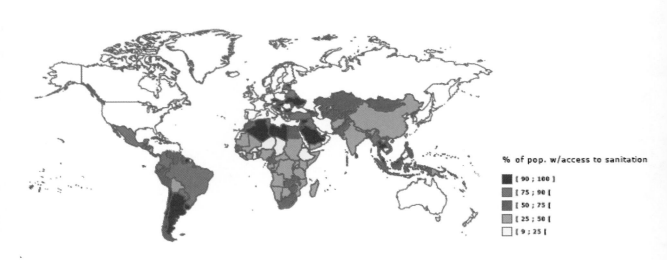

% of pop. w/access to sanitation

- ■ [90 ; 100]
- [75 ; 90 [
- [50 ; 75 [
- [25 ; 50 [
- □ [9 ; 25 [

CRS AID ACTIVITIES IN SUPPORT OF WATER SUPPLY AND SANITATION – ISBN 978-92-64-05172-0 – © OECD/WWC 2008

Map 3. Aid to water supply and sanitation per capita,
average commitments for 2005-06, USD

USD per capita

[10 ; 90.3]

[3 ; 10 [

[1 ; 3 [

[0 ; 1 [

The levels of access to water supply and sanitation among the top recipients of aid to these sectors can be summarised as follows:

- Nigeria has the lowest levels of access for both water supply and sanitation, 48% and 44% respectively.
- Bangladesh, Benin, China, Indonesia, India, Lesotho and Viet Nam face poor access to sanitation facilities, with less than 40% of the population having access [exept for Indonesia (55%) and Viet Nam (61%)]. However these countries have relatively high levels of access to water supply supply (from 67 to 86%).
- The remaining countries, Albania, Costa Rica, Iraq, Jordan, Lebanon, Malaysia, Morocco and Tunisia, have very high levels of access to both water supply and sanitation (above 90%, except Iraq and Morocco with slightly lower levels).

Charts 6a. and 6b. Aid to water supply and sanitation per capita in relation to the degree of access to water supply and sanitation facilities by recipient countries

Chart 6a. Water supply

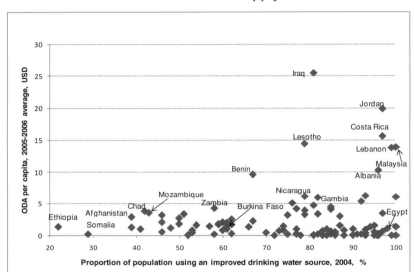

Chart 6b. Sanitation facilities

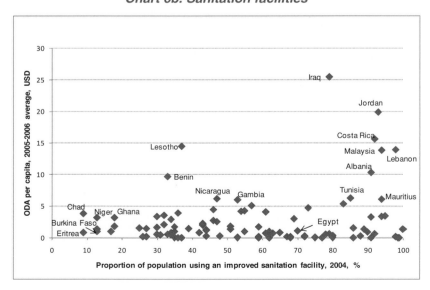

CRS AID ACTIVITIES IN SUPPORT OF WATER SUPPLY AND SANITATION – ISBN 978-92-64-05172-0 – © OECD/WWC 2008

Charts 6a and 6b confirm that a significant portion of aid is allocated to countries in difficult situations. Afghanistan, Chad, Laos, Madagascar, Mali, Mozambique, Niger, Zambia all suffer from poor access to water supply (less than 60%) and to sanitation (less than 55%) and received at least on average USD 2.5 per capita over 2005-2006.

However, numerous countries with low levels of access to water supply and/or sanitation received very little during the same period (e.g. Angola, Central African Republic, Republic of Congo, Somalia, Togo received less than USD 0.5 per capita, see also Tables 6a. and 6b.) while the countries with higher levels of access received more (e.g. Albania, Costa Rica, Iraq, Jordan, Lebanon, Malaysia received at least USD 13 per capita).

Tables 6a. and 6b. Countries most in need and levels of ODA per capita

Access: % coverage (2004); ODA per capita: USD (average 2005-2006)

Table 6a. Lowest levels of access to water supply

Water supply	Water supply coverage, %	ODA for WSS per capita, USD
Ethiopia	22	1.3
Somalia	29	0.2
Papua New Guinea	39	1.3
Afghanistan	39	2.9
Cambodia	41	1.0
Chad	42	3.8
Mozambique	43	3.5
Niger	46	3.2
Madagascar	46	2.1
Congo, Dem. Rep.	46	0.6

Table 6b. Lowest levels of access to sanitation

Sanitation	Sanitation coverage, %	ODA for WSS per capita, USD
Chad	9	3.8
Eritrea	9	0.8
Ethiopia	13	1.3
Niger	13	3.2
Burkina Faso	13	1.0
Cambodia	17	1.0
Guinea	18	1.8
Ghana	18	3.2
Namibia	25	1.6
Somalia	26	0.2

III. Methodological note: challenges in collecting data on aid to water

Budget support not allocated to sectors

A number of donors claim to increasingly channel their aid to developing countries through general budget support instead of undertaking specific projects in identified sectors. Donors' budget support, once integrated in developing countries' domestic budgets, will contribute to the development of the water sector, but this contribution is not specified and not taken into account in CRS/DAC statistics on aid for water.

General budget support contributions are not earmarked for any specific use but are accompanied by various understandings and agreements on the government's development strategy. This implies that an individual donor cannot control the extent to which its contribution focuses on a particular sector. Furthermore, the contributions are spent through the recipient government's own financial management system. Donors do not control the spending but monitor the implementation of the recipient's strategy as a whole on the basis of an agreed set of indicators.

The United Kingdom estimates its contribution to the water sector through general support based on the assumption that recipient governments spend the same proportion of their budget support on the water sector as the percentage of their total spending on water (set out in their PRSPs). This methodology was discussed within the *DAC Working Party on Statistics (WP-STAT)* but there was no consensus to use it in standard DAC statistical presentations. The discussion will continue but, in any case, such data would be considered as estimates, and would need to be presented separately from standard statistics on aid flows by sector (as for imputed multilateral contributions).

Separating aid to water supply from aid to sanitation

The UN have declared 2008 the *International Year of Sanitation* to highlight the specific challenges involved in the sanitation area. For UN Water and the EUWI African Working Group, it would be desirable to distinguish between aid flows for the purpose of water supply and those for sanitation only. The situation of countries vis-à-vis water supply or sanitation can vary a lot, with sanitation being a more acute problem than water supply in a number of developing countries.

The WP-STAT addressed this methodological issue at its meeting in June 2008, with UN-Water presenting a proposal for introducing separate purpose codes for "water supply" and "sanitation". Members favoured having disaggregated data on water and sanitation in principle, but felt that, for practical reasons, the existing combined code should be retained. Indeed, the differentiation was not always possible for projects mixing both aspects. A revised proposal will be discussed by the WP-STAT at its next meeting (probably February 2009).[11]

In the meantime, in view of testing the feasibility of introducing separate purpose codes, members were asked to try and allocate amounts of their largest projects over 2005-2006 against "water supply" and "sanitation" components. Members were able to allocate 42% of their total bilateral aid against the two components, and stated that a further 20% addressed both aspects but components could not be estimated. No information was provided for the remaining 38% of aid.

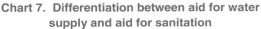

Chart 7. Differentiation between aid for water supply and aid for sanitation

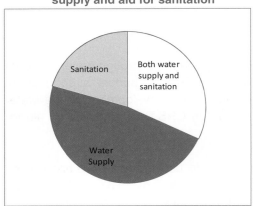

Among the projects examined (62% of total aid), the allocation suggests that almost half of the amount examined goes to water supply aspects, and 21% to sanitation, whereas the remaining 32% address both aspects in unknown proportions.

---

[11] Only the United States currently have codes that separately identify water supply and sanitation, but also have codes that cover both. Austria will shortly introduce separate codes.

 CRS AID ACTIVITIES IN SUPPORT OF WATER SUPPLY AND SANITATION – ISBN 978-92-64-05172-0 – © OECD/WWC 2008

Monitoring outputs

CRS/DAC statistics monitor aid inputs (ODA volumes) but experts in the water sector are also generally interested in the outputs of aid (i.e. number of water connections etc.). The complexity of evaluating aid outputs makes it impossible to monitor these on a systematic basis at project level through the CRS. However, the Development Evaluation Resource Centre (DEReC) provides access to a wide array of evaluation publications and reports, including on the water sector (see www.oecd.org/dac/evaluationnetwork/derec).

Also, in their policy briefs, in the "performance evaluation" section, a number of members mention a few figures on outputs reached or targeted in terms of number of people gaining access to water supply and sanitation services.

PART 2

Donor profiles
on aid to the water supply and sanitation sector

Statistics and policy briefs for each DAC member

www.oecd.org/dac/stats/crs/water

Donor profiles on aid to the water sector supply and sanitation sector

The following profiles cover both statistical and policy aspects of DAC members' aid to the water supply and sanitation sector. For each member, summary statistics on aid to water supply and sanitation are presented in the form of charts and tables, and textual information describes the policy/strategy for development co-operation in the water sector.

Statistics are based on members' reporting to the CRS and DAC, and comply with the definition of aid to water supply and definition given in section I.1 of this report. Information shown includes the total volume of aid to water and trends over time, the share of aid to water in total bilateral sector-allocable aid, the top ten recipients, the geographical distribution and the sub-sector distribution. One table also shows statistics based on a wider definition of aid to water that includes water transport, hydro-electric power plants and agricultural water resources.

The **policy briefs** were prepared by the World Water Council in collaboration with members. They provide a short description of members' development co-operation policy in the water supply and sanitation sector. The information covers main policy statements (including possible commitments to increase aid for water supply and sanitation), geographical and sub-sectoral focus, aid modalities, rules and practices, performance evaluation, coordination with other actors. Web references are also provided for further reading.

 CRS AID ACTIVITIES IN SUPPORT OF WATER SUPPLY AND SANITATION – ISBN 978-92-64-05172-0 – © OECD/WWC 2008

AUSTRALIA
Aid at a glance - Water supply and sanitation

Summary statistics

Unless otherwise stated, figures refer to 2005-2006 annual average commitments expressed in 2006 constant prices.

Total aid to water supply and sanitation			
	USD million	Aid to water by Australia as a share of total aid by Australia	Aid to water by Australia as a share of total DAC members' aid to water
Australia	**6.2**	**0%**	**0%**
For reference, total DAC	4974.0	8%	100%

Top ten recipients of aid to water supply and sanitation			
		Aid to water by Australia to that recipient as a share of	
	USD million	total aid by Australia to that recipient*	total DAC members' aid to water to that recipient
Indonesia	2.6	2%	2%
Timor-Leste	1.0	2%	15%
Sri Lanka	0.5	4%	1%
China	0.4	2%	0%
Philippines	0.2	0%	2%
Nauru	0.1	1%	97%
Viet Nam	0.1	0%	0%
Cook Islands	0.1	6%	33%
Tonga	0.0	1%	3%
India	0.04	0%	0%

Aid to all water-related sectors	
	USD million
Water supply and sanitation	6.2
Water transport	1.9
Hydro-electric power plants	0.0
Agricultural water resources	3.7
Total water-related aid	11.8

Annual aid to water supply and sanitation, 1993-2006 (2006 constant prices)

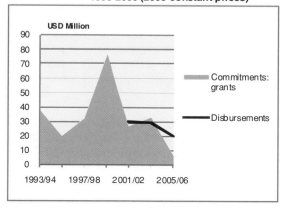

Regional distribution of aid to water supply and sanitation,

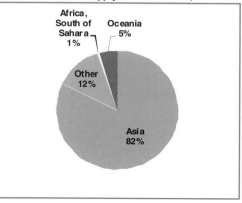

Water supply and sanitation aid by subsector

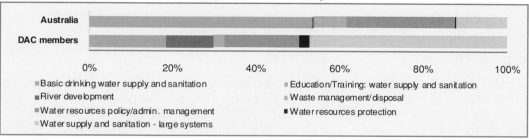

* % of sector allocable aid

AUSTRALIA
Development Co-operation Policy in the water supply and sanitation sector

POLICY STATEMENT

Australia's aid and foreign policy objectives with respect to Official Development Assistance for Water (ODAW) recognise the importance of water to regional growth and stability in the Pacific and Asia, and will increase development assistance in this sector by A$300 million from 2008/09 to 2010/11. Australian experience and expertise in ODAW aim at strengthening sustainable access to water and helping secure regional and national prosperity.

The objectives of AusAID's water-related policies align with broad international goals, and in particular with the United Nations Millennium Development Goals. Australia's water assistance builds on lessons learned from domestic and international experience and tailors activities to individual country circumstances. It also builds on existing policies on gender and development, poverty reduction and governance, and identifies strategic interventions that maximise the impact of Australia's aid. Australia's ODAW encapsulates the sustainable management of water resources as much as the supply of drinking water and provision of sanitation services.

GEOGRAPHICAL AND SUB-SECTORAL FOCUS

Australia's water assistance focuses on the Asia-Pacific region. A significant number of countries in Australia's region, including Papua New Guinea, Laos, Cambodia, Indonesia and East Timor appear unlikely to achieve Millennium Development Goal 7. Australia's aid initiatives will target partner countries in the Asia-Pacific region that experience serious deficits in access to safe drinking water and basic sanitation, or face critical challenges in protecting freshwater sources. Partnerships will strengthen water planning and management, enhance access to safe drinking water and improved sanitation, increase water conservation and storage capacity, and enhance solid and hazardous waste management to prevent contamination of water.

Australia's water assistance will focus on: 1) ensuring access for the poor; 2) increasing access to improved sanitation 3) increasing coverage in urban areas; 4) implementing sustainable pricing policies; 5) increasing aid effectiveness; 6) address environmental and use practices that result in poor water quality and availability; 7) maintenance of existing infrastructure; and 8) participation of user groups in the development and operation of water services.

Current programs which focus on two central themes - water governance and delivery systems - will continue and will be scaled up in areas affected by climate change. Within water governance, emphasis will be placed on strengthening water policy as well as institutional arrangements and legal frameworks, building water resource planning and management capacities, and promoting better allocation of water that reflects the true cost of water. The improvement of delivery systems will be achieved through additional finance for capital investment; building capacity in partner government systems to deliver services and make them more sustainable; building partnerships with domestic private sector providers and improvements in sub-national government funding of basic operations and maintenance related to water supply and sanitation infrastructures; improving water quality; improving water efficiencies especially in the agricultural sector; and supporting better flood and drought management.

RULES AND PRACTICES

To implement Australia's aid program, specific emphasis will be placed on three guiding principles: 1) promote attention to water issues in Australia's aid program by ensuring active community participation in program design to ensure new water facilities meet community demands and receive local support; 2) enhance access and disseminate Australian knowledge and expertise by developing a dedicated water section on the Australian Development Gateway, an internet portal for developing countries, and funding

 CRS AID ACTIVITIES IN SUPPORT OF WATER SUPPLY AND SANITATION – ISBN 978-92-64-05172-0 – © OECD/WWC 2008

research to tackle practical and policy level issues in developing countries; 3) build strategic partnerships with global and local organisations by selecting and working with partners that complement Australian efforts and help maximise the overall impact of Australian aid.

AusAID's approach for managing water activities is outlined in the *Safe water guide*. AusAID's framework for managing water projects incorporates best practice principles for safe water from the *Australian drinking water guidelines 2004*, the WHO *Guidelines for drinking-water quality* (3rd edn), and AusAID's *Environmental management guide for Australia's aid program 2003*. The framework identifies five steps that may need to be taken when designing, implementing, monitoring and evaluating water-related activities in the Australian aid program, particularly the provision of safe drinking water supplies: 1) understanding the policy and legal setting; 2) assessing water quality and outline the management plan; 3) implementing the provisions for managing water quality; 4) monitoring and evaluate water quality management; 5) reviewing water quality management in AusAID.

PERFORMANCE EVALUATION

Australian ODA framework does not provide a specific performance evaluation process for the water and sanitation sector. For selected activities, AusAID evaluates the effectiveness of the aid activity and draws lessons about the activity design and the suitability of the management and the principles and practices that were followed. Overall, this evaluation process serves to assess the development effectiveness and value for money return for AusAID-funded activities.

COORDINATION WITH OTHER ACTORS

Australia has a strong track record of forging partnerships on water issues. The partnership between the *Murray Darling Basin Commission* and the *Mekong River Commission (MRC)* has drawn on Australian experience in transboundary water management to strengthen the MRC's capacity. Australia also has a long history of collaboration with the World Bank and Asian Development Bank. This brings Australian expertise to bear on issues that Australia, as a mid-sized donor, may not otherwise be in a position to tackle. The *Australia-World Bank Partnership for Water and Coastal Resource Management* will also facilitate the application of Australian knowledge and expertise to World Bank policy formulation and water-related investments.

Australia will continue to focus on strategic alliances to help partner governments secure finance from the World Bank and Asian Development Bank, Asia-Pacific's major water investors, and the private sector. Partnerships with multilateral agencies, such as the World Health Organisation, UNICEF, the Commonwealth Scientific and Industrial Research Organisation (CSIRO) and other partners will also be continued to develop innovative solutions for water security. Australia will extend its successful collaboration with NGOs and also forge partnerships with global and regional water organisations, the private sector and civil society.

Web References:

Australian Government Overseas Aid Water Program: http://www.ausaid.gov.au/keyaid/water.cfm

AUSTRIA
Aid at a glance - Water supply and sanitation

Summary statistics

Unless otherwise stated, figures refer to 2005-2006 annual average commitments expressed in 2006 constant prices.

Total aid to water supply and sanitation

	USD million	Aid to water by Austria as a share of total aid by Austria	Aid to water by Austria as a share of total DAC members' aid to water
Austria	**18.8**	**7%**	**0%**
For reference, total DAC	*4974.0*	*8%*	*100%*

Top ten recipients of aid to water supply and sanitation

		Aid to water by Austria to that recipient as a share of	
	USD million	total aid by Austria to that recipient*	total DAC members' aid to water to that recipient
Uganda	4.5	41%	26%
China	2.3	9%	1%
Albania	2.2	38%	10%
Macedonia (TFYR)	2.0	36%	26%
Cape Verde	0.9	29%	6%
Bosnia-Herzegovina	0.7	3%	8%
Moldova	0.6	34%	85%
Montenegro	0.6	14%	66%
Serbia	0.5	2%	2%
Mozambique	0.3	5%	0%

Aid to all water-related sectors

	USD million
Water supply and sanitation	18.8
Water transport	0.0
Hydro-electric power plants	1.1
Agricultural water resources	0.6
Total water-related aid	*20.6*

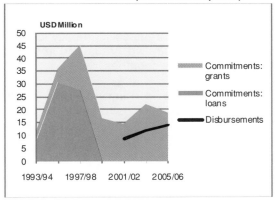

Annual aid to water supply and sanitation, 1993-2006 (2006 constant prices)

Commitments: grants
Commitments: loans
Disbursements

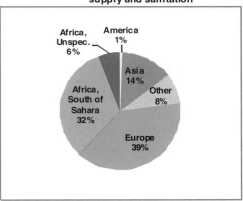

Regional distribution of aid to water supply and sanitation

Africa, Unspec. 6%
America 1%
Asia 14%
Other 8%
Europe 39%
Africa, South of Sahara 32%

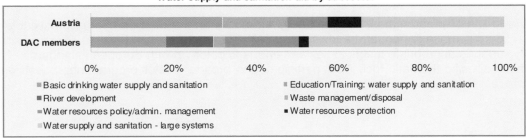

Water supply and sanitation aid by subsector

Austria
DAC members

0%   20%   40%   60%   80%   100%

- Basic drinking water supply and sanitation
- Education/Training: water supply and sanitation
- River development
- Waste management/disposal
- Water resources policy/admin. management
- Water resources protection
- Water supply and sanitation - large systems

* % of sector allocable aid

AUSTRIA
Development Co-operation Policy in the water supply and sanitation sector

POLICY STATEMENT

Water and sanitation is one of six thematic priorities of Austrian Development Co-operation (ADC); a revised policy document for the water sector has been adopted in 2008.
In the water sector, Austria aims at contributing to the Millennium Development Goals by improving access to safe water supply and basic sanitation. Priority is given to the satisfaction of the basic needs of all population groups.

To be sustainable, improvements in access to water supply and sanitation services need to be embedded in a comprehensive and participatory development process. ADC is therefore committed to fostering nationally-owned sector development through long-term programmes. This implies support of and alignment to national structures, targets and development strategies, in line with the Paris Declaration agenda. It also implies the need for a holistic approach, where infrastructure investments must go hand in hand with the creation or strengthening of the necessary institutional, organisational and regulatory structures; building capacities and knowledge; and raising awareness for hygiene, operation and maintenance sustainability and the issue of resource protection. The principles of Integrated Water Resources Management (IWRM) are seen as the best way to ensure equitable access for all user groups, to preserve the water resources and the environment, and to avoid conflicts.

To implement this approach, ADC supports and actively participates in sector coordination and financing, as well as in the development of national strategies, programmes and instruments. A mix of different intervention levels and instruments are used, from piloting of innovative approaches to the support of joint sector programmes and the promotion of regional networking.

GEOGRAPHICAL AND SUB-SECTORAL FOCUS

ADC is generally focusing on bilateral programmes in 14 priority countries (5 located in Africa, 7 in South East Europe, and 2 in Asia and Central America, respectively), whereas bilateral engagement in other countries is being replaced by regional programmes. Currently there are specific (mostly long-term) water sector programmes in 5 African countries, in the Palestinian Administrative Areas and in the South East European priority countries. NGO co-financing is not restricted to these countries.

ADC's traditional sub-sectoral focus is water supply and sanitation (WS&S) in rural areas, including small towns. Specific support is provided to the development of appropriate instruments, technologies and institutional structures to finance and to ensure the sustainability of WS&S infrastructure in the context of decentralisation.

Sanitation receives increasing attention within the sector programme and a sanitation strategy is being prepared at the moment. Austria has been supporting sustainable and ecological sanitation approaches for years, and is now planning to increase targeted investments in sanitation, including dedicated programmes aiming to raise sanitation coverage and hygiene awareness.

With regard to IWRM programmes, Austria places emphasis on the development of practicable solutions for water resources management at the lowest appropriate (e.g. sub-basin) level. In Eastern Europe, ADC specifically supports appropriate technologies and efforts to catch up with European standards.
Beyond these specific thematic foci, ADC participates in broader sector programmes based on national planning processes.

CRS AID ACTIVITIES IN SUPPORT OF WATER SUPPLY AND SANITATION – ISBN 978-92-64-05172-0 – © OECD/WWC 2008

AID MODALITIES

The main modality of bilateral aid is currently funding of jointly agreed sector programmes, implemented either through partner government institutions or – decreasingly – by Austrian organisations selected by tendering, but in line with national rules, processes and reporting modalities as much as possible. As a rule, the programmes are agreed and established at the highest possible level, depending on the status of institutional and sector development in the partner country. Austria actively participates in coordinated approaches, including Sector-Wide Approaches (SWAp) and joint financing arrangements (basket funding) where these emerge. Sector budget support is not yet done but envisaged for the near future. Pilot projects are supported where a clear strategy for evaluation and scaling - up is available.

In addition, there are two aid modalities besides bilateral programming: one is co-financing of NGO programmes and projects based on their own initiatives. The other one is support to Austrian companies in developing countries where positive development impacts or synergies may be expected. The share of Austrian aid to the water sector channelled through these two modalities is relatively small.

RULES AND PRACTICES

The Federal Ministry for European and International Affairs, responsible for the coordination of Austrian development policy, formulates strategies and a rolling Three-Year Programme. The Austrian Development Agency (ADA) is the operational unit in charge of implementing programmes and projects, including all the above-mentioned aid modalities. The contributions of other government institutions are mostly non-sector specific or minor in volume.

There is no general sector programme for water and sanitation. Detailed programmes are formulated in regional or country dialogues where coordination with partners is maintained.

All ADC water sector programmes and projects are screened for policy compliance by ADA's Quality Assurance and Knowledge Management division, with particular attention to the cross-cutting issues of gender and environment which are of outstanding importance for water sector interventions.

PERFORMANCE EVALUATION

Apart from regular project cycle management ADA operates an evaluation unit responsible for both strategic and programme/project-level evaluations. Evaluations are carried out by teams of independent, interdisciplinary experts with the involvement of local specialists. The last external evaluation of Austria's development co-operation in the water sector was carried out in 2003.

ADC strives to strengthen and use national monitoring and performance measurement systems. For instance, in Uganda it is closely involved in joint performance evaluation mechanisms and contributes to their enhancement. In this context, joint evaluations will increasingly become a regular element of joint programmes.

In view of fostering results-based management and quantifying the effectiveness of the programmes, the options to strengthen outcome and impact analysis for the water sector are currently being studied. Again, these analyses will be conducted in collaboration with the programme partners as far as possible.

COORDINATION WITH OTHER ACTORS

At the international level, Austria participates actively in sector relevant CSD and OECD processes as well as in the EU Water Initiative (EUWI). It leads one out of six work packages of SPLASH, the EUWI's European Research Area Network (Era-Net) aiming at improving the effectiveness of European funded research on water for development and at developing the networking capacities of local research organisations. Support to other multilateral sector institutions and initiatives includes the World Bank's Water and Sanitation Program (WSP - Africa Region) and the African Water Facility. Austria further contributes to the EU-Africa Infrastructure Trust Fund and to the Private Infrastructure Development Group

(PIDG), a multi-donor organisation aiming to encourage private infrastructure investment in developing countries.

Austria strives to enhance collaboration and division of labour with other donors. Examples in the water sector are the co-operation with the Swiss Development Co-operation in the Republic of Moldova and co-financing with the ACP-EU Water Facility in Uganda and Rwanda.

Austria recognises the important, complementary role of non-state partners in water sector development, encourages a collaborative relationship and provides appropriate aid delivery channels. Collaboration with NGOs is implemented through various co-financing programmes and through calls for proposals, as well as by direct collaboration with NGOs in the South. ADC systematically supports the development of the local private sector within its programmes and facilitates private-sector projects of Austrian and European enterprises that serve the development interests of the host country.

Web References:

Portal of Austrian Official Development Co-operation: http://www.entwicklung.at/en.html
Austrian Development Agency - http://www.entwicklung.at/en/akteure/ada.html

 CRS AID ACTIVITIES IN SUPPORT OF WATER SUPPLY AND SANITATION – ISBN 978-92-64-05172-0 – © OECD/WWC 2008

BELGIUM
Aid at a glance - Water supply and sanitation

Summary statistics

Unless otherwise stated, figures refer to 2005-2006 annual average commitments expressed in 2006 constant prices.

Total aid to water supply and sanitation			
	USD million	Aid to water by Belgium as a share of total aid by Belgium	Aid to water by Belgium as a share of total DAC members' aid to water
Belgium	**63.4**	**8%**	**1%**
For reference, total DAC	*4974.0*	*8%*	*100%*

Top ten recipients of aid to water supply and sanitation				Aid to all water-related sectors	
		Aid to water by Belgium to that recipient as a share of			
	USD million	total aid by Belgium to that recipient*	total DAC members' aid to water to that recipient		USD million
Ghana	10.0	52%	27%	Water supply and sanitation	63.4
Tunisia	9.5	77%	15%	Water transport	7.4
Morocco	6.1	46%	4%	Hydro-electric power plants	0.0
Senegal	5.5	29%	21%	Agricultural water resources	2.3
Niger	5.1	17%	16%	*Total water-related aid*	*73.1*
Congo, Dem. Rep.	4.3	4%	13%		
Viet Nam	3.9	16%	4%		
Mali	2.8	15%	7%		
Palestinian Adm. Areas	2.6	15%	3%		
Benin	2.1	13%	3%		

Annual aid to water supply and sanitation, 1993-2006 (2006 constant prices)

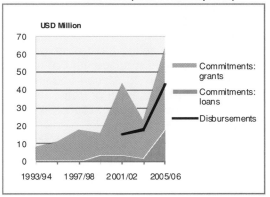

Regional distribution of aid to water supply and sanitation

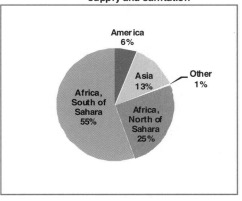

Water supply and sanitation aid by subsector

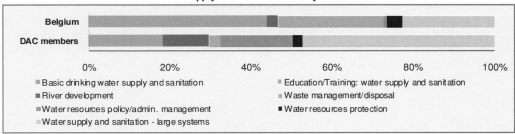

* % of sector allocable aid

BELGIUM
Development Co-operation Policy in the water supply and sanitation sector

POLICY STATEMENT

Belgian ODA for water and sanitation fosters Millennium Development Goal 7. One of the strategic pillars of Belgium Development Co-operation (BDC) is integrated water and sanitation initiatives, including awareness and education to hygiene. According to the 1999 Law on International Co-operation (LIC), Belgian assistance is concentrated in five key areas to promote sustainable human development, among which food security and basic infrastructures. Basic infrastructures are essential to meeting fundamental requirements in areas such as water supply and sanitation. The connection between strategic planning for water and sanitation and the basic infrastructure focus area is deemed crucial to ensure the financing of improved drinking water or sanitation service operations through innovative fiscal approaches linking services cost recovery with inter-sector subsidy.

Belgian Technical Co-operation (BTC) is currently carrying out over a hundred projects linked to the basic infrastructure sector in co-operation with local and international partners. Over half of these are intended to meet basic requirements in the areas of water supply and sanitation. Support for the water sector is one of Belgium's development priorities in partner countries such as Senegal and Rwanda. In Senegal, where Belgium has more than 20 years of experience on the ground in the provision of drinking water and water purification, support is guaranteed until 2012: a budget of €11 million was set aside in 2007 for work on the infrastructure that will provide almost half a million Senegalese with access to potable water. In Rwanda, almost a half of the €140 million pledged until 2010 is going to rural development in the form of support for drinking water, among others.

GEOGRAPHICAL AND SUB-SECTORAL FOCUS

At present, 18 countries have been selected for direct bilateral co-operation: Algeria, Benin, Bolivia, Burundi, DR Congo, Ecuador, Mali, Morocco, Mozambique, Niger, Palestine, Peru, Rwanda, Senegal, South Africa, Tanzania, Uganda and Vietnam. More than 80% of Belgian ODA for water and sanitation target Africa. Numerous Belgian projects set out to provide potable water to these countries. At the request of the governments in Morocco, Algeria and Senegal, Belgium is investing most of its governmental aid there in the provision of drinking water and the improvement and solid management of the water sector. The Belgian Survival Fund (BSF) also helps to ensure food security in 20 African countries by attacking food-related problems, and a reliable water supply is considered an essential prerequisite for developing arable and livestock farming. Finally, BDC funds emergency aid which includes actions taken during crises to meet vital needs for food and water.

AID MODALITIES

The framework for Belgian ODA does not provide specific aid modalities for water and sanitation. Co-operation between governments is prepared and financed by the Directorate-General for Development Co-operation (DGDC), but it is implemented by the BTC. The DGDC finances and coordinates other programmes, among which co-financing programmes with recognised Belgian NGOs and their partners from developing countries (up to 85% of expenditures). NGOs that have received special programme recognition may submit a programme application. NGOs who have not received programme recognition will only be eligible for the co-financing of smaller-scale projects. A programme that has been approved can rely on a pre-arranged amount for 3 years, or 2 years in the case of a project.

RULES AND PRACTICES

Belgian ODA for water and sanitation is not subject to specific rules and practices. According to the LIC, development actions must consider 5 topics: gender equality, the empowerment of women, protecting the environment, promoting a social economy, and children's rights. The Belgian government is obliged by

 CRS AID ACTIVITIES IN SUPPORT OF WATER SUPPLY AND SANITATION – ISBN 978-92-64-05172-0 – © OECD/WWC 2008

parliament to report annually on the results of efforts towards the MDGs. Development policy is defined by the Belgian federal government and its minister for Development Co-operation, but shadows the European and international development agenda in accordance with the Paris Declaration on Aid Effectiveness, EU's Code of Conduct on Complementarity and Division of Labour in Development Policy, and the EU's European Consensus on Development. Belgium is committed to the following policy formulations, among others: 1) long-term strategic programming; 2) division of labour and complementarity between donors; 3) alignment of implementation modalities; 4) predictability of ODA.

PERFORMANCE EVALUATION

Project evaluations are the responsibility of various departments within the DGDC and their implementation partners, such as the BTC, NGOs, universities and others. The DGDC's Monitoring and Evaluation Department (MED) participates in evaluations, controls the quality of evaluation processes and enhances the integration of evaluation results. The MED cooperates with the Special Evaluation Unit (SEU), an external unit that evaluates strategies and policies relating to all types of co-operation projects and programmes carried out within the framework of Belgian ODA. Evaluations are tailored to the demands and procedures of the partner countries, in line with the Paris Declaration on aid effectiveness.

BDC assesses the relevance of development in the water and sanitation sector using the criteria defined by the OECD – DAC, which help to ascertain whether the actions in question take sufficient account of the following underlying principles: 1) strengthening institutional and management capabilities; 2) social and economic impact; 3) technical and financial viability; 4) effectiveness of the planned implementation procedure; 5) focus on equality between men and women; 6) respect for protection and safeguarding of the environment (see "Rapport de la Cour des comptes transmis à la Chambre des représentants, *Tirer des enseignements des évaluations de projets de développement*", Brussels, November 2006, p.33).

COORDINATION WITH OTHER ACTORS

BDC cooperates today with 21 international development organisations. Belgium engages in multilateral ODA for water and sanitation through contributions to activities and programmes carried out by international organisations such as the World Bank (WB) and the United Nations Development Program (UNDP). In particular, the Belgian contribution to the WB Water and Sanitation Programme reached €400 000 in 2005. At the humanitarian level, BDC works closely together with NGOs, humanitarian UN organisations and the International Red Cross to foster food security and access to water.

Web References:

Belgian development co-operation agency: http://www.btcctb.org/
Statistics: http://www.dgdc.be/en/statistics/index.html
Other relevant document: www.dgcd.be/documents/fr/themes/omd/rapport_parlement_omd_2007.doc

CANADA
Aid at a glance - Water supply and sanitation

Summary statistics

Unless otherwise stated, figures refer to 2005-2006 annual average commitments expressed in 2006 constant prices.

Total aid to water supply and sanitation			
	USD million	Aid to water by Canada as a share of total aid by Canada	Aid to water by Canada as a share of total DAC members' aid to water
Canada	**31.7**	**2%**	**1%**
For reference, total DAC	*4974.0*	*8%*	*100%*

Top ten recipients of aid to water supply and sanitation			
		Aid to water by Canada to that recipient as a share of	
	USD million	total aid by Canada to that recipient*	total DAC members' aid to water to that recipient
Indonesia	3.7	5%	2%
Sri Lanka	2.9	9%	5%
Honduras	2.5	45%	18%
Maldives	2.2	96%	99%
Bolivia	1.7	11%	6%
Ghana	1.4	14%	4%
Bangladesh	1.4	2%	1%
Algeria	0.9	20%	5%
Tajikistan	0.8	12%	36%
Senegal	0.7	9%	3%

Aid to all water-related sectors	
	USD million
Water supply and sanitation	31.7
Water transport	0.0
Hydro-electric power plants	1.0
Agricultural water resources	7.9
Total water-related aid	*40.7*

Annual aid to water supply and sanitation, 1993-2006 (2006 constant prices)

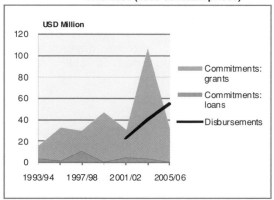

Regional distribution of aid to water supply and sanitation **

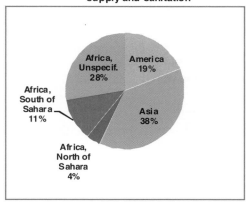

Water supply and sanitation aid by subsector

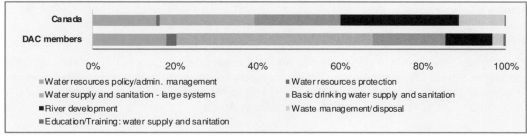

* % of sector allocable aid

** Commitments in Asia in 2005/06 are mostly attributed to the aftermath of the Tsunami in december 2004.

CANADA
Development Co-operation Policy in the water supply and sanitation sector

POLICY STATEMENT

CIDA's development strategy is based on 3 pillars: the Millennium Development Goals (2000), the Monterrey Consensus (2002) and the Paris Declaration on Aid Effectiveness (2005). In this framework, CIDA does not have a specific policy on water, but programs and projects demonstrated CIDA's support for water and sanitation, Integrated Water Resources Management (IWRM) and transboundary water resources management. CIDA recognizes that sustainable results can be best achieved within the context of an IWRM approach that considers and plans for a wide range of water uses, consults users and affected stakeholders at the community level, and integrates the principles of equity, environmental sustainability, and local ownership.

Canada has committed to align its country programs to the priorities identified within the National Poverty Reduction Strategies Papers (PRSP) and other national planning processes put in place by developing countries. In applying this model, pre-determined investment targets for specific sectors or sub-sectors are inappropriate in CIDA's view, in order to permit the fullest expression and implementation of the Paris principles.

In 2006, Canada committed to doubling international assistance by 2010-2011 from the 2001-2002 level, and has subsequently increased aid budget allocations by 8% annually. Though, no sectoral commitments are available about water supply and sanitation specifically.

GEOGRAPHICAL AND SUB-SECTORAL FOCUS

In the last 20 years, CIDA has focused aid on 30 core countries and regions. Despite this focus, Canadian aid for water can be characterized as being widely dispersed with at least some bilateral programming in approximately 80 countries. In 2003, CIDA undertook to concentrate aid in fewer countries. Among them, CIDA selected a limited number of the world's poorest countries for an enhanced partnership relationship. The criteria used to select such a list of countries include a high level of poverty and a commitment to development effectiveness, as demonstrated through efforts to improve governance, ensure local ownership of PRSPs, end corruption and make effective use of aid funds. Special consideration is also given to countries with the potential to exercise regional leadership.

AID MODALITIES

In CIDA's practices, the most important instrument is the PRSP. Often working in support of PRSPs are Sector-Wide Approaches (SWAp). CIDA maintains programming in a large number of countries which are now involved in the PRSP process and increasingly aligns its programming orientations along the priorities identified by the governments in their PRSPs. The agency continues to use different sorts of program-based approaches (PBAs), including several SWAps (mostly in Africa), and numerous non-sectoral program-based initiatives. In addition, the growing use of coordinated approaches, such as joint assistance strategies with other donors in Ghana, Tanzania, Kenya and Zambia, enables to strengthen donor harmonization.

RULES AND PRACTICES

Gender equity and women's empowerment are included among Canada's significant objectives in all engaged investments. Water projects often include equality training and a requirement for a fair representation of women in water committees and sub-committees. Women's groups are particularly taken into consideration by CIDA's programmes, especially in Africa.

PERFORMANCE EVALUATION

CIDA applies a robust results-based approach. All programs and projects have their own Logical Framework Analysis (LFA) with relevant indicators that are used for monitoring, as well as mid-term and end-of-project evaluations.

COORDINATION WITH OTHER ACTORS

In 2006, Canada has raised its share contribution to the Global Environment Facility (GEF) replenishment from 4.28% to 6.39%, and agreed to provide USD 158.9 million over the period 2007-2010. CIDA also supports regional institutions, such as the Nile Basin Initiative. In addition, CIDA works in partnership with new donors and former CIDA recipients which bring additional knowledge, expertise, financial and technical resources to development programming.

Canada promotes recognition of the role that non-state partners can play, especially for aid effectiveness, encourages a more collaborative relationship between state and non-state partners, and advocates better usage of non-state partners in the selection of aid delivery mechanisms and channels;

Web References:

CIDA official website http://www.acdi-cida.gc.ca

DENMARK
Aid at a glance - Water supply and sanitation

Summary statistics

Unless otherwise stated, figures refer to 2005-2006 annual average commitments expressed in 2006 constant prices.

Total aid to water supply and sanitation

	USD million	Aid to water by Denmark as a share of total aid by Denmark	Aid to water by Denmark as a share of total DAC members' aid to water
Denmark	**123.3**	**11%**	**2%**
For reference, total DAC	*4974.0*	*8%*	*100%*

Top recipients of aid to water supply and sanitation

		Aid to water by Denmark to that recipient as a share of	
	USD million	total aid by Denmark to that recipient*	total DAC members' aid to water to that recipient
Viet Nam	36.4	46%	33%
Bangladesh	30.1	29%	18%
Zambia	18.5	58%	49%
Sri Lanka	14.8	88%	23%
Niger	9.9	87%	32%
Mali	5.0	79%	13%
South Africa	0.7	2%	2%
Bhutan	0.3	7%	53%
Nepal	0.2	1%	1%

Aid to all water-related sectors

	USD million
Water supply and sanitation	123.3
Water transport	0.4
Hydro-electric power plants	0.0
Agricultural water resources	0.0
Total water-related aid	*123.7*

Annual aid to water supply and sanitation, 1993-2006 (2006 constant prices)

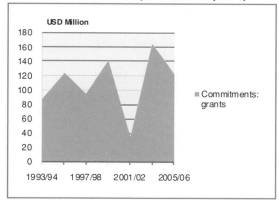

Regional distribution of aid to water supply and sanitation

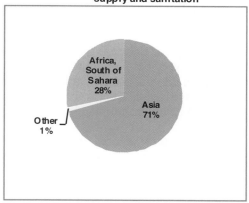

Water supply and sanitation aid by subsector

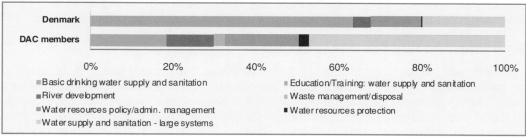

Denmark
DAC members

- Basic drinking water supply and sanitation
- Education/Training: water supply and sanitation
- River development
- Waste management/disposal
- Water resources policy/admin. management
- Water resources protection
- Water supply and sanitation - large systems

* % of sector allocable aid

DENMARK
Development Co-operation Policy in the water supply and sanitation sector

POLICY STATEMENT

The water sector is expected to remain a key sector for Denmark's development assistance, but the number of focus countries may be further reduced, and direct Danish interventions may also be reduced as a result of the introduction of Joint Assistance Strategies and the principle of Lead Donor. Denmark subscribes to the harmonization and alignment agenda, and this will impact on how the support will be managed in the future.

Denmark will no longer develop its own global strategy papers, as this would be in conflict with the alignment and harmonization process: instead, Denmark will support the development of national sector strategies, but will continue to produce specific Best practices Papers on issues relevant for the sector. Danish Sector Policy Guidelines for Water Supply, Sanitation and Integrated Water Resources Management (IWRM) have been therefore replaced by overall Aid Management Guidelines, which give the direction for all Danish bilateral development assistance. Responding to the changing international aid environment (Monterrey Conference, Paris Declaration, Johannesburg World Summit on Sustainable Development), the Danish Ministry of Foreign Affairs issued the latest in 2005.

Water supply and sanitation is part of a global climate change policy led by Danish government. Denmark takes a holistic approach to the water sector, meaning that support to water sector will include water supply and sanitation as well as IWRM in all program countries with water sector support programs.

The core of Danish development policy is long-term and binding co-operation with the developing countries, in the framework of their national Poverty Reduction Strategies Papers (PRSP). The Sector Programme Support (SPS) concept and the Sector Wide Approach to planning (SWAp) principles are widely used within the sector. Danida drew up a second generation of SPS, aimed at facilitating harmonisation with other parties and simplifying management and monitoring arrangements. Through this, Denmark shows its will to implement the UN Joint Monitoring Framework.

Support to IWRM has been on the increase over the last decade, partly due to additional funding being available under the special environmental support programs originally provided outside the ODA regime, but now being an integral part of Danish development assistance. In 2005, the Danish government defined IWRM as its main target. Efforts have been focused on policy and legislative framework, water actions plans, water and environment, and regional conflict prevention.

Water governance, both at local and central level, is a key issue in Danish assistance to the water sector. Danida substantially contributes to the development of planning and implementation capacities at national and sub-national levels, e.g. through supporting decentralisation and deconcentration processes. Danida is leading capacity-building programmes, including institutional strengthening, organisational development and human resources development.

Danida fully agrees that there is a strong need to pay more attention to sanitation issues. The international sanitation year (2008) poses an opportunity to make a serious push forward in this respect.

GEOGRAPHICAL AND SUB-SECTORAL FOCUS

Danish Government has committed to consolidate Africa as the primary focus of its development policy priorities, because this is where challenges are the greatest. Denmark has recently included 3 new water sector programmes to its global bilateral portfolio in the water sector: in 2005 for Zambia and in 2006 for Mali and Niger. Consequently, Denmark provides at present sector programme support to the water sector in the following countries: Bangladesh, Benin, Bhutan, Burkina Faso, Ghana, Kenya, Mali, Niger, Sri

Lanka, Uganda, Vietnam, Zambia. Danida is active in some of the poorest countries of the world and in the poorer regions in these countries, with a particular focus on rural areas.

AID MODALITIES

Danish development co-operation is mobilizing a very wide range of financial flows such as sector budget support, multi-donor financing, basket funding, pooling of technical assistance and other forms of multi-partner co-operation. The water programmes are most often linked to national development plans, PRSPs and medium-term expenditure frameworks (MTEF).

Denmark is moving from second generation bilateral Sector Program Support (SPS) to new aid modalities based on the Rome and Paris declaration on harmonization and alignment. New programs are being prepared with closer integration into national systems and in closer co-operation with development partners. Programs will be closer linked to national sector strategies, and implementation will be by national institutions. Financial flows will be through earmarked sector budget support (if possible through common basket arrangements). In some countries funding will be channeled through discretionary funding through local government institutions (eventually without earmarking to sector specific activities).

RULES AND PRACTICES

Danida interventions in the water sector are mainly based upon the principle of water and sanitation for all. But some individual Danida interventions can give special attention to the poor, mainly through subsidies to the poorest or by charging low fees for basic service levels. To improve water governance, Danida pays special attention to democratic, participatory and transparent decision-making processes, involving women, children, community-based organisations and civil society. Gender empowerment is a crucial issue for Denmark. Equitable access to water resources, water services and improved sanitation, as well as equitable participation in the governance of the water sector, are areas where women benefit directly from water sector development. At the local level, gender mainstreaming approaches are implemented within the community structures as well as within the private sector.

Water and sanitation programmes normally include hygiene promotion and information about HIV/AIDS, notably preventive measures. Access to safe water and sanitation is an important part of health care and the fight against HIV/AIDS.

Assistance to water sector activities are planned and implemented with a view to securing that interventions are managed at the lowest appropriate levels, and a demand-responsive approach is applied in the design of programmes. Consumers' contributions to the investment and contributions to operation and maintenance costs are key principles in the planning and implementation of activities.

PERFORMANCE EVALUATION

It is estimated that between 1999 and 2005 some 5.7 million people were reached with improved water supply services through the support of Danida, whereas some 3.8 million people were reached with improved sanitation facilities (Source: Evaluation of Danida's support by Danish Ministry of Foreign Affairs, 1999-2005). Compared to Danish disbursements during the same period, this means that about USD 21 were necessary to improve access to water or sanitation for one person.

COORDINATION WITH OTHER ACTORS

Denmark is actively involved, partly through financial assistance, in a number of global and regional water and sanitation initiatives and programmes, including: Water and Sanitation Programme (WSP – USD 5.8 million in 2002-2004), Global Water Partnership (GWP), UNEP Water Initiative (UCC-Water), World Water Assessment Programme (WWAP), African Water Facility, Rural Water Supply and Sanitation Initiative, Centre for Regional Water Supply and Sanitation in West Africa (CREPA). Denmark supports with substantial amounts various transboundary river basin programmes, such as the Nile Basin Initiative, the Regional Water Resources Management Initiative in Southern Africa, and the Mekong River Commission, to which Denmark has been the largest single donor since the early 1990s.

Through its membership in the GWP, Denmark intends to ensure that poverty orientation and cross-cutting issues such as environment and gender are given well-defined and prominent roles. Denmark hopes thereby to foster political will, knowledge transfer, and policy dialogue regarding IWRM.

The government has emphasized new multi-partner co-operation in the concrete implementation of the development interventions, including with civil society, the business community and through the civilian-military co-operation.

Danish NGOs are key partners, especially in Bangladesh and Uganda. The government will work in close collaboration with them and other stakeholders to revise its Civil Society Strategy: emphasis will be placed on strengthening NGO initiatives in Africa. User organisations are also supported vigorously, although the support is focused on the construction period more than on maintenance and post-intervention sustainability.

Danida support to the private sector is more limited. Danish assistance seeks to empower the local and emerging private sector involved in maintenance, operation and installation of equipment. The supported private sector includes informal service providers, who are often dominant in peri-urban areas, and must be included as stakeholders.

Finally, Danish Water Forum is a network for water and development consisting of almost all the key players in the Danish water sector, i.e. Danish ministries, authorities, companies, universities and NGOs.

Web References:

Evaluation of Danish support to Water Supply and Sanitation 1999-2005
http://www.um.dk/NR/rdonlyres/05AF05E6-D08C-467C-ACE2-1AAB27889CF0/0/20075WSS_EnglishSummary.pdf

Harmonisation and Alignment in Water Sector Programmes and Initiatives 2006
http://www.odi.org.uk/wpp/resources/project-reports/06-harmonisation-alignment.pdf

Denmark Development Co-operation in Water and Sanitation 2004
http://www.danidadevforum.um.dk/NR/rdonlyres/16107513-4921-4CB3-A352-E34462C35E1B/0/DenmarksDevelopmentCo-operationinWaterandSanitation.pdf

EUROPEAN COMMISSION
Aid at a glance - Water supply and sanitation

Summary statistics

Unless otherwise stated, figures refer to 2005-2006 annual average commitments expressed in 2006 constant prices.

Total aid to water supply and sanitation			
	USD million	Aid to water by EC as a share of total aid by EC	Aid to water by EC as a share of total DAC members' aid to water
EC	**729.7**	**8%**	**15%**
For reference, total DAC	*4974.0*	*8%*	*100%*

Top ten recipients of aid to water supply and sanitation			
		Aid to water by EC to that recipient as a share of	
	USD million	total aid by EC to that recipient*	total DAC members' aid to water to that recipient
Egypt	51.2	34%	62%
India	46.8	37%	10%
Mozambique	40.5	33%	58%
Ethiopia	31.8	14%	63%
Morocco	26.7	14%	19%
Tanzania	25.6	22%	43%
Lesotho	22.9	65%	88%
Chad	19.2	19%	70%
Turkey	18.4	4%	99%
Croatia	18.0	12%	96%

Aid to all water-related sectors	
	USD million
Water supply and sanitation	729.7
Water transport	18.6
Hydro-electric power plants	57.5
Agricultural water resources	9.7
Total water-related aid	*815.6*

Annual aid to water supply and sanitation, 1993-2006 (2006 constant prices)

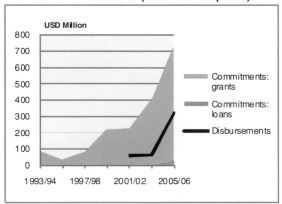

Regional distribution of aid to water supply and sanitation

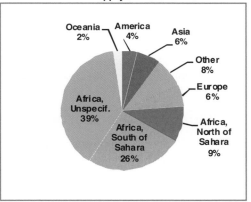

Water supply and sanitation aid by subsector

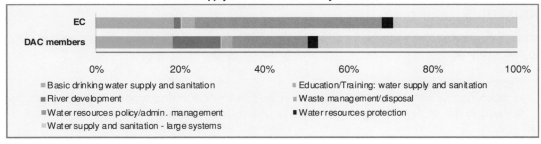

* % of sector allocable aid

EUROPEAN COMMISSION
Development Co-operation Policy in the water supply and sanitation sector

POLICY STATEMENT

The EU's development policy promotes a framework for integrated water resources management, based on three priorities:

- Provide access to safe drinking water and adequate sanitation to all people to reduce poverty, improve public health and increase livelihood opportunities.
- Establish and strengthen organizations and infrastructure for the sustainable and equitable management of transboundary rivers, lakes and groundwater.
- Coordinate the equitable, sustainable and appropriate distribution of water between various users.

The European Commission responds with an integrated framework combining political processes and complementary financial instruments to strengthen the capacity of developing countries to meet the global water and sanitation crisis. These instruments are:

- *The European Water Initiative:* this is an area of political dialogue and consultation among governments, decentralised government structures, civil society and the private sector in partner countries, the Member States of the European Union and the European Commission. The common aim is to achieve the Millennium Goals.
- *The Support for National and Regional Programmes*: These programmes, which cover all development aid sectors, are in line with the co-operation between the European Union and third countries and regions. In 2006, the support accounted for 271 million Euros (excluding emergency aid). 60 countries will receive a total of almost 700 million Euros (excluding emergency aid) for the period 2008-2013.
- *The ACP-EU Water Facility*: the *facility* financially supports the *initiative* through positive actions. The *facility* has financed projects in countries in Africa, the Caribbean and the Pacific (ACP) with a total of 500 million Euros between 2004 and 2007.
- *The EU-Africa Partnership for Infrastructures*: this programme was created in 2006 to support the development of grand trans-African networks. The management of cross-border catchment areas, flood defense programmes, knowledge acquisition and water resource monitoring with a view to better management and sustainable regional infrastructures are the particular focus of this partnership for the water sector.

GEOGRAPHICAL AND SUB-SECTORAL FOCUS

The European Commission covers 60 developing countries in the water sector, with an important focus on ACP countries (32 countries in Africa, 11 in Asia, 5 in Latin America, 5 in the Pacific, 4 in the Neighbourhood and 3 in the Caribbean. Indicatively, the sub-sector water supply, sanitation and integrated water resources management (IWRM) respectively represent 43 %, 32 % and 19 % of European Commission aid, the remaining 6 % targeting emergency actions.

AID MODALITIES

Commission support in the sector is based on three financing modalities, i.e. sector budget support, pooled funding and projects.

- *Sector budget support*
  In accordance with the Paris declaration and in the countries where it is relevant, sector budget support is the preferred financing modality of the Commission.

 CRS AID ACTIVITIES IN SUPPORT OF WATER SUPPLY AND SANITATION – ISBN 978-92-64-05172-0 – © OECD/WWC 2008

- *Pooled funding*
  The different types of pooled funding will depend on the choice of institution to manage it and whether the accounting and reporting procedures are modelled on those of the government (decentralised management), of an international organization (joint management), of an international or national public sector body (indirect decentralised management) or of the EC.

- *Projects (including calls for proposals)*
  This modality is principally used in all countries where use of sector budget support and pooled funding is not possible, but also for contracts for major works, identification and formulation studies, technical assistance and pilot activities.

RULES AND PRACTICES

The European Commission development policy is formulated by the DG Development, the implementation of centralised co-operation and the sector support is ensured by DG Europeaid. On the field, Delegations are present in almost all the countries and deal with national and regional co-operation.

COORDINATION WITH OTHER ACTORS

The European Commission and Member States are working together to increase the effectiveness of their development assistance and are cooperating with other donors and stakeholders. Depending on the countries benefiting from EC aid, the Delegation can play the role of leader of the donors in the water sector.

The European Commission also belongs to the coordination committee of the EU water initiative.

Web References:

European Commission Directorate General for Development:
http://ec.europa.eu/development/index_en.cfm

European Commission Co-operation office Europeaid: http://ec.europa.eu/europeaid/index_en.htm

European Union Water Initiative: http://www.euwi.net/index.php?main=1

FINLAND
Aid at a glance - Water supply and sanitation

Summary statistics

Unless otherwise stated, figures refer to 2005-2006 annual average commitments expressed in 2006 constant prices.

Total aid to water supply and sanitation			
	USD million	Aid to water by Finland as a share of total aid by Finland	Aid to water by Finland as a share of total DAC members' aid to water
Finland	**44.2**	**10%**	**1%**
For reference, total DAC	4974.0	8%	100%

Top ten recipients of aid to water supply and sanitation			
		Aid to water by Finland to that recipient as a share of	
	USD million	total aid by Finland to that recipient*	total DAC members' aid to water to that recipient
Viet Nam	9.1	26%	8%
Nepal	8.6	65%	47%
Ethiopia	6.1	42%	12%
Sri Lanka	4.6	26%	7%
Palestinian Adm. Areas	3.1	66%	4%
Philippines	2.5	78%	18%
China	1.1	7%	0%
Kenya	0.5	2%	1%
Georgia	0.4	34%	3%
Egypt	0.3	30%	0%

Aid to all water-related sectors	
	USD million
Water supply and sanitation	44.2
Water transport	0.0
Hydro-electric power plants	0.7
Agricultural water resources	0.0
Total water-related aid	45.0

Annual aid to water supply and sanitation, 1993-2006 (2006 constant prices)

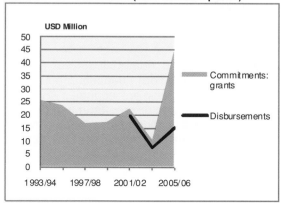

Regional distribution of aid to water supply and sanitation

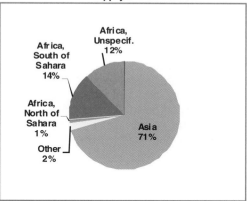

Water supply and sanitation aid by subsector

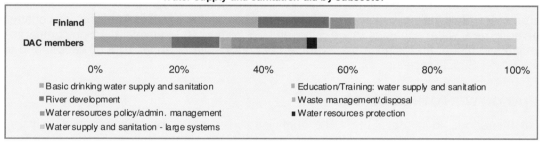

* % of sector allocable aid

FINLAND
Development Co-operation Policy in the water supply and sanitation sector

POLICY STATEMENT

Development co-operation is planned jointly with the partner country on the basis of its development plans and ownership. Finnish development co-operation builds on the partner countries' own poverty reduction strategies. The content of development co-operation is decided jointly with partner countries every 2 to 3 years in bilateral consultations. As part of programme-based co-operation, Finland uses budget support as one instrument in countries where this is feasible. Improved public sector management, transparency and public monitoring are both preconditions for and consequences of programme co-operation. Whenever the level of governance permits, Finland uses the partner countries' own administrative systems and helps to strengthen the management of public-sector finances.

The new development policy (2008-2012) does not provide clear quantitative objective for the water and sanitation sector.

GEOGRAPHICAL AND SUB-SECTORAL FOCUS

Finland currently supports water supply and sanitation projects in 8 countries. Of the total of some 30 projects, 60% are rural water supply and sanitation (RWSS) interventions, the rest are urban water projects. In the urban water supply projects, the progress made in developing the capacity of the respective water authorities is a key component of Finnish development assistance.

AID MODALITIES

The Finnish government stresses the importance of traditional official development co-operation, but at the same time contributes actively to the international debate on the new mechanisms (innovative financing mechanisms). These should complement official development co-operation, and funding from innovative mechanisms should be channelled primarily using existing aid delivery channels.

RULES AND PRACTICES

In addition to the water supply and sanitation project funding and related technical assistance (TA) activities, DIDC (Department for International Development Co-operation) finances direct TA activities to support sector policy and respective strategy development in the recipient countries. Human resource development has also been an important activity in the DIDC water sector programme in the form of post-graduate courses for African engineering students. Finally, the participation of local people in dialogue and decision-making concerning their own living conditions is an essential objective in the DIDC projects.

PERFORMANCE EVALUATION

Rough estimates of population covered show that, between 1968 and 2007, in the order of some 6 million people have received new safe and reliable water service or are now being served by a rehabilitated system that provides an improved quality of service. Achievements in sanitation are less significant due to lesser targets set, thus falling behind the achievements in water service. The cost of construction per capita served in rural water schemes range between US$ 20 and 40 (equivalent) which compares well with any international criteria, for instance the average cost per capita of about US$ 53 reported in the World Bank Rural Water Supply projects.

COORDINATION WITH OTHER ACTORS

As part of the efforts to improve the UN's administrative system for environmental matters, Finland considers it important to strengthen and intensify the work of the United Nations Environment Programme, with the aim of transforming the UNEP into the United Nations Environmental Organisation (UNEO).

Finland is also in favour of increasing ways for the poorest developing countries to be able to exert an influence on financing institutions' decision-making. A global development partnership should be created involving the public and private sectors in both developing and developed countries.

Since 1994, Finland has supported the Global Environmental Facility (GEF) with approximately 70 million dollars. Finland also supports the Nile Basin Initiative (NBI – 2 million Euros between 2004 and 2006) and the Mekong River Commission.

A new government policy on non-governmental organisations (NGO Development Co-operation Guidelines) was approved in 2006. NGOs' work complements official development co-operation on a bilateral, multilateral and EU basis. In development co-operation with NGOs, the Government strives to boost the effectiveness of operations and the NGOs' general capacity, notably by providing training.

The government simplified the procurement rules in order to facilitate co-operation between Finnish and partner countries' institutions; one particular aim is to strengthen co-operation between universities and research institutes.

Finnfund (Finnish Fund for Industrial Co-operation Ltd.) is a development finance company that provides long-term risk capital for private projects in developing countries, including in the water and sanitation sector.

Web References

Development Policy Programme 2007 http://formin.finland.fi/public/default.aspx?contentid=107497

FRANCE
Aid at a glance - Water supply and sanitation

Summary statistics

Unless otherwise stated, figures refer to 2005-2006 annual average commitments expressed in 2006 constant prices.

Total aid to water supply and sanitation			
	USD million	Aid to water by France as a share of total aid by France	Aid to water by France as a share of total DAC members' aid to water
France	**185.7**	**4%**	**4%**
For reference, total DAC	*4974.0*	*8%*	*100%*

Top ten recipients of aid to water supply and sanitation			
		Aid to water by France to that recipient as a share of	
	USD million	total aid by France to that recipient*	total DAC members' aid to water to that recipient
Morocco	39.1	10%	27%
Kenya	29.5	31%	43%
South Africa	25.1	21%	67%
Lebanon	19.3	28%	38%
Chad	8.0	30%	29%
Palestinian Adm. Areas	7.5	35%	9%
Cambodia	7.0	26%	51%
Niger	6.9	11%	22%
Benin	6.4	10%	10%
Sri Lanka	6.4	76%	10%

Aid to all water-related sectors	
	USD million
Water supply and sanitation	185.7
Water transport	8.6
Hydro-electric power plants	10.9
Agricultural water resources	17.9
Total water-related aid	*223.0*

Annual aid to water supply and sanitation, 1993-2006 (2006 constant prices)

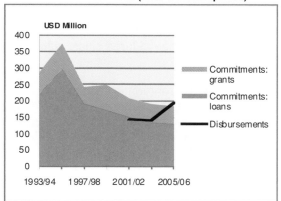

Regional distribution of aid to water supply and sanitation

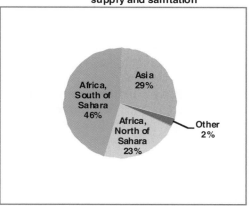

Water supply and sanitation aid by subsector

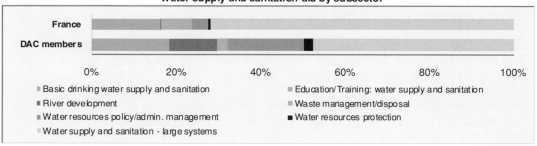

- Basic drinking water supply and sanitation
- River development
- Water resources policy/admin. management
- Water supply and sanitation - large systems
- Education/Training: water supply and sanitation
- Waste management/disposal
- Water resources protection

* % of sector allocable aid

FRANCE
Development Co-operation Policy in the water supply and sanitation sector

POLICY STATEMENT

A reform of the French co-operation framework occurred in 2004. Since then, the Inter-ministerial Committee for International Co-operation and Development (in French CICID) defines the national co-operation strategy, co-ordinates and evaluates the French public development aid policy and international development co-operation. Chaired by the Prime Minister, the CICID brings together representatives of the ministries most directly concerned with development issues, among others the Ministry of Foreign Affairs (MFA) and the Ministry of Economy (including Finance and Industry).

Under the MFA, the State Secretary for Co-operation and its **Directorate General for International Co-operation and Development (DGCID)** have the leading role in co-ordinating the implementation of the strategies by the different actors, whether public or private, central or local, including local governments, water syndicates, water agencies, NGOs, research institutes, etc. The French development agency (**Agence Française de Développement, AFD**) has been mandated by the CICID to be the key operator of the French bilateral co-operation and is therefore in charge of all ranges of operation, from institutional support to implementation of projects and programmes in beneficiary countries.

"**Water and Sanitation**" is one of the 7 priority intervention sectors of French ODA defined in 2004 to support the achievement of the Millennium Development Goals (MDGs). In 2005, the CICID validated a **Long-term Framework Strategy for Water Supply and Sanitation** which still is the reference document for any intervention. (The French definition of the water sector includes water supply, sanitation, IWRM, irrigation and pastoral hydraulics.). It is based on strong principles such as implementing participatory approaches, developing adapted technologies, and training partners' human resources. The Strategy sets quantitative objectives (see below), prioritises additional aid to Africa and the following themes:

- **Sanitation & Hygiene**, because positive impacts on health, environment and even economy are not effective with efforts on access to drinking water only;

- **access to basic services (both sanitation and water supply) for the poorest people,** particularly in the rural areas, small towns and low income areas in large cities, without neglecting rehabilitation and extension projects of existing networks;

- **water resources management** in order to meet large environmental, health and food security challenges; this includes improving the knowledge on water resources and building decision making tools, protecting water quality - notably through domestic and industrial wastewater collection and treatment – and preserving water resources in quantity, through water savings in irrigation and urban water distribution networks.

France's overall strategy is complemented by **Strategic papers (Documents d'orientation stratégique, DOS)**, elaborated by multi-stakeholders working groups. They lay down the principles of French co-operation and define strategic framework and guidance for the operational interventions. Four strategic papers deal with water: **Trans-boundary Integrated Water Resources Management; Efficient Water for Agriculture; Sanitation; Professional Training** in the water sector.

AFD's water strategy or Sectoral intervention framework validated in 2007 is aligned on the Long-term Framework Strategy for Water Supply and Sanitation. With the overall objective of water resources sustainability, it is based on an integrated approach as follows:

- **Integrated Water Resources Management (IWRM)** is considered as the framework through which equilibrium of the large water cycle can be reached. Efforts are concentrated on co-operative management and knowledge improvement in (trans-boundary) river basins.

- **Ensuring access to water and sanitation for all, and quality of the service**. AFD focuses on the very poor and its subsidies are allocated in priority to access to sanitation (hygiene education, sanitary equipments, collection systems). AFD supports amongst all reduction of commercial losses, strengthening of exploitation and maintenance and implementation of affordable / economical tariffs.

- **Good governance**. AFD provides assistance to partner countries in formulating their Poverty Reduction Strategic Plans (PRSP) and strives to channel financing towards poor and under-equipped areas. It supports the development of programming tools that help implementing national policies, and values programme approaches and budget support. To enhance management efficiency, investments are consistently and systematically associated with an improvement of the institutional and management frameworks.

- **Water for food**, to meet the challenge of feeding the world population and reach a greater efficiency of water use for agriculture. AFD supports policies aiming at food security through agriculture value chains rather than policies of food self-sufficiency.

In the 2005 Framework Strategy, **France committed to doubling its ODA to the water sector** by 2009, reaching an additional aid of 180 million Euros annually with reference to 2001-2003 annual ODA average (2008 USD 278.6 million), and to channelling a significant part of this increase to Africa. This amount totals an average of 75 million Euros for increasing multilateral contributions and of 105 million Euros for increasing bilateral commitments. The doubling of French ODA for water will be focused on: access to sanitation; targeting of the poorest populations, especially those without access to water supply and sanitation, the objective being to give access to water supply and sanitation to 9 million Africans by 2015; water resources management, particularly water treatment and water savings in irrigation.

France also committed to increasing ODA for water channelled via NGOs, for instance through calls for proposal, and to developing financial mechanisms facilitating additional ODA flows, for instance potential resources from local governments and water agencies. **AFD** also committed to a **strong contribution to the French ODA** doubling by increasing its assistance from an average of 145 million Euros for 2001-2005 up to 290 million Euros in 2009 (2008 USD 448.6 million), half of which being allocated to Africa – based on the critical assumption of the continuing increase of the total AFD budget and a constant share of the water sector against other sectors. With the development of new financial instruments, the 2009 targets are as follows: 100 million Euros of grants, 140 million Euros of sovereign loans, 50 million Euros of non-sovereign loans. AFD also announced additional objectives: support of investments in the water for food sub-sector and improvement of aid efficiency in terms of number of recipients. In particular, AFD wishes to provide wastewater treatment to at least as much people as the number to which it provides access to freshwater by 2009.

In 2007, AFD exceeded its objectives with 404 million Euros of new commitments to the water sector in developing countries (2008 USD 621 million). Regarding the WSS, half is allocated to water supply and half to sanitation. According to AFD's estimates, these amounts should give access to freshwater to 3 million people and sanitation to 1 million people. With 180 million Euros of new commitments in 2007 (57% of the total), Africa has been the main recipient of this effort.

GEOGRAPHICAL AND SUB-SECTORAL FOCUS

The French government has set a Priority Solidarity Zone (PSZ) with most countries located in sub-Saharan Africa, others in Near East, North Africa, South-East Asia and Caribbean. Efforts in trans-boundary IWRM are focused on priority river basins, especially Niger, Senegal, Nile, Mekong, Volta and Congo.

AID MODALITIES

Aid grants can be allocated through several mechanisms from different Ministries. One of them is the Priority Solidarity Fund, the purpose of which being to finance the French support for institutional, social, cultural and research development to PSZ countries. Another one is the co-funding of IWRM projects (with the French Fund for World Environment - FFEM), water and sanitation projects (decentralised co-operation - MFA, AFD), etc.

AFD offers a wide range of financial tools to meet its partners' needs: loans to states for large infrastructures, loans to public or private operators with subsidy according to the social activities, support to local authorities, underwritings to raise local savings, grants to states to support sector programmes, grants for NGOs' programmes, co-financing of decentralised co-operation programmes, etc... About 90% of AFD loans are sovereign, and the remaining ones are allocated to both public and private sectors.

RULES AND PRACTICES

French ODA to the water sector is now totally untied. It does not promote a unique water management model but focuses on optimizing service quality for all users, enhancing participative and demand-based approaches, promoting support public as well as private operators depending on the context and public-private partnerships, strengthening capacity building of all levels of beneficiaries, from States to local operators, with special attention to professional training.

If the state is responsible for sectoral policy formulation, economic efficiency and respect of public rules, water resources and water supply management are more and more devolved upon local authorities. That is why France promotes decentralised co-operation and partnerships with local actors. AFD intends to increase its support through direct financing (sub-sovereign loans) and to facilitate access to financing for local actors (both public and private), notably by mobilising local savings, providing more loans in local currencies, and lengthening their duration.

In 2005 the Oudin-Santini law made it possible for local authorities and the water agencies to apportion up to 1% of their water and sanitation budget specifically to water and sanitation projects in developing countries. This law is a major tool for decentralised co-operation and enables local authorities to support a part of the effort to improve access to water and sanitation. Out of an estimated potential of 100 million Euros per year, more than 10% per year is now being allocated this way.

PERFORMANCE EVALUATION

AFD estimates that, between 2002 and 2006: more than 20 million people benefited from its intervention in the water and sanitation sector; 3.5 million people gained access to water and 12.5 million benefited from an improvement of the water supply; 1 million people gained access to sanitation and 3.5 million benefited from an improvement of the sanitation system. AFD's projects enabled to save 80 million m^3 of water per year thanks to technical improvements in the supply networks, and to treat sewages corresponding to the water consumption of 1.6 million people. The annual objectives until 2009 are to give access to water to 1.6 million people, to improve water supply for 5.6 million people, to give access to sanitation to 600 000 people, and to improve sanitation systems for 2 million people.

COORDINATION WITH OTHER ACTORS

France provides an increased support to various multilateral funds and institutions involved in the water sector. In addition to its contribution to the Water and Sanitation Program, France is particularly active on the European stage: it supports the EU Water Initiative and the ACP-EU Facility implemented by the European Commission (DGCID or AFD co-finance several projects). France provides a significant share of the EU-ACP Water Facility (121.5 million Euros over 3 years), and of a number of multilateral organisations' budgets (25% for the European Development Fund, 6% for IDA). In collaboration with the New Partnership for Africa's Development (NEPAD), France supports the Rural Water Supply and Sanitation Initiative of the African Development Bank. One fifth of its contribution to the African Development Fund will be allocated to this Initiative (25 million Euros annually over 3 years). DGCID also committed 12 million Euros over 5 years to the African Water Facility.

 CRS AID ACTIVITIES IN SUPPORT OF WATER SUPPLY AND SANITATION – ISBN 978-92-64-05172-0 – © OECD/WWC 2008

Web References:

Ministère des Affaires Etrangères (MAE) : Stratégie sectorielle : eau et assainissement - mai 2005
http://www.diplomatie.gouv.fr/fr/actions-france_830/aide-au-developpement_1060/politique-francaise_3024/strategies-gouvernementales_5156/strategies-sectorielles-cicid_4570/strategie-sectorielle-eau-assainissement-mai-2005_11834.html

Agence Française de Développement (AFD) : Cadre d'intervention sectoriel Eau 2007-2009
http://www.afd.fr/jahia/webdav/site/myjahiasite/users/administrateur/public/Portail%20Eau%20et%20Assainissement/pdf/CIS-eau-2007-2009.pdf

GERMANY
Aid at a glance - Water supply and sanitation

Summary statistics

Unless otherwise stated, figures refer to 2005-2006 annual average commitments expressed in 2006 constant prices.

Total aid to water supply and sanitation			
	USD million	Aid to water by Germany as a share of total aid by Germany	Aid to water by Germany as a share of total DAC members' aid to water
Germany	**452.7**	**9%**	**9%**
For reference, total DAC	*4974.0*	*8%*	*100%*

Top ten recipients of aid to water supply and sanitation			
		Aid to water by Germany to that recipient as a share of	
	USD million	total aid by Germany to that recipient*	total DAC members' aid to water to that recipient
Morocco	48.9	40%	34%
Jordan	43.5	73%	40%
Yemen	29.6	45%	63%
Congo, Dem. Rep.	18.9	29%	55%
Burundi	18.2	93%	93%
Tunisia	18.0	47%	29%
Azerbaijan	16.2	53%	71%
Peru	15.0	28%	69%
Serbia	14.9	12%	45%
Kenya	14.6	26%	21%

Aid to all water-related sectors	
	USD million
Water supply and sanitation	452.7
Water transport	6.4
Hydro-electric power plants	42.4
Agricultural water resources	113.0
Total water-related aid	*614.4*

Annual aid to water supply and sanitation, 1993-2006 (2006 constant prices)

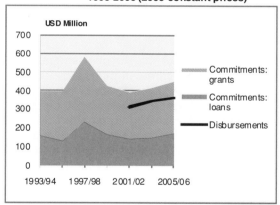

Regional distribution of aid to water supply and sanitation

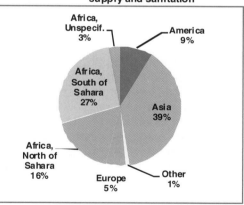

Water supply and sanitation aid by subsector

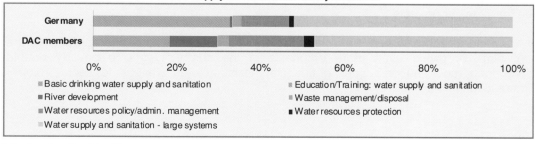

- Basic drinking water supply and sanitation
- River development
- Water resources policy/admin. management
- Water supply and sanitation - large systems
- Education/Training: water supply and sanitation
- Waste management/disposal
- Water resources protection

* % of sector allocable aid

GERMANY
Development Co-operation Policy in the water supply and sanitation sector

POLICY STATEMENT

The German development co-operation in the water sector is formulated in the Water Sector Strategy 2006. Its main guidelines are the Millennium Development Goals (MDG), the Millennium Declaration and the Programme of Action 2015, which sets out how the German government intends to make its contribution to achieving the MDG. In its Development policy action plan on human rights, Germany makes concrete commitments on how to promote economic, social and cultural rights. This also includes supporting the right to water, which spells out non-discriminatory access for all.

German development co-operation in water is based on the concept of Integrated Water Resources Management (IWRM). The aim is to help translate people's right of access to drinking water and basic sanitation into reality. IWRM allows taking a flexible, process-oriented, holistic approach towards making optimum use of water resources and the eco-system. Its fields of action comprise:

- Sustainable management of water resources
- Giving water sector reforms a poverty dimension
- Efficient and sustainable water supply and sanitation
- Using water efficiently for food production
- Enhancing effectiveness through co-operation

One top priority of German development co-operation is to make hygienic water and sanitation services accessible as rapidly as possible to previously undersupplied population groups. Germany stresses the importance of ensuring that this supply is sustainable and can be maintained by the people locally. Another priority area of support is **transboundary water co-operation**. Support is provided in Africa for regional institutions that are concerned with the joint management of water resources by riparian states. Germany is also notably assisting the African Ministers' Council on Water (AMCOW) on transboundary co-operation. German development co-operation is also an important player in the **sanitation** field, including in the promotion of ecological sanitation concepts (*ecosan*).

To improve the absorption capacity and implementation capability of the national structures, Germany recommends combining projects' implementation with water sector reforms in many partner countries. By improving the political, legal and organisational framework, it enables to establish or develop transparent and efficient institutions and regulations in the water sector. German development co-operation also assists its partner countries in formulating their sector policies and promotes decentralisation, modernization and entrepreneurial management as well as private sector involvement. German development programmes and projects are then embedded in national strategies with a long-term commitment. German development co-operation is active at all levels, macro-, meso- and micro level and is further characterised by a strong long-term presence of its advisers.

Germany is one of the world's three biggest bilateral donors in the water sector, providing an annual average volume of around 350 million Euros in development assistance each year. In 2006, commitments were made for 52 water projects.

Faced by the need to enhance further the effectiveness of its contributions and approaches the German government aims at giving even more emphasis in the coming years to poverty orientation and participation, donor coordination, coupling financial assistance as well as to sector reform and the mobilisation of local funds.

 CRS AID ACTIVITIES IN SUPPORT OF WATER SUPPLY AND SANITATION – ISBN 978-92-64-05172-0 – © OECD/WWC 2008

GEOGRAPHICAL AND SUB-SECTORAL FOCUS

Currently, Germany supports water programs in 70 countries worldwide, and has made water a priority sector in 28 of these countries. Funding concentrated particularly on sub-Saharan Africa and countries in the Near and Middle East.

Sectorally, most spending in the water sector is on water supply and sanitation/waste water management. Around 11 % of funds go to water resources policy, administration and protection. The promotion of water-related regional co-operation and the assistance to regional organisations have become increasingly important.

AID MODALITIES

In order to enhance the effectiveness of development co-operation and simplify and unify the tendering, appraisal and reporting procedures, Germany is intensively increasing its involvement in budget financing, joint programmes and programme-oriented joint financing with other donors (Sector wide Approaches, SWAPs).

RULES AND PRACTICES

German development policy is formulated by the Federal Ministry for Economic Co-operation and Development (BMZ). German bilateral development co-operation is carried out by implementing agencies. Those agencies work hand in hand on financial and technical co-operation.

COORDINATION WITH OTHER ACTORS

Germany is involved in various international and multilateral initiatives aimed at improving global access to water and sanitation, providing around 173 million Euros in core contributions. Apart from this, Germany is engaged in the European Water Initiative (EUWI), whose Africa Working Group was chaired by Germany in 2006; BMZ also supports a number of specific programs of international organisations such as the UN Water Decade Programme on Capacity Development (UNW-DPC), which receives support of 1 million Euros per year for a period of 3 years (2007-2009); the Joint Monitoring Programme (JMP) with a pledge of 400,000 Euros for 2007 and 2008; and the UN Secretary General's Advisory Board on Water and Sanitation (UNSGAB) where the German parliamentarian, Dr. Uschi Eid, is the acting Vice-President.

Other international initiatives aimed at addressing the challenges in the water sector include the support of the work of the World Commission on Dams and support to the co-operation between river basin commissions including support to the African Ministers' Council on Water (AMCOW).

Germany is a co-initiator and supports the Water Dialogues, a multi-stakeholder dialogue on water and the private sector, which provide a forum for a balanced, independent assessment of private sector participation in the water sector. The Global Water Partnership (GWP) also receives financial support from the German government for its work. Germany is also supporting the Water Integrity Network (WIN), which is attached to Transparency International in Berlin, with 200,000 Euros in 2007 and continues to support this network in 2008 in its work of fighting corruption in the water sector.

Finally, the BMZ supports development measures conducted by German NGOs. The participatory rights of NGOs have been strengthened, and they play an important role as competent partners in dialogue and in reminding governments to deliver on their promises.

Web References:

The Water Sector in German development co-operation, 2006:
http://www.bmz.de/en/service/infothek/fach/materialien/Materialie153.pdf
Water Sector Strategy 2006: http://www.bmz.de/en/service/infothek/fach/konzepte/konzept152

GREECE
Aid at a glance - Water supply and sanitation

Summary statistics

Unless otherwise stated, figures refer to 2005-2006 annual average commitments expressed in 2006 constant prices.

Total aid to water supply and sanitation			
	USD million	Aid to water by Greece as a share of total aid by Greece	Aid to water by Greece as a share of total DAC members' aid to water
Greece	**0.8**	**1%**	**0%**
For reference, total DAC	*4974.0*	*8%*	*100%*

Top ten recipients of aid to water supply and sanitation			
		Aid to water by Greece to that recipient as a share of	
	USD million	total aid by Greece to that recipient*	total DAC members' aid to water to that recipient
Albania	0.1	1%	1%
Palestinian Adm. Areas	0.1	6%	0%
Armenia	0.1	5%	13%
Syria	0.1	2%	0%
Iran	0.1	12%	3%
Moldova	0.05	2%	6%
Korea, Dem. Rep.	0.04	27%	69%
Ethiopia	0.03	3%	0%
Sudan	0.03	4%	0%
Serbia	0.03	0%	0%

Aid to all water-related sectors	
	USD million
Water supply and sanitation	0.8
Water transport	0.0
Hydro-electric power plants	0.0
Agricultural water resources	0.1
Total water-related aid	*0.9*

Annual aid to water supply and sanitation, 1993-2006 (2006 constant prices)

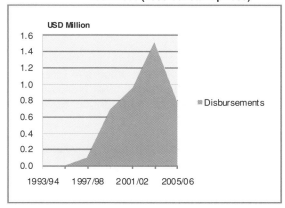

Regional distribution of aid to water supply and sanitation

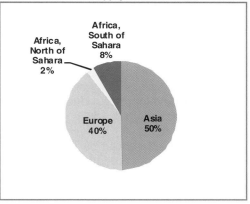

Water supply and sanitation aid by subsector

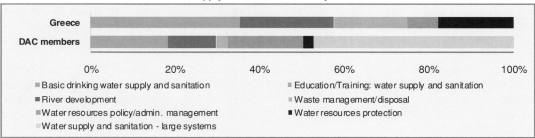

* % of sector allocable aid

GREECE
Development Co-operation Policy in the water supply and sanitation sector

POLICY STATEMENT

The development co-operation policy framework of the Hellenic Republic supports the undertaking of co-ordination process to achieve the MDGs. In particular, the development co-operation in water is schematically shaped by three intersecting circles: the MDGs, as set by the UN; the EU priorities, as defined by the Council of Development Co-operation Ministers; and the Greek foreign policy priorities. The common area between the three circles specifies the range of Greek activities.

The political dimension of water supply, sanitation and integrated water resources becomes highly important under the framework of sustainable development and international co-operation. Ensuring people's right of access to drinking water, wise governance of water supply, and efficiency of water supply and sanitation constitute priorities of the Greek Development Co-operation policy.

The Hellenic republic, as a donor country, encourages bilateral co-operation with developing countries, and is committed to align its country programmes to the priorities of the national Poverty Reduction Strategies Papers (PRSP) put in place by partner countries. The development co-operation is implemented both through public and private sectors as well as with the assistance of Greek NGOs having relevant expertise and experience.

GEOGAPHICAL AND SUB-SECTORAL FOCUS

During the last four years Greece has implemented projects in a number of countries in the Mediterranean (Egypt, Jordan, Syria, Palestinian Administrative Areas) and according to the new 5-year planning will increase funding to the water sector in general, especially in Sub-Saharan Africa and the Near and Middle East.

AID MODALITIES

Greek development aid in the water sector is provided in the form of grants.

NGO projects must go through Hellenic Aid call for proposals to be adopted and funded. The proposals received are first reviewed by Hellenic Aid, in collaboration with development officers in Greek embassies, and then assessed by a nine-member Committee for certification and evaluation of NGOs. Rehabilitation, development, and development education programmes/projects can be co-financed up to 50% of the costs, and in some cases up to 75%.

Priority sectors are basic social services (including water supply and sanitation), environment and rural development, income generation, small infrastructure and local business initiatives for combating unemployment. They also include cross-cutting issues such as human rights, gender equality and establishment of democratic institutions.

Once approved, projects are usually funded in three stages: 50% when the contract is signed, 30% when the project is half completed and 20% at the end. Disbursements are subject to an assessment of the project implementation. In order to ensure the public funds are efficiently spent, strengthened regulations have been adopted when co-financing NGO projects.

Over the last five years, Hellenic Aid has co-financed an increasing number of NGO projects through the call-for-proposals process, with a view to strengthen the role of civil society and promote systematic co-operation with it. To improve the call-for-proposals process in terms of transparency and efficiency, since 2005, an electronic form has been available on the web site of the Ministry of foreign Affairs (replacing the former Special guide).

CRS AID ACTIVITIES IN SUPPORT OF WATER SUPPLY AND SANITATION – ISBN 978-92-64-05172-0 – © OECD/WWC 2008

In order to ensure that the money is efficiently spent and meets international criteria, Hellenic aid has tightened the rules and set up extensive ex-ante assessment process covering the technical, management and financial capacity of the NGO, its knowledge of the partner country and its ability to work with local partners. Hellenic Aid assesses the validity of the projects submitted, for which tougher conditions are required. NGOs are required to provide: a document of agreement with a credible local NGO or a local authority; a document from an official authority of the recipient country referring to the need for implementing the proposed programme/project, and a document of agreement from the closest Greek diplomatic authority.

At the same time new contracts between Hellenic Aid and NGOs are more demanding.

RULES AND PRACTICES

Co-financed ODA projects for water must be in line with the Greek national development co-operation strategy. They must also take into account partner country development priorities, respond to basic needs, improve livelihoods and promote local community ownership.

COORDINATION WITH OTHER ACTORS

Greek ODA projects are implemented either by Hellenic Aid and other Greek ministries, or in co-ordination with Greek NGOs and Universities and the private sector.

IRELAND
Aid at a glance - Water supply and sanitation

Summary statistics

Unless otherwise stated, figures refer to 2005-2006 annual average commitments expressed in 2006 constant prices.

Total aid to water supply and sanitation			
	USD million	Aid to water by Ireland as a share of total aid by Ireland	Aid to water by Ireland as a share of total DAC members' aid to water
Ireland	**17.2**	**5%**	**0%**
For reference, total DAC	*4974.0*	*8%*	*100%*

Top ten recipients of aid to water supply and sanitation			
		Aid to water by Ireland to that recipient as a share of	
	USD million	total aid by Ireland to that recipient*	total DAC members' aid to water to that recipient
Zambia	4.4	16%	12%
South Africa	4.2	25%	11%
Lesotho	3.0	24%	12%
Ethiopia	0.9	3%	2%
Kenya	0.6	8%	1%
Mozambique	0.5	1%	1%
Afghanistan	0.4	21%	2%
Sudan	0.3	5%	2%
India	0.3	6%	0%
Pakistan	0.2	23%	1%

Aid to all water-related sectors	
	USD million
Water supply and sanitation	17.2
Water transport	0.0
Hydro-electric power plants	0.0
Agricultural water resources	0.1
Total water-related aid	*17.2*

Annual aid to water supply and sanitation, 1993-2006 (2006 constant prices)

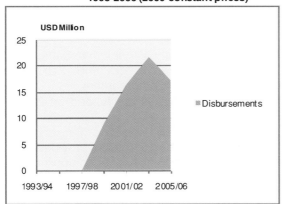

Regional distribution of aid to water supply and sanitation

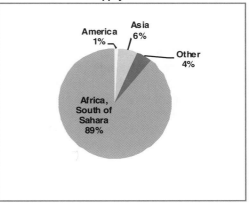

Water supply and sanitation aid by subsector

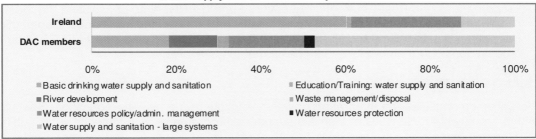

- Basic drinking water supply and sanitation
- River development
- Water resources policy/admin. management
- Water supply and sanitation - large systems
- Education/Training: water supply and sanitation
- Waste management/disposal
- Water resources protection

* % of sector allocable aid

IRELAND
Development Co-operation Policy in the water supply and sanitation sector

POLICY STATEMENT

Irish Aid recognises that access to adequate sanitation and safe drinking water is a basic human need and is also one of the key determinants of sustainable development. Irish Aid policy is to support the development of sanitation, hygiene promotion and water supply in developing countries in a manner that conforms to national policies and that facilitates access to these resources by those who are disadvantaged. The policy recognises the inter-dependence of the MDGs and particularly the need for integrated programming in HIV & AIDS, health, education and local development.

Two specific objectives shape Irish Aid's support to the sector: (i) Strengthened national and sub-national policies, processes and systems for equitable access to adequate sanitation, hygienic practices and safe water; (ii) Integrated and balanced approaches to sanitation, water and hygiene adopted across the sector.

GEOGRAPHICAL AND SUB-SECTORAL FOCUS

Sector level support is channelled through Government systems in South Africa (Masimbambane SWAp, earmarked for Limpopo Province), Zambia (SWAp development and Northern Province) and Lesotho (SWAp development and Department of Rural Water Supply). In addition to these priority countries, civil society funding schemes support specific projects in other developing countries.

AID MODALITIES

Irish Aid's development assistance involves various grant types, each clearly focused on different aspects of development:

- Budget support in Tanzania and Mozambique and support to the Poverty Action Fund in Uganda cover programmes aimed at achieving the sanitation and water MDGs. .
- Sector Wide Approach in South Africa, direct government support in Zambia and Lesotho; local government support in Northern Province Zambia
- Silent partnership in local government budget support through CIDA in Inhambane, Mozambique
- Multi-donor trust funds: support to the Water and Sanitation Programme (EUR 1.2 million per year); core support to UNICEF under a multi-annual partnership agreement
- Under the Multi Annual Programming Scheme, four Irish NGOs are funded (in 2007, EUR 2.7m was spent on sanitation and water-related projects).
- Under the Civil Society Fund, support for NGOs and other civil society organisations from Ireland to respond to the development needs of poor communities overseas is provided to a maximum ceiling of EUR 200 000 per year and per application, and up to 75% of each application's budget.
- The Micro-Projects Scheme supports small projects for up to 75% of eligible costs. The maximum grant is EUR 20 000.
- The Emergency Humanitarian Assistance Fund (EHAF) aims at saving and protecting lives in emergency situations through activities such as the provision of water, while the Emergency Preparedness and Post-Emergency Recovery Fund (EPPR) aims at assisting the most vulnerable people in post-emergency societies to re-establish their lives and livelihoods. Contributions sought from either the EHAF or EPPR budget are between EUR 75 000 and EUR 250 000, with each request considered on merits and in the light of funds available.

RULES AND PRACTICES

Irish Aid policy emphasises the integration of water, sanitation and hygiene education. Sustainability and partnership are key characteristics of Irish Aid supported programmes with financial sustainability carefully weighed against exclusion of the poorest and marginalised. Gender and environmental issues are mainstreamed across all Irish Aid programmes but receive particular attention within the water and sanitation arena.

PERFORMANCE EVALUATION

Within Irish Aid, evaluation is managed by the Evaluation and Audit Unit. The Unit's mandate covers evaluation of Official Development Assistance for Water funds administered by the Department of Foreign Affairs. The Unit reports to an independent Audit Committee. In order to enhance the independence of the evaluation arrangements for the Irish Aid programme, the Advisory Board for Irish Aid might review and comment on the reports produced by the Evaluation and Audit Unit.

Evaluation in Irish Aid is guided by the OECD-DAC Principles for Evaluation, which include partnership, impartiality, transparency, credibility and independence. Irish Aid also uses the OECD-DAC's 'Criteria for Evaluating Development Assistance', among which relevance, effectiveness, efficiency, impact and sustainability.

CORDINATION WITH OTHER ACTORS

Irish Aid works in close partnership with other donors and multilateral organisations and with non-governmental organisations and missionaries. Coordination with international organisations such as the World Bank, IMF and the UN Funds and Programmes is important to Irish Aid as a means of enhancing the value of the programme. The EU is a critical partner for Irish Aid in maximising the effectiveness of its development assistance. Irish Aid participates in the Africa Working Group of the EU Water Initiative.

Web References:

Irish Aid: http://www.irishaid.gov.ie/

ITALY
Aid at a glance - Water supply and sanitation

Summary statistics

Unless otherwise stated, figures refer to 2005-2006 annual average commitments expressed in 2006 constant prices.

Total aid to water supply and sanitation			
	USD million	Aid to water by Italy as a share of total aid by Italy	Aid to water by Italy as a share of total DAC members' aid to water
Italy	**63.4**	**9%**	**1%**
For reference, total DAC	*4974.0*	*8%*	*100%*

Top ten recipients of aid to water supply and sanitation			
		Aid to water by Italy to that recipient as a share of	
	USD million	total aid by Italy to that recipient*	total DAC members' aid to water to that recipient
India	16.1	79%	3%
Lebanon	13.9	31%	27%
Bangladesh	8.3	45%	5%
Viet nam	6.7	75%	6%
Jordan	4.6	81%	4%
Albania	3.9	4%	17%
Honduras	2.5	16%	18%
Nicaragua	2.4	5%	14%
Morocco	0.8	2%	1%
Mali	0.8	65%	2%

Aid to all water-related sectors	
	USD million
Water supply and sanitation	63.4
Water transport	-
Hydro-electric power plants	-
Agricultural water resources	-
Total water-related aid	-

Annual aid to water supply and sanitation, 1993-2006 (2006 constant prices)

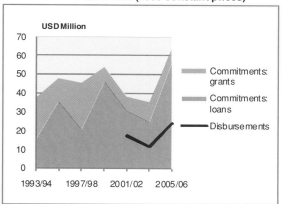

Regional distribution of aid to water supply and sanitation

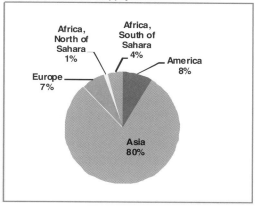

Water supply and sanitation aid by subsector

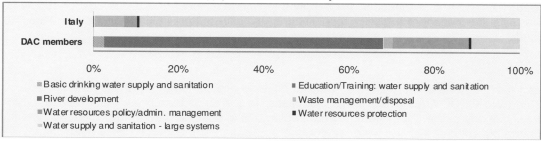

* % of sector allocable aid

ITALY
Development Co-operation Policy in the water supply and sanitation sector

POLICY STATEMENT

The Italian co-operation with developing countries regards the Millennium Development Goals as a reference policy framework. Italy is also committed to the Monterrey Consensus (2002) and adopts the Paris Declaration on Aid Effectiveness (2005) guidelines in its official development assistance. Consistently, the Italian Development Co-operation does not apply a distinctive co-operation policy on water. In general, programmes and projects in the water sector which represent a significant share of both the Italian bilateral aid and the Italian contributions to multilateral institutions are either shaped on the relevant policies and strategies of target countries or on relevant regional bodies. In particular, whenever possible, Italy aligns its country programmes to the priorities and modalities identified within the national Poverty Reduction Strategies Papers (PRSP) of recipient countries.

As from 2006, the Italian aid in the water sector has been delivered as part of a broader environmental policy framework, based on the MDG 7 "environmental sustainability' and MDG 1 "poverty eradication" as well as on the Johannesburg resolution on Sustainable Development (2002). Moreover, the directives outlined in the Compendium of Actions adopted in 2006 by UNSGAB (the so called "Hashimoto Action Plan") are considered of great importance by the Italian Government in its co-operation policy for the water sector. In this perspective, Italy considers "access to water as a human right" as a theme of strategic importance together with climate change, desertification and sustainable forestry. Italian initiatives promoting access to water, in addition to environmental sustainability, also encompass the principles of partnership, at local and global levels. It also supports equitable sharing of water resources among interested stakeholders and across different geographical and administrative boundaries. Similarly as in any other sector of development co-operation, Italy actively promotes mainstreaming of women's empowerment into its water programmes and projects.

No specific commitments for the coming years are available for the water sector, but Italy, along with EU countries, pledged to reach 0.51% of GNI as total official development assistance in 2010 and 0.7 of GNI in 2015.

GEOGRAPHICAL AND SUB-SECTORAL FOCUS

The geographical focus of the Italian ODA in general also applies to the water sector and includes Sub-Saharan Africa, conflict and post-conflicts countries together with others of historical importance for the Italian assistance. Particular attention is given to the Mediterranean and Middle East Region in the framework of the Barcelona Process (Italy, together with France and Spain is a founder of the Euro-Mediterranean Information System on the know-how in the Water Sector - EMWIS). The above geographical priorities should contribute to reducing the fragmentation of Italian fund flows in general so as for the water sector.

The sub-sectoral focus of Italian programmes and projects normally aligns with national or regional priorities of recipients, with a slight predominance of water and sanitation initiatives. It is to be remarked that "water supply" and "sanitation" are deeply intertwined in most operations. Water supply is also often found as a sub-component in many food security initiatives, thus linking "access to water" to "way into livelihoods".

AID MODALITIES

The long term trend is to increase the share of bilateral assistance of the Italian aid, since this modality is lower than the OECD-DAC average. Nevertheless, in the short term, Italy will increase its commitments with organisations of the UN system, toward implementing specific programmatic decisions and increasing the general aid volume. For the water sector, the application of the above policy guidelines, will expectedly

CRS AID ACTIVITIES IN SUPPORT OF WATER SUPPLY AND SANITATION – ISBN 978-92-64-05172-0 – © OECD/WWC 2008

lead to increased aid volumes through multilateral organisations specialised in environmental sustainability and food sector security.

In the water and sanitation sector, as for other basic services sectors, Italy is increasing its involvements in joint programmes and programme-oriented joint financing with other donors, in a bid to enhancing the effectiveness of aid flows and increasing the recipients' ownership of development initiatives.

A significant share of Italian development co-operation is channelled through non-governmental organisations. Water and sanitation is a typical sector of involvement of many of them. The added value, in this case, is the ability to link effectively and flexibly with local actors, so as to deliver aid at grassroots levels. In a similar perspective, Italy strongly promotes the "decentralised" co-operation between Italian local institutions and their peers in recipient countries.

COORDINATION WITH OTHER ACTORS

The coordination with other European Union, OECD-DAC and UN donors is highly valued by Italy for the role it plays in ensuring effective development co-operation. Consistently, Italy assures an active participation to global decisions while contributing to their timely implementation through international agreements and commitments. Particular attention is given by Italy to the Water Information Sector for which it implements specific initiatives including through UNDESA and FAO.

The Italian aid objectives are consistent with the European Union development policies. One third of Italian official development assistance is delivered through the European Commission, i.e. in full coordination with EU members and partners. In this context, the active Italian role in the EU Water Initiative stands out together with the Italian support to the EU Water facility, in particular in their Monitoring and Reporting components.

Italy supports a range of initiatives in the water and environment sectors through its voluntary contributions to the main global and regional multilateral financing institutions (World Bank, European Investment Bank and African Development Bank being the most obvious examples).

JAPAN
Aid at a glance - Water supply and sanitation

Summary statistics

Unless otherwise stated, figures refer to 2005-2006 annual average commitments expressed in 2006 constant prices.

Total aid to water supply and sanitation			
	USD million	Aid to water by Japan as a share of total aid by Japan	Aid to water by Japan as a share of total DAC members' aid to water
Japan	**1626.0**	**20%**	**33%**
For reference, total DAC	*4974.0*	*8%*	*100%*

Top ten recipients of aid to water supply and sanitation			
		Aid to water by Japan to that recipient as a share of	
	USD million	total aid by Japan to that recipient*	total DAC members' aid to water to that recipient
India	401.4	32%	82%
China	369.1	38%	95%
Malaysia	349.5	85%	100%
Indonesia	114.6	12%	73%
Costa Rica	64.6	90%	97%
Bangladesh	53.1	27%	31%
Viet Nam	26.9	3%	25%
Tunisia	25.6	35%	41%
Pakistan	19.3	7%	57%
Morocco	18.5	14%	13%

Aid to all water-related sectors	
	USD million
Water supply and sanitation	1626.0
Water transport	255.9
Hydro-electric power plants	296.3
Agricultural water resources	246.4
Total water-related aid	*2424.5*

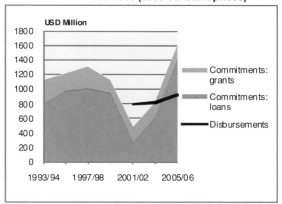

Annual aid to water supply and sanitation, 1993-2006 (2006 constant prices)

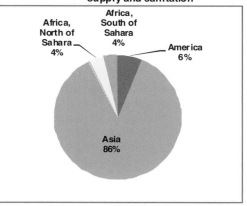

Regional distribution of aid to water supply and sanitation

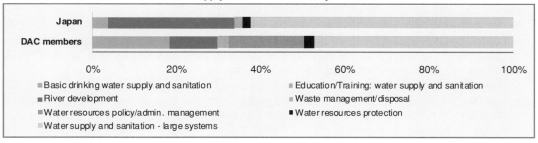

Water supply and sanitation aid by subsector

Legend:
- Basic drinking water supply and sanitation
- River development
- Water resources policy/admin. management
- Water supply and sanitation - large systems
- Education/Training: water supply and sanitation
- Waste management/disposal
- Water resources protection

* % of sector allocable aid

JAPAN
Development Co-operation Policy in the water supply and sanitation sector

POLICY STATEMENT

Water is one of the top-priority sectors of Japanese assistance. In order to improve its effectiveness, Japan launched the "Water and Sanitation Broad Partnership Initiative" (WASABI) on the occasion of the 4th World Water Forum in Mexico in 2006. This initiative sets the guidelines for Japanese assistance to water and sanitation for the coming years. Its basic principles are:

- Pursuing water use sustainability: promoting water resources management according to the Integrated Water Resources Management (IWRM) approach and transboundary management, ensuring project harmonization, supporting institutional development and efforts to monitor and forecast long-term trends related to water in developing countries, ensuring equitable access to water.
- Emphasizing the human security perspective, encouraging participation of inhabitants, taking gender issues into consideration.
- Emphasizing capacity development: promoting development of the organizations, policies, systems and human resources of the governments of developing countries, in order to maximise the effects of infrastructure development. Promoting development of technological and managerial capacities at the local level.
- Pursuing synergy through cross-sectoral measures: promoting coordination with closely related sectors, such as health, education, disaster reduction, urban and rural development, industrial development, environmental and ecological conservation and gender equity.
- Considering local conditions and appropriate technology.

WASABI's comprehensive measures are:

- Promotion of IWRM approaches, focusing on regions in which aggravation of structural water shortage is predicted.
- Provision of safe drinking water and sanitation: In rural areas, Japan supports the development of water supply facilities tailored to local conditions and the capacity development of local communities (including women). Japan also supports implementation of ecological sanitation (EcoSan) systems. In urban areas, Japan promotes the use of private funds in addition to ODA in order to respond to the large-scale financial needs. Japan provides both software and hardware support. In situations where developing a household water and sewage system is difficult due to financial limitations, Japan provides support for transitional measures such as collection and disposal of human waste. Japan supports the capacity development of the operational entities, such as response to privatization issues and cost recovery, and methodologies, such as monitoring water leakage and water quality.
- Support regarding water use for food and other purposes: Japan promotes multi-faceted use of water including agricultural water (irrigation facilities and water-saving agricultural technologies), electricity generation, industrial water, and water transport..
- Water pollution prevention and ecosystem conservation: Japan promotes sanitary education. As a countermeasure against industrial wastewater from factories, Japan strives to transfer its expertise in identifying the source of pollution and establishing effluent regulations, as well as technologies for effective use of water and wastewater treatment.
- Mitigation of damage from water-related disaster: Japan supports the development of information systems concerning surface and ground water, and infrastructures development including flood control, erosion control, drought management and sewage facilities. In addition, to strengthen the disaster response capacities of individuals and local communities, Japan supports the

development of hazard maps and the self-help efforts of inhabitants towards disaster reduction and crisis management, such as disaster drills.

TICAD IV and G8 Hokkaido Toyako Summit
The year 2008 has given an unique opportunity for Japan to host both the forth Tokyo International Conference on African Development (TICAD IV) and the Group of Eight (G8) Summit in Toyako, Hokkaido, bringing together key challenges, including water and sanitation, and acting as a catalyst to facilitate global co-operation towards resolution.

Given that water has a cross-sectoral nature as it is inextricably linked to development, poverty reduction, environment, agriculture, education, gender, security and so on, the participants of TICAD IV acknowledged the importance of water in light of addressing environmental issues and climate change. In this regard, Japan announced to dispatch a newly organized technical assistance corps of water specialists, "W-SAT," or the "Water Security Action Team," to improve water resources management on the ground of African nations.

At the G8 Hokkaido Toyako Summit, advocating Japan's strong stand on water and sanitation, Japan, together with other G8 members, addressed the crucial importance of good water cycle management in order to utilize the limited resource in an effective way, and urged focus on the concept of "Good Water Governance," with particular focus on Sub-Saharan Africa and Asia-Pacific. In addition, in the International Year of Sanitation, the G8 members called upon national governments to prioritize access to sanitation to meet the Millennium Development Goal on water and sanitation.

GEOGRAPHICAL AND SUB-SECTORAL FOCUS

Based on the strategic orientation of overall foreign policy and the deliberations of the Overseas Economic Co-operation Council, the Ministry of Foreign Affairs (MOFA) deliberates over assistance policies for each region and methods to address individual sectors and challenges. Japan clearly prioritizes its area of regional influence; all its major partner countries are its most populated neighbours. Therefore, Japan brings a substantial support to several countries where the absolute amounts of poor people are the highest.

AID MODALITIES

The Poverty Reduction Strategy Paper (PRSP) has been positioned as the basic, country-focused strategy for achieving the MDGs. But Japan's ODA for water up to now has remained mainly focused on project-type assistance. This kind of co-operation is based on Japan's principles, which place emphasis on the self-help efforts of the recipient country; aid is provided based on a premise that after completion of projects, recipient countries will secure budgets to autonomously operate and maintain the achieved results. Many developing countries, particularly in sub-Saharan Africa, are requesting that Japan provide budget support, which the country intends to do, as long as it supports the self-help efforts of the recipients. In 2007, Japan introduced a new instrument of grant aid for assisting PRSP through budget support.

Loans remain predominant as well, as MOFA considers them as consistent with its self-help efforts strategy towards recipient countries.

Japan provides a part of its assistance directly to projects led by local governments and NGOs, through the Grant Assistance for Grass-roots Human Security Projects system. This aid can be expected to further contribute to strengthen civil society of the recipient countries.

RULES AND PRACTICES

Japan supports the capacity development of developing countries based on methods such as comprehensive flood control measures, integrated lake basin management (ILBM), and groundwater regulation and management. The philosophy that Japan has developed seeks not merely to transfer technology, but also to empower the people. By transferring Japanese knowledge and technology, Japan can contribute to self-help efforts of developing countries: its strategy places emphasis on having developing countries play a leading role in solving their development problems, and on improving the capacity not only of individuals but of organizations and institutions. Thus, Japan's technical co-operation also consists in transferring Japan's expertise to the people who play leadership roles in developing countries, in order to establish efficient legal institutions and an economic and social foundation.

Japan promotes human resource development for water projects through technical co-operation in coordination with the development of infrastructure, in pursuit of efficiency in operation, maintenance and management of the facilities. Japan emphasizes having Japanese people work closely together with locals. Japanese experts teach skills on-site and formulate manuals together with local officials. In 2006, Japan accepted 633 trainees from developing countries and dispatched 64 experts to them. Considering the goal of the Initiative for Japan's ODA on Water (1000 people over the next 5 years from 2003), 541 experts have been dispatched and 4001 trainees have been accepted between 2003 and 2006.

PERFORMANCE EVALUATION

It is estimated that Japan's ODA has provided access to safe and stable drinking water and sanitary sewerage system to more than 40 million people between 1996 and 2000 (source: MOFA's latest available evaluation).

COORDINATION WITH OTHER ACTORS

WASABI Initiative is designed to bolster co-operation with international organizations, other donor countries, domestic and overseas NGOs and other concerned parties. Japan is conducting several projects in collaboration with international organisations such as the African Development Bank or UNICEF. Japan also supported the establishment of systems such as the Global Earth Observation System of Systems (GEOSS).

Moreover, Japan launched in 2002 the "United States-Japan Clean Water for People Initiative". Japan and the United States are currently exploring how to attract private funds for the development of regional water and sewage infrastructure by combining Japanese ODA loan from JBIC with USAID investment guarantees. Indonesia, India, Philippines and Jamaica have been chosen as pilot countries.

Japanese assistance is employed with the efforts of a wide range of participants: Japanese NGOs are important partners as they play a leading role at the grassroots level; advisors from consulting firms are active during the definition of ODA projects; academics and public institutions contribute to the evaluation of the projects. Several initiatives, such as the Japan Water Forum (JWF), enable this general synergy between Japanese government, NGOs, private sector and academia. The Japan Platform (JPF) is a framework whereby each sector cooperates during a crisis or disaster to deliver emergency aid in a rapid and effective manner. JICA has also undertaken the JICA Partnership Program (JPP) with all these stakeholders to jointly implement projects in developing countries.

Web references:

Water and Sanitation Broad Partnership Initiative (WASABI) 2006
http://www.mofa.go.jp//policy/oda/category/water/wasabi0603-2.pdf

LUXEMBOURG
Aid at a glance - Water supply and sanitation

Summary statistics

Unless otherwise stated, figures refer to 2005-2006 annual average commitments expressed in 2006 constant prices.

Total aid to water supply and sanitation			
	USD million	Aid to water by Luxembourg as a share of total aid by Luxembourg	Aid to water by Luxembourg as a share of total DAC members' aid to water
Luxembourg	**12.2**	**9%**	**0%**
For reference, total DAC	*4974.0*	*8%*	*100%*

Top ten recipients of aid to water supply and sanitation			
		Aid to water by Luxembourg to that recipient as a share of	
	USD million	total aid by Luxembourg to that recipient*	total DAC members' aid to water to that recipient
Nicaragua	2.0	23%	12%
El Salvador	1.6	20%	23%
Morocco	1.2	83%	1%
Namibia	1.1	34%	35%
Senegal	1.1	10%	4%
Mali	0.9	11%	3%
Albania	0.8	97%	4%
States Ex-Yugoslavia	0.8	100%	57%
Viet Nam	0.4	3%	0%
Niger	0.2	2%	0%

Aid to all water-related sectors	
	USD million
Water supply and sanitation	12.2
Water transport	0.0
Hydro-electric power plants	0.0
Agricultural water resources	0.5
Total water-related aid	*12.7*

Annual aid to water supply and sanitation, 1993-2006 (2006 constant prices)

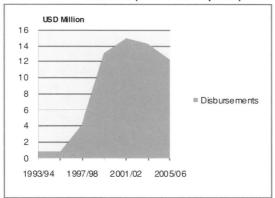

Regional distribution of aid to water supply and sanitation

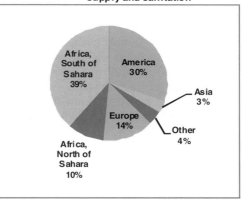

Water supply and sanitation aid by subsector

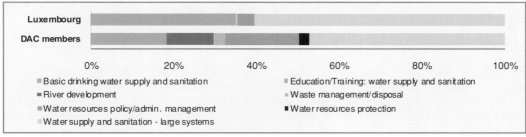

* % of sector allocable aid

LUXEMBOURG
Development Co-operation Policy in the water supply and sanitation sector

POLICY STATEMENT

Responsibility for co-operation policy lies with the Minister for Co-operation and Humanitarian Action. Within this Ministry, the management of development policy is entrusted to the Development Co-operation Directorate (DCD). The DCD prepares PICs (*Programmes indicatifs de coopération – Indicative co-operation programmes*), identifies programmes and projects and is also responsible for multilateral financing (apart from the Bretton Woods institutions), for humanitarian aid, and for grants to NGOs. The DCD uses an executing agency, Lux-Development, which designs and implements bilateral projects.

Luxembourg's ODA aims at eliminating poverty in conformity with the MDGs. Meeting basic needs for water is identified as a traditional priority focus for Luxembourg Development Co-operation. This priority stems from the following: 1) water is essential to life; 2) all recipient countries in which Luxembourg is active suffer from water stress; 3) improvements in health levels remain inconsistent without proper access to quality drinking water and adequate sanitation; 4) water is essential to agriculture, which is the primary means of survival in developing countries; 5) insufficient access to water and environmental flows provoke social insecurity and political instability which undermine development.

The pattern of Luxembourg's assistance for water and sanitation is varied and tailored to the recipient countries' needs. Recently, Luxembourg has initiated a progressive shift from an emphasis on building water and sanitation infrastructures to development and reinforcement of local management and implementation capacities.

Luxembourg's commitments in ODA for water and sanitation are on a programme / country basis. For example, Luxembourg is specifically committed to support water and sanitation programmes from 2007-08 to 2011 for a total budget of €15 M in Senegal, €10.5 M in Mali, €8 M in Nicaragua, and €5 M in Salvador.

GEOGRAPHICAL AND SUB-SECTORAL FOCUS

Concentrating its assistance geographically on the least developed countries is a strategic element of Luxembourg's co-operation policy, which targets 10 priority partner countries, 6 of them in sub-Saharan Africa. Initiatives with respect to water and sanitation are implemented in all partner countries, often as components of larger rural or urban integrated development. The essential points of sectoral focus are found in general government statements or specific operational documents (PICs, project formulation mandates given to Lux-Development, and Lux-Development's methodological manual).

RULES AND PRACTICES

Luxembourg's ODA for water and sanitation is not subject to specific rules and practices. The Development Co-operation Act of 1996 establishes the legal and regulatory framework for Luxembourg's co-operation. The strategy is based on the benchmark framework established by the Millennium Declaration, the Monterrey Consensus on development finance, and the Paris Declaration on aid effectiveness. Luxembourg's "Strategies and Principles" statement guides activities in the spirit of sustainable development in all aspects - social, economic and environmental.

PERFORMANCE EVALUATION

Luxembourg's ODA for water and sanitation is not subject to a specific performance evaluation process. The various evaluation modalities generally applicable also apply to the water sector. (For an overview, see OECD – DAC *Peer review of Luxembourg* 2008, p.46-47.) Luxembourg's involvement in the Diber Water and Sanitation Project, Albania, which gave access to drinking water to 70% of the 8,700 inhabitant target population, is a successful example of ODA for water since it was implemented in the context of water services decentralization at the national level, and was executed in a low density area with minimal potential financial contribution from beneficiaries. In this context, the high initial costs for infrastructure emphasised the importance of ODA, with Luxembourg contributing close to €1 M in 2007. From 2005 to 2008, a €5.5 M Lux-Development project in Salvador resulted in access to drinking water and sanitation services for a population of 12,500 in 12 rural communities. Finally, Luxembourg's involvement in Morocco is another successful endeavour, as LDC will soon complete the last of 4 successive projects within the Programme d'Alimentation Groupée en Eau Potable des Populations Rurales (PAGER), for which the Moroccan State Secretariat for Water received the UN Award for Improvement of Public Service Results in 2004.

COORDINATION WITH OTHER ACTORS

Luxembourg's co-operation has become more strategic thanks to new framework agreements reached with the WHO, UNDP and UNICEF. These agencies were selected by Luxembourg because of the close alignment between their activities and Luxembourg's priority sectors and cross-cutting issues. This is shown by the projects that received earmarked financing from Luxembourg, such as UNDP's water and sanitation activities in Mali.

Web References:

Ministry of Foreign Affairs: http://www.mae.lu/MAE.taf?IdNav=7&IdLang=FR
Lux-Development: http://www.lux-development.lu/
2007 Annual report of Lux-development: http://www.lux-development.lu/publication/rapann/rapann_uk.pdf

NETHERLANDS
Aid at a glance - Water supply and sanitation

Summary statistics

Unless otherwise stated, figures refer to 2005-2006 annual average commitments expressed in 2006 constant prices.

Total aid to water supply and sanitation			
	USD million	Aid to water by Netherlands as a share of total aid by Netherlands	Aid to water by Netherlands as a share of total DAC members' aid to water
Netherlands	**334.1**	**10%**	**7%**
For reference, total DAC	*4974.0*	*8%*	*100%*

Top ten recipients of aid to water supply and sanitation

		Aid to water by Netherlands to that recipient as a share of	
	USD million	total aid by Netherlands to that recipient*	total DAC members' aid to water to that recipient
Bangladesh	52.4	47%	31%
Benin	33.7	75%	52%
Indonesia	30.0	19%	19%
Viet Nam	22.3	39%	20%
Mozambique	17.0	19%	24%
Egypt	15.1	37%	18%
Yemen	13.5	45%	29%
Pakistan	12.9	22%	38%
Bolivia	8.4	23%	27%
Senegal	6.7	14%	26%

Aid to all water-related sectors

	USD million
Water supply and sanitation	334.1
Water transport	0.3
Hydro-electric power plants	0.0
Agricultural water resources	21.4
Total water-related aid	*355.9*

Annual aid to water supply and sanitation, 1993-2006 (2006 constant prices)

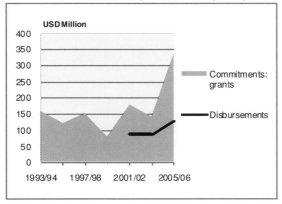

Regional distribution of aid to water supply and sanitation

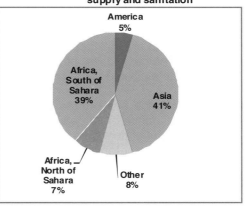

Water supply and sanitation aid by subsector

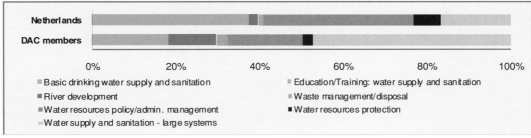

- Basic drinking water supply and sanitation
- Education/Training: water supply and sanitation
- River development
- Waste management/disposal
- Water resources policy/admin. management
- Water resources protection
- Water supply and sanitation - large systems

* % of sector allocable aid

NETHERLANDS
Development Co-operation Policy in the water supply and sanitation sector

POLICY STATEMENT

Water has been an important component of Dutch development co-operation since the 1970s, and Integrated Water Resources Management (IWRM) forms the core of Dutch water development policy: it takes into account the entire hydrological system, including balancing the needs of different groups in society, and deals with flooding, water shortages and water pollution, while at the same time paying attention to water use in households, agriculture, industry and nature. Access to drinking water and sanitation is another key element. The Netherlands supports developing countries in achieving MDG 7 by providing access to safe drinking water and improved sanitation to at least a half of those who had no such access. Finally, Dutch co-operation helps countries to carry out the necessary reforms in the water sector and to increase their knowledge of water management.

The Netherlands is committed to providing 50 million people with sustainable access to safe drinking water and improved sanitation services by 2015. The sustainability of the results is of particular importance. Each co-operation contract will come with a strict monitoring protocol to ensure that the services provided will still be functional in 2015.

Efforts to reach this output target will focus on poor people, especially in Sub-Saharan Africa. The programmes will be based in rural and peri-urban areas and focus on improving the position of women, promoting sustainability and creating leverage for additional investments.

The Netherlands wants also to ensure that: developing countries are given help in charting the impact of climate change and in offsetting the risks associated with such change; and international environmental protection agreements are concluded and international co-operation is improved.

GEOGRAPHICAL AND SUB-SECTORAL FOCUS

The Netherlands is co-funding the installation of drinking water, sanitation and irrigation systems in 18 countries, 12 of which are in Africa. The country is supporting the drafting of IWRM plans in 10 countries, 6 of which are in Africa. These plans will serve as blueprints for the sustainable consumption and management of the available water sources.

A substantial effort has been made since 1998 to reduce the dispersion of Dutch assistance by more clearly defining its "partnership" countries. The Netherlands consciously distinguishes between 3 different types of partner country: countries where the focus is on accelerating MDG achievement, countries with a focus on security and development, and countries with a broad-based relationship.

AID MODALITIES

In its bilateral programmes the Netherlands provides budget support and assistance to sector-wide programmes where possible and supports projects when conditions for such sector-wide programmes are not yet in place. A substantial part of the Netherlands' development co-operation programme on water is implemented through multilateral organisations, civil society organisations and the private sector. For each of these categories specific funding modalities and facilities exist.

RULES AND PRACTICES

Dutch approach to development coordination is based on 3 important principles: ownership [promoting the use of country-owned strategies, in particular the Poverty Reduction Strategy Papers (PRSP) as a framework for implementation, monitoring and evaluation, donor coordination and policy dialogue in priority countries]; utilisation of domestic resources; poverty reduction.

 CRS AID ACTIVITIES IN SUPPORT OF WATER SUPPLY AND SANITATION – ISBN 978-92-64-05172-0 – © OECD/WWC 2008

Gender equality, availability of qualified personnel in even the poorest areas and public access to the information needed to call government and service providers to account are also important.

The Netherlands supports governments that want to improve and expand their water policies and water management, notably through a "water partnership" with the development banks and through the UNESCO-IHE (Institute for Water Education), WaterNet and CapNet capacity building programmes. The Netherlands also help developing countries build their capacity to adapt to climate change in vulnerable sectors like land use, food production, water and health.

But Dutch focus will not be exclusively on central governments. The policy dialogue will examine how local authorities and NGOs can play their role in collaboration with the government or to supplement its activities. Adequate financing of decentralised government bodies and other service providers will be a natural part of this dialogue.

The Netherlands has systematically drawn up output targets for its water and sanitation programmes. In 2008, contracts have been signed which will provide safe drinking water for 26 million people and improved sanitation facilities for 29 million.

COORDINATION WITH OTHER ACTORS

The Netherlands is actively involved in coordinated action with its major multilateral partners, in particular in European policy dialogue and coordination.

Many of the targets are implemented through multilateral organisations like UNICEF, UN-Habitat, GWP, and WASH-CC. Thus, the Dutch government contributes 2.4 million Euros to the UN-Habitat Lake Victoria Initiative. This multi-donor programme will enable local authorities in Kenya, Tanzania and Uganda to provide drinking water and sanitation services to more than 600 000 people around Lake Victoria, and it is expected that the actual amount of people served by the programme towards the end of its implementation (in 2009) will increase to at least 1.4 million. Co-operation with regional development banks and the World Bank takes 2 forms: promotion of IWRM and specific support to raise the level of investment (loans) in the water sector and MDG 7 realisation in particular.

The Netherlands supports water-related development in 6 transboundary river basins: Mekong, Senegal, Incomati, Maputo, Zambezi and Nile. Dutch contribution to Nile Basin Initiative is currently 19.2 million Euros, aimed at 3 programmes: Applied Training, Nile Transboundary Environmental Action and Efficient Water Use for Agriculture. The projects entail concrete actions in the field of capacity building, agricultural productivity and environmental management. For the next 3 years the Netherlands will contribute an additional 12 million Euros to Subsidiary Action Programme investment projects.

The development assistance policy is implemented in alliance with Dutch civil society, with which the government signed the Pact of Schokland for an improved collaboration in achieving the MDGs. Indeed, Dutch development assistance makes an extensive use of private and non-governmental organisations (NGOs) for implementing its programmes, and the Minister for Development Co-operation maintains an active dialogue with NGOs (both Dutch and international), academic and research institutions and the business sector. The Dutch government also supports multi-stakeholder partnerships between partners from different sectors, like governments, civil society organisations, the business sector and knowledge institutes, in rich and poor countries alike.

Public-private partnerships have been signed with several national and international banks and companies with the aim of mobilising private capital and knowledge. Traditionally, the Netherlands has been home to many businesses skilled in constructing drinking water facilities and treating wastewater. Within the new ORIO programme for public-private partnership for investement projects in key sectors, water management will be one of the eligible sectors.

Web References:

http://www.waterland.net/showdownload.cfm?objecttype=mark.hive.contentobjects.download.pdf&objectid
=055D3485-F199-606C-08016F78500D067D
http://www.vewin.nl/SiteCollectionDocuments/Publicaties/Waterspiegel/2006/Holland%20Water%20Aid%2
0brochure%202006.pdf

CRS AID ACTIVITIES IN SUPPORT OF WATER SUPPLY AND SANITATION – ISBN 978-92-64-05172-0 – © OECD/WWC 2008

NEW ZEALAND
Aid at a glance - Water supply and sanitation

Summary statistics

Unless otherwise stated, figures refer to 2005-2006 annual average commitments expressed in 2006 constant prices.

Total aid to water supply and sanitation			
	USD million	Aid to water by New Zealand as a share of total aid by New Zealand	Aid to water by New Zealand as a share of total DAC members' aid to water
New Zealand	**4.1**	**2%**	**0%**
For reference, total DAC	*4974.0*	*8%*	*100%*

Top ten recipients of aid to water supply and sanitation		Aid to water by New Zealand to that recipient as a share of	
	USD million	total aid by New Zealand to that recipient*	total DAC members' aid to water to that recipient
Tonga	1.1	11%	93%
Fiji	0.4	5%	87%
Sri Lanka	0.4	65%	1%
Vanuatu	0.3	3%	60%
Cook Islands	0.2	3%	61%
Myanmar	0.2	23%	35%
Cambodia	0.2	2%	1%
Niue	0.2	2%	100%
Uzbekistan	0.1	95%	1%
Uganda	0.04	27%	0%

Aid to all water-related sectors	
	USD million
Water supply and sanitation	4.1
Water transport	1.4
Hydro-electric power plants	0.1
Agricultural water resources	0.0
Total water-related aid	*5.6*

Annual aid to water supply and sanitation, 1993-2006 (2006 constant prices)

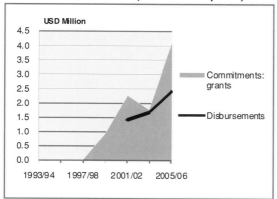

Regional distribution of aid to water supply and sanitation

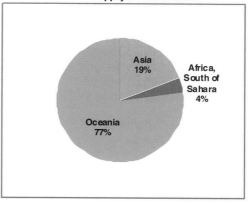

Water supply and sanitation aid by subsector

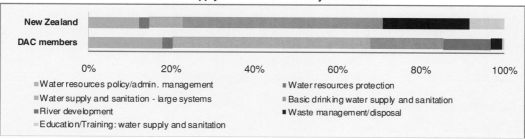

* % of sector allocable aid

NEW ZEALAND
Development Co-operation Policy in the water supply and sanitation sector

POLICY STATEMENT

New Zealand has no specific ODA policy statement on water and sanitation but its assistance has a direct impact on development in that sector. New Zealand's ODA contributes to the elimination of poverty in accordance with the MDGs. Meeting basic needs for water is identified as an intermediate target outcome within the global developmental impact design to lift people out of poverty. Health, sustainable livelihood and environment, which are foci of key strategies in New Zealand's ODA, imply assistance for water and sanitation that aims at improving access to, and quality of, appropriate and affordable services and infrastructures, as well as strengthening the relevant national or local policy and regulatory environment.

GEOGRAPHICAL AND SUB-SECTORAL FOCUS

New Zealand ODA's geographic focus is primarily on the Pacific. New Zealand Agency for International Development (NZAID) has 11 core bilateral country partners (Papua New Guinea, Solomon Islands, Vanuatu, Fiji, Tuvalu, Kiribati, Tonga, Samoa, Cook Islands, Niue and Tokelau) defined as countries in which there is a government-to-government programme determined by a jointly agreed strategy and a 3-year funding commitment. NZAID is engaged in a number of other countries through its regional programmes for South Asia, Africa and Latin America for specific activities.

With respect to sectoral focus, NZAID supports activities under a regional partnership for improving access to water and sanitation coordinated by the Secretariat for the Pacific Applied Geoscience Commission (SOPAC) under which Pacific countries have identified environmental pressure caused by urban water supply systems as a priority concern. With NZAID support, SOPAC and the regional partners are helping Pacific countries to develop urban water management plans, and to repair and maintain reticulation systems over the long-term.

AID MODALITIES

The framework for New Zealand ODA does not provide specific aid modalities for water and sanitation. New Zealand generally supports water, sanitation and human settlement projects through NZAID regional and bilateral programmes. Support for civil society organisations overseas takes a variety of forms ranging from core funding of regional and national umbrella bodies, to long-term programme support for local NGOs and small project funding for grassroots projects.

RULES AND PRACTICES

New Zealand ODA for water and sanitation is not subject to specific rules and practices. The vision, outcomes and operating principles of New Zealand's ODA programme are detailed in NZAID's Policy Statement, *Towards a Safe and Just World Free of Poverty*. This policy statement is supported by other more detailed strategy and policy documents, including: the *Five Year Strategy (2004/05-2009/10);* annual business plans; regional, country and sectoral strategies; and draft strategies for cross-cutting issues such as human rights, governance, gender and environment that need to be mainstreamed. Core operating principles reflected throughout NZAID development co-operation are: protecting and promoting human rights, strategic approach to poverty elimination, sustainability, equity, partnerships, participation, coordination, access and accountability. In general, NZAID is responsible for managing New Zealand's ODA programme overseas as a semi-autonomous body within the Ministry of Foreign Affairs and Trade. It works in accordance with the Government's strategic directions for foreign policy and ODA.

PERFORMANCE EVALUATION

New Zealand ODA for water and sanitation is not subject to a specific evaluation process. NZAID undertakes reviews and evaluations to assess the performance and effectiveness of programme activities. NZAID has established an Evaluation Committee to provide oversight of evaluation planning and ensure close links between evaluation and programme development. Evaluations are expected to provide information for three key purposes: accountability, learning and improvement. Relevant OECD - DAC Guidelines contributing directly to the performance evaluation process in New Zealand assistance policy are *DAC Principles for Evaluation of Development Assistance* and *Harmonising Donor Practices for Effective Aid Delivery*.

COORDINATION WITH OTHER ACTORS

New Zealand ODA has no specific coordination pattern in the water and sanitation sector. NZAID notably engages at the global level through multinational agencies and international development banks.

Web References:

New Zealand's International Aid and Development Agency: http://www.nzaid.govt.nz/

NORWAY
Aid at a glance - Water supply and sanitation

Summary statistics

Unless otherwise stated, figures refer to 2005-2006 annual average commitments expressed in 2006 constant prices.

Total aid to water supply and sanitation

	USD million	Aid to water by Norway as a share of total aid by Norway	Aid to water by Norway as a share of total DAC members' aid to water
Norway	**38.9**	**2%**	**1%**
For reference, total DAC	*4974.0*	*8%*	*100%*

Top ten recipients of aid to water supply and sanitation

	USD million	Aid to water by Norway to that recipient as a share of	
		total aid by Norway to that recipient*	total DAC members' aid to water to that recipient
Serbia	2.6	7%	8%
Albania	2.4	39%	10%
Bosnia-Herzegovina	2.3	13%	24%
Sudan	2.0	2%	14%
Iraq	1.9	17%	0%
Viet Nam	1.8	13%	2%
Ethiopia	1.2	5%	2%
Laos	1.0	13%	5%
Eritrea	0.7	10%	23%
Palestinian Adm. Areas	0.6	1%	1%

Aid to all water-related sectors

	USD million
Water supply and sanitation	38.9
Water transport	1.4
Hydro-electric power plants	9.8
Agricultural water resources	3.0
Total water-related aid	*53.1*

Annual aid to water supply and sanitation, 1993-2006 (2006 constant prices)

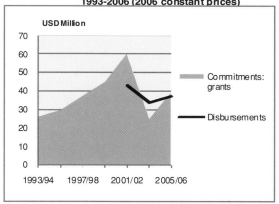

Regional distribution of aid to water supply and sanitation

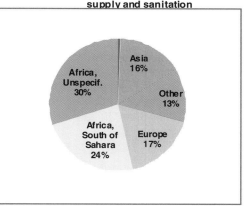

Water supply and sanitation aid by subsector

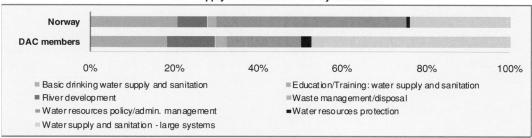

* % of sector allocable aid

NORWAY
Development Co-operation Policy in the water supply and sanitation sector

POLICY STATEMENT

Improved water supply, sanitary conditions and hygiene are crucial in the fight against poverty and for the achievement of the MDGs. The Norwegian Action Plan for Environment in Development Co-operation identifies water resources management and water and sanitation as a thematic priority. According to the plan from 2006, Norway intends to:

- support the development and implementation of plans for integrated water resources management, including for transboundary watercourses. Particular emphasis will be placed on promoting the ecosystem approach and supporting institutions that are mandated to ensure sustainable management and use of water resources;
- promote efficient water use, particularly in agriculture;
- focus attention on the importance of sanitation and hygiene, and of reducing contamination of water resources;
- support the improvement of water supply and sanitary conditions in other sectors, for example by supporting the installation of satisfactory water supply and sanitary and hygiene facilities in schools and health institutions;
- assist priority countries in achieving water and sanitation targets, focusing particularly on sanitation;
- promote community-based management of catchment areas, including support for rainwater harvesting and other small-scale water projects;
- increase awareness of and promote research on how water resources are affected by climate change; and
- work to secure all people the right to water and promote acceptance of the principle that water resources are a common good.

GEOGRAPHICAL AND SUB-SECTORAL FOCUS

Some years ago, Norway reduced the number of sectors in which it was to be involved on a bilateral basis in a particular country. Norway's geographical and sub-sectoral focus in ODA for water and sanitation are on a programme/country basis. Norway supports infrastructure or institutional development in Laos, Vietnam, Afghanistan; Palestinian Administrative Areas, and transboundary water resources management through support to co-operation on the Zambezi, through ZACPRO, and on the Nile, through the multi-donor supported Nile Basin Initiative, including programmes in the sub-basins.

AID MODALITIES

Norway's development co-operation with many of its Partner countries is based on a general Memorandum of Understanding identifying the areas of concentration for Norwegian support and/or on a Main Agreement/Framework Agreement. When several donors are involved, agreement for donor co-operation is through a Joint Financing Arrangement or an Arrangement on Delegated Co-operation.

RULES AND PRACTICES

Norway's ODA for water and sanitation is not subject to specific rules and practices. The MFA's Development Co-operation Manual describes the key principles, procedures and standard working methods in different phases of a programme cycle, from the initial planning to implementation and completion.

PERFORMANCE EVALUATION

NORAD's Evaluation Department is responsible for planning and implementing independent evaluations of activities financed through Norwegian Development Co-operation. Bilateral aid is largely evaluated based on OECD – DAC criteria, whereas activities carried out by UN bodies are evaluated using the criteria and standards of the UN Evaluation Group. Each evaluation sheds light on the relevance, the effectiveness, the sustainability, the efficiency and the impact of a specific activity.

An evaluation of Norway's ODA for water and sanitation was carried out in 2008: NORAD Report 16/2008, *Norad's Assistance to Water Supply and Sanitation Development in Tanzania and Kenya during the 70s, 80s and 90s.*

COORDINATION WITH OTHER ACTORS

At the multilateral level, Norway's ODA gives priority to co-operation with and supports UNDP, UNICEF, UN-HABITAT and the international financial institutions, including GEF, on transboundary waters, but also cooperates with other international initiatives and organisations such as the Water and Sanitation Program, the Global Water Partnership and the Water Supply and Sanitation Collaborative Council. Norway has supported the African Water Facility, since it was launched. Norway will continue to support these organisations' key programmes on water resources management and on water and sanitation. At the bilateral level, NGOs are an important channel for carrying out water- and sanitation-related measures in selected countries.

Web References:

The Norwegian Agency for Development Co-operation (NORAD): http://www.norad.no/
The Norway Ministry of Foreign Affairs: http://www.regjeringen.no/en/dep/ud.html?id=833

PORTUGAL
Aid at a glance - Water supply and sanitation

Summary statistics

Unless otherwise stated, figures refer to 2005-2006 annual average commitments expressed in 2006 constant prices.

Total aid to water supply and sanitation			
	USD million	Aid to water by Portugal as a share of total aid by Portugal	Aid to water by Portugal as a share of total DAC members' aid to water
Portugal	**1.6**	**1%**	**0%**
For reference, total DAC	*4974.0*	*8%*	*100%*

Top recipients of aid to water supply and sanitation		Aid to water by Portugal to that recipient as a share of	
	USD million	total aid by Portugal to that recipient*	total DAC members' aid to water to that recipient
Angola	0.6	3%	9%
Timor-Leste	0.5	2%	8%
Mozambique	0.3	2%	0%
Guinea-Bissau	0.1	1%	16%
Cape Verde	0.01	0%	0%
Indonesia	0.001	24%	0%

Aid to all water-related sectors	
	USD million
Water supply and sanitation	1.6
Water transport	0.2
Hydro-electric power plants	0.0
Agricultural water resources	0.0
Total water-related aid	*1.8*

Annual aid to water supply and sanitation, 1993-2006 (2006 constant prices)

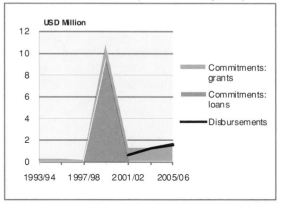

Regional distribution of aid to water supply and sanitation

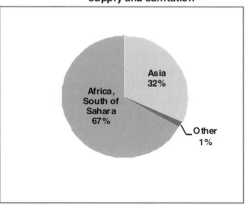

Water supply and sanitation aid by subsector

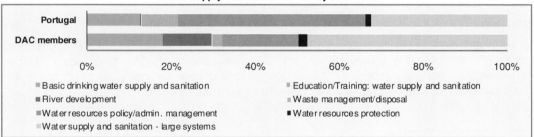

* % of sector allocable aid

PORTUGAL
Development Co-operation Policy in the water supply and sanitation sector

POLICY STATEMENT

Portugal's development co-operation strategy and policy is based on the following documents: Main Planning Options (GOP) 2005-2008; National Strategy for sustainable Development (ENDS) and Strategic Vision for Portuguese Co-operation (*Cabinet Resolutions Nos. 52/2005; 109/2007 and 196/2005*).

The guiding principles are to: i) contribute to the achievement of the Millennium Development Goals; ii) contribute to the reinforcement of human security particularly in 'Fragile States' or post conflict situations; iii) promote economic development while insuring social and environmental sustainability; and iv) actively participate in the international debates in support to the principal of international convergence and around common objectives. The Portuguese development co-operation priorities do not highlight a specific policy on water and sanitation or Integrated Water Resources Management (IWRM).

Portugal has a highly decentralised aid programme spread over 15 different ministries plus universities, other public institutions and 308 municipal governments. The bulk of ODA is administered by the Ministry of Foreign Affairs and the Ministry of Finance. The Portuguese Institute for Development Support (IPAD) – a part of the MFA – is responsible for co-ordination. Water sector and sanitation co-operation activities are mostly in the scope of Ministry of Environment.

GEOGRAPHICAL AND SUB-SECTORAL FOCUS

The Strategic Vision document mentioned above defined priorities for the Portuguese development co-operation at two levels, geographical and sectoral. According to the geographical priorities, Portuguese speaking countries - most especially PALOP[12] and East-Timor - are priority intervention areas. The sectoral priorities are the following: good governance, participation and democracy, sustainable development and fight against poverty (that includes education, health, rural development, environment protection and natural resources sustainable management, economic growth, private sector development and training and job creation) and education for development.

In terms of development co-operation at a bilateral level, Portugal has been developing diverse projects in the sector of the environment, in PALOP countries (African countries having the Portuguese as its official language) and in East Timor, which focus, above all, on the areas of water and sanitation, combating climate change and institutional capacity building for this sector. Examples include the *Water Supply and Sanitation programme for Luanda* (Angola) - *Water treatment plant and technical assistance to the EPAL -* EP Luanda (Water Public Company); *The Project of Infrastructures of water supply, sanitation and solid waste sectors in Lumbo* – Mozambique; The Projects *of Rehabilitation of the Ataúro Aqueduct and of Water supply and sanitation, conveyance and distribution systems in Dili* - Timor-Leste. In Guinea-Bissau Portugal supports NGO projects like *The Communitarian management of water, sanitation and sanitary education on the islands of Uno and Formosa* (NGO CIC), and supports since 2001 an important Sanitation System and Waste Collecting Project in Bissau. In Cape Verde, an island with severe lack of water problems, Portugal finances the *Project of Substitution of the desalination for inverse osmosis of the Mindelo.*

AID MODALITIES

Portugal's main aid modality in its bilateral programme is Technical Co-operation (TC), which is primarily targeted at local training and empowerment. This TC takes various shapes, particularly teacher training, sending co-operation staff, awarding scholarships (including internal ones, for training in the partner country), and technical assistance to partner's government administration.

---

[12] PALOP: African countries having the Portuguese as its official language (Angola, Mozambique, Guinée-Bissau, Equatorial Guinea, Cap Verde and Sao Tome & Principe).

CRS AID ACTIVITIES IN SUPPORT OF WATER SUPPLY AND SANITATION – ISBN 978-92-64-05172-0 – © OECD/WWC 2008

Whenever possible, depending on the country, IPAD tries to actively participate in the SWAP processes for the water and sanitation sector. Water projects are also supported in aid to NGOs.

RULES AND PRACTICES

Regarding the programmes and projects financing, IPAD follows, as far as possible, the rules and practices of the EU and the OECD guidelines.

PERFORMANCE EVALUATION

IPAD applies a result-based project management. The project application forms are harmonised with the European Commission's forms. All the programmes and projects integrate relevant indicators for monitoring and evaluation.

COORDINATION WITH OTHER ACTORS

At a multilateral level, Portugal has contributed financially towards the main conventions, protocols and funds dedicated to environmental protection in general and water in particular, namely the *Global Environment Facility* (GEF), the *Multilateral Fund for the Montreal Protocol*, or the *UN Framework-Convention on Climate Change*. Portugal has also joined the EU *Water Facility Initiative*.

Through the Ministry of Finance and Public Administration, Portugal supports water sector activities indirectly by contributing to the resource replenishment of Multilateral Development Financial Institutions, such as the International Development Association (IDA), the African Development Fund (AfDF) and the Asian Development Fund (ADF), which have developed extensive programmes in support of water sector development.

SPAIN
Aid at a glance - Water supply and sanitation

Summary statistics

Unless otherwise stated, figures refer to 2005-2006 annual average commitments expressed in 2006 constant prices.

Total aid to water supply and sanitation			
	USD million	Aid to water by Spain as a share of total aid by Spain	Aid to water by Spain as a share of total DAC members' aid to water
Spain	58.1	4%	1%
For reference, total DAC	4974.0	8%	100%

Top ten recipients of aid to water supply and sanitation			
		Aid to water by Spain to that recipient as a share of	
	USD million	total aid by Spain to that recipient*	total DAC members' aid to water to that recipient
Honduras	7.3	18%	52%
Peru	5.5	8%	26%
Nicaragua	5.4	10%	31%
Guatemala	4.3	13%	22%
El Salvador	4.2	11%	62%
Mongolia	3.0	100%	68%
Morocco	2.3	4%	2%
China	1.9	3%	1%
Ecuador	1.4	4%	7%
Bolivia	1.3	2%	4%

Aid to all water-related sectors	
	USD million
Water supply and sanitation	58.1
Water transport	15.0
Hydro-electric power plants	0.0
Agricultural water resources	4.5
Total water-related aid	77.6

Annual aid to water supply and sanitation, 1993-2006 (2006 constant prices)

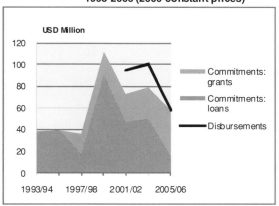

Regional distribution of aid to water supply and sanitation

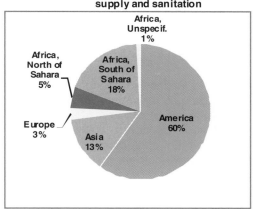

Water supply and sanitation aid by subsector

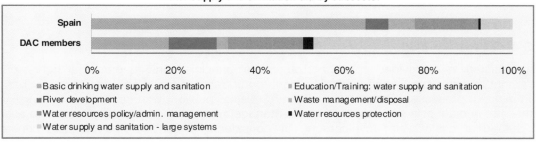

* % of sector allocable aid

SPAIN
Development Co-operation Policy in the water supply and sanitation sector

POLICY STATEMENT

Spanish Environment and Sustainable Development Strategy Paper has been drawn up in compliance to the Master Plan's 2005-2008 priorities on environment, and those commitments assumed by Spain when ratifying the international agreements stated in the UN Global Conferences. The European Commission's resolutions and communications, working on development co-operation issues, and OECD-DAC guidelines and recommendations for Environment stemmed, have been also taken into account.

The Spanish strategy in the water sector is based on the concept of Integrated Water Resources Management (IWRM). Its main objectives are to:

* Ensure the sustainable use of water resources;
* Strengthen institutional and human capacities within environmental management, as well as the processes of action and social participation, in order to reduce socio-environmental vulnerability and help an efficient and sustainable environmental management; and
* Support countries integration at regional level in order to tackle the water challenges which overdo the local level, border clashes resolution and economic integration.

The Spanish Agency for International Co-operation and for the Development (AECID) strives to integrate environmental issues in the development strategies of the recipient countries, including satisfaction of basic social needs, emergency plans and aspects of the climate change to promote the sustainable development.

GEOGRAPHICAL AND SUB-SECTORAL FOCUS

The geographic framework is divided in regions and sub-regions. In order to determine the priorities at regional level, a global analysis of both the environmental situation and the characteristics and circumstances has been developed. There is a classification of the countries taking into account the per capita income and the criteria for geographic allocation fixed by the Master Plan 2005-2008.

The level of per capita income may guide the priority actions fixed by the strategic lines, although other factors must be also taken into account in order to have a wider view since per capita income indicators could conceal levels of poverty. Actions must also take some of the regional factor into account: people's cultural and socio-economic characteristics, social inequities when referring to the access to water, energetic resources, sustainable development policies, potential partnerships in the region, civil society involvement in development policies, institutional capacities, etc... Special attention is given to the Latin American region due to historical reasons and because of the easiness of the Spanish Co-operation for working there, and to the Least Developed Countries (LDCs).

In the water and sanitation sector, high priority is given to Sub-Saharan Africa, Middle-East, Maghreb, Central and Eastern Europe, while medium priority is given to Central and South America, Asia and the Pacific.

RULES AND PRACTICES

Spanish assistance is provided in respect of several principles and fundamental rights: promotion of people's right to water supply and sanitation, right to information; social and private sector participation in decision-making processes; gender equality (in raising public awareness and training, in capacity strengthening, in diagnosis and research and in management); taking cultural diversity into account within actions aimed at sustainable development, by considering difficulties and people's concrete needs; systematizing good practices (notably the precautionary principle); healthy and equitable environment; common responsibilities; environmental responsibilities and sovereignty on natural resources.

PERFORMANCE EVALUATION

The integration of the Paris Declaration within the Environment and Sustainable Development Strategy Paper turns it into the appropriate instrument for harmonization and coordination amongst the actors of the Spanish Co-operation and an element for coherence policy of the Spanish Administration with respect to Environmental issues. The monitoring and evaluation system of the Strategy Paper is based on 3 criteria: ownership, internal and external coherence.

COORDINATION WITH OTHER ACTORS

In order to achieve the Millennium Development Goals (MDGs), Spain contributed to the creation of the Millennium Fund Spain-UNDP. Spain also supports the Global Environmental Facility (GEF), and strives to foster sectoral coordination of the co-operation institutions with the different national actors (ministries, autonomous communities, NGOs, trade unions, universities, research centres, etc.).

Web References:

AECID (Environment and Development Strategy Paper)
http://www.aecid.es/
http://www.maec.es/es/MenuPpal/Cooperacion%20Internacional/Publicaciones%20y%20documentacin/Documents/DES%20Medio%20Ambiente%20RE%20inglés.pdf (executive summary; English version)

Annual Plans for the Spanish Co-operation:
http://www.maec.es/es/MenuPpal/Cooperacion%20Internacional/Publicaciones%20y%20documentacin/Paginas/publlicaciones_cooperacion.aspx

SWEDEN
Aid at a glance - Water supply and sanitation

Summary statistics

Unless otherwise stated, figures refer to 2005-2006 annual average commitments expressed in 2006 constant prices.

Total aid to water supply and sanitation			
	USD million	Aid to water by Sweden as a share of total aid by Sweden	Aid to water by Sweden as a share of total DAC members' aid to water
Sweden	**97.0**	**5%**	**2%**
For reference, total DAC	*4974.0*	*8%*	*100%*

Top ten recipients of aid to water supply and sanitation			
		Aid to water by Sweden to that recipient as a share of	
	USD million	total aid by Sweden to that recipient*	total DAC members' aid to water to that recipient
Sri Lanka	19.5	51%	30%
Kenya	13.5	17%	20%
Palestinian Adm. Areas	3.2	11%	4%
Georgia	2.8	22%	19%
Philippines	2.6	35%	19%
Bolivia	2.4	10%	8%
Serbia	2.2	6%	7%
Indonesia	2.2	32%	1%
Honduras	1.6	14%	12%
Albania	1.1	7%	5%

Aid to all water-related sectors	
	USD million
Water supply and sanitation	97.0
Water transport	1.2
Hydro-electric power plants	1.3
Agricultural water resources	0.0
Total water-related aid	*99.5*

Annual aid to water supply and sanitation, 1993-2006 (2006 constant prices)

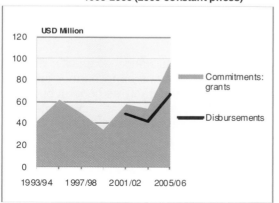

Regional distribution of aid to water supply and sanitation

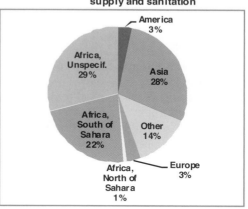

Water supply and sanitation aid by subsector

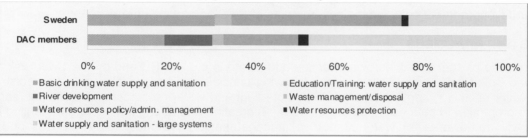

* Basic drinking water supply and sanitation
* Education/Training: water supply and sanitation
* River development
* Waste management/disposal
* Water resources policy/admin. management
* Water resources protection
* Water supply and sanitation - large systems

* % of sector allocable aid

SWEDEN
Development Co-operation Policy in the water supply and sanitation sector

POLICY STATEMENT

The Swedish International Development Agency's (Sida) position paper on *Management and Use of Water resources* and *Pure water – Strategy for water supply and Sanitation*, apply to all Swedish development co-operation in the areas of water resources management and water supply and sanitation (this covers integrated water resources management; transboundary waters; water supply; sanitation and hygiene promotion for households; as well as industrial water and wastewater management).

The main objective for Swedish development co-operation is to improve the livelihoods of poor people. In order to meet this objective, the following sub-objectives are considered the most important: improved health; environmental sustainability through protection and more efficient use of water resources as well as treatment of human waste and wastewater; enhanced incomes and economic development at household and macro levels and improved democratic governance.

In order to address the objectives strategically, Swedish assistance focuses on the following areas: 1) integrated water resources management 2) water supply, sanitation and hygiene promotion in urban and peri-urban slums; 3) waste and wastewater management in urban areas; 4) water supply, sanitation and hygiene promotion in rural areas; 5) capacity building; 6) water supply, sanitation and hygiene promotion in emergency situations.

GEOGRAPHICAL AND SUB-SECTORAL FOCUS

Sida's assistance in the water sector primarily targets Sida's programme countries in Africa South of Sahara but also countries in South Asia, South East Asia and Latin America. Support is also given to key regional and global actors and programmes. Sida's support to the water sector is based on programme-based approaches and includes support to reformation and development of key water and sanitation institutions including water governance.

AID MODALITIES

Sida supports programme-based approaches with technical assistance, capacity development and investments in water infrastructure through grants, credits and guarantees to partner countries. Swedish bilateral ODA is mainly channelled through government systems preferably as general budget support and sector budget programmes. Support is also provided to local and regional governmental institutions, independent actors such as CBOs and NGOs and promotes public-private partnerships.

Sida primarily uses partner country systems for procurement when such are available and are in accordance with internationally accepted principles and good procurement practices. If country systems are not deemed efficient and transparent Sida Procurement Guidelines (SPG) are applied. SPG are based on the World Banks procedural rules and conform to the legal principles stated in relevant European Community treaties and directives, provide policies and rules to be applied by co-operation partners or by Sida in Sida-financed operations. Applicable procurement guidelines are stipulated in the general co-operation agreement between the governments of Sweden and the partner country.

RULES AND PRACTICES

Based on the partner countries own priorities addressed in e.g. a PRS and on the Swedish governments approved Country Assistance Strategy the following principles and approaches guide Swedish assistance to the water sector: 1) application of principles of integrated water resources management, sanitation and hygiene promotion in both rural and urban areas; 2) creation and strengthening of an enabling institutional and policy frameworks for water resources management; 3) promotion of awareness and advocacy; 4) participation of consumers in all relevant stages of water supply and sanitation i.e. service design, delivery, maintenance, management and monitoring in order to recognise the poor as active citizens and to counteract practices of social exclusion; 5) good governance in water sector; 6) reform and refinement of water and sanitation utilities to address the diverse service needs of all citizens; 7) development and use of water and sanitation systems that are socially, economically and environmentally appropriate and sustainable, including rainwater harvesting and ecological sanitation; 8) strengthening effective cost recovery for operation, repairs and maintenance, and adequate reservation for depreciation and future investments, while at the same time developing mechanisms for appropriate cross-subsidisation in favour of the poor; 9) development and strengthening of appropriate financing mechanisms that link the domestic capital market to poor people's own resources; 10) development of mixed public/private sector solutions for sector investment, service delivery and operations and maintenance; 11) development of user-friendly environmental management and monitoring techniques for all levels of water supply and sanitation service provision and wastewater management and recycling, including capacity building for improved industrial water management.

An additional principle to the above with special regard to humanitarian assistance in emergency is the coordination of water supply and sanitation planning and inputs from all sources as well as the integration of services into existing, local institutional frameworks.

PERFORMANCE EVALUATION

Evaluations of ODA for specific water programmes have been concluded. The general framework for performance assessment provides that evaluations of development assistance through Sida are conducted at three levels:

1) by the sectorial and regional departments. Each department carries out evaluations within its own area of responsibility.
2) by the Department for Evaluation (UTV), which takes a broader view, focusing on thematic and strategic issues of wide relevance. UTV has an independent status and report directly to Sida's Director General. In addition to carrying out evaluations on its own, or in co-operation with other organisations, UTV supports the evaluation activities of Sida's other departments.
3) by the Swedish Agency for Development Evaluation, which is an independent agency mandated to evaluate all Swedish development co-operation.

COORDINATION WITH OTHER ACTORS

Sida actively seeks collaboration with other actors:

1) Co-operation with most UN actors – such as UNICEF, UNDP, Habitat, and WHO among others – takes place at national level wherever this is regarded as the most efficient way of supporting development in the sector. Sida supports regional and global initiatives through these actors.
2) Sida cooperates with various international organisations in the humanitarian sector, for example the Red Cross/Crescent Movement and international NGOs.
3) Sida strives to coordinate efforts with development banks such as the World Bank, African Development Bank, the European Development Bank, the European Investment Bank and other regional development banks, in order to link up with major credit schemes as well as with new sector thinking and reforms advocated by the banks. In some cases joint funding mechanisms are established (e.g. trust funds such as the Africa Water Facility, investment credit funds and investment guarantees).
4) Through its membership in the EU, Sweden contributes to the extensive development activities administered by the European Commission.
5) Sida actively seeks coordination with other bilateral agencies in the water resources, water supply and sanitation sector.
6) Sida takes an active part in various regional coordination mechanisms (e.g. the SADEC, River Basin Organisations, AMCOW, and the Nile Basin Initiative) aimed at capacity development, advocacy, conflict prevention and shared investment programmes. Water and sanitation as well as industrial pollution are key issues in these initiatives.

7) Sida participates in organisations and networks such as the Global Water Partnership, Water and Sanitation Programme and the Water Supply and Sanitation Collaborative Council to promote policy development and advocacy mainly on a global level. Networking and exchange of experience constitute important features of this co-operation.

8) The Swedish Resource Base: this includes central, regional and local government authorities, agencies and utilities, universities, consultancy and construction companies and non-governmental organisations.

Web References:

Swedish International Development Co-operation Agency:
http://www.sida.se/sida/jsp/sida.jsp?d=121&language=en_US

SWITZERLAND
Aid at a glance - Water supply and sanitation

Summary statistics

Unless otherwise stated, figures refer to 2005-2006 annual average commitments expressed in 2006 constant prices.

Total aid to water supply and sanitation			
	USD million	Aid to water by Switzerland as a share of total aid by Switzerland	Aid to water by Switzerland as a share of total DAC members' aid to water
Switzerland	**50.1**	**7%**	**1%**
For reference, total DAC	*4974.0*	*8%*	*100%*

Top ten recipients of aid to water supply and sanitation			
		Aid to water by Switzerland to that recipient as a share of	
	USD million	total aid by Switzerland to that recipient*	total DAC members' aid to water to that recipient
Tanzania	7.7	27%	13%
Azerbaijan	6.5	90%	28%
Serbia	5.6	13%	17%
Bosnia-Herzegovina	5.1	31%	54%
Macedonia (TFYR)	4.4	44%	58%
Mozambique	4.1	39%	6%
India	1.8	17%	0%
Bangladesh	1.2	15%	1%
Mali	1.2	11%	3%
El Salvador	0.9	46%	14%

Aid to all water-related sectors	
	USD million
Water supply and sanitation	50.1
Water transport	0.0
Hydro-electric power plants	0.0
Agricultural water resources	0.0
Total water-related aid	*50.1*

Annual aid to water supply and sanitation, 1993-2006 (2006 constant prices)

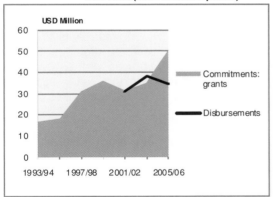

Regional distribution of aid to water supply and sanitation

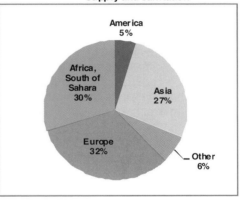

Water supply and sanitation aid by subsector

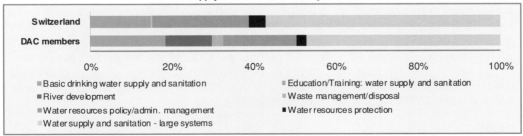

* % of sector allocable aid

SWITZERLAND
Development Co-operation Policy in the water supply and sanitation sector

POLICY STATEMENT

The Orientation Paper Water 2015 outlines how the Swiss Agency for Development and Co-operation (SDC) aims to contribute in the water sector towards the achievement of the MDGs in the period leading up to 2015. The international reference framework is provided by the various Rio Conventions as well as environments agreements such as the Climate Convention, the Kyoto Protocol, the Convention of Biological Diversity, Desertification and Preservation of Wetlands (Ramsar). SDC promotes a holistic view of water issues that considers the entire water cycle through the concept of Integrated Water Resources Management (IWRM). In this framework, access to services for the poor is the entry point for the Swiss engagement. SDC's priorities are:

- Satisfying basic needs: ensuring equitable access to water and basic water services for all without discrimination;
- Ensuring the efficient, non-polluting and sustainable use of water and water infrastructure in agriculture;
- Ensuring water cycle, through sustainable management and protection of the water resources;
- Reducing natural hazards and risks: measures to cope with droughts and floods are an integral part of all water programmes;
- Participation, including gender equity, protection of cultural values and local knowledge; and
- Good governance, in both private and public sectors and in the civil society.

SDC differentiates 4 water uses (According to the categorization proposed by the Global Water Partnership - GWP), that are all linked in the frame of IWRM. Water for people (including sanitation) and water for food are the main targets for SDC investments.

- Water for people: SDC programmes comprise water supply, sanitation, hygiene, wastewater and solid waste management and recycling. The provision of water for drinking purposes has priority over all other uses. To improve welfare in a sustainable way, investments in water supply and sanitation are linked to income generation (horticulture, livestock keeping). SDC supports the HCES approach (Household Centred Environmental Sanitation) as well as ecological sanitation (EcoSan). In addition to a long-standing focus on rural drinking water supply issues, SDC increasingly supports the improvement of water supply and sanitation services in small towns and peri-urban areas.

- Water for food: implementation of the "more crops and jobs per drop" strategy, as long as this is planned in an environmentally friendly and socially acceptable way; promotion of cost-effective and environmentally friendly technologies and user systems for irrigation (drip-irrigation, treadle pumps) as well as improved farming practices and soil and water conservation (water efficient crop varieties, slope agriculture technologies, rainwater harvesting). SDC concentrates its efforts at the village level, with an important focus on hillside farming and watershed development. SDC does not support new construction of large-scale irrigation schemes. Where such schemes already exist, SDC supports improved management and rehabilitation measures.

- Water for nature: Across all water use categories, SDC promotes water savings and conservation to preserve the functionality and the regeneration capacity of ecosystems upstream and downstream. SDC supports the concept of Environmental Flows, as well as permanent ground and surface water monitoring, mapping and modelling systems.

- Water for other uses (industry, energy, transportation): SDC insists on compliance with the guidelines of the World Commission on Dams and defends the interests of the poorest through

 CRS AID ACTIVITIES IN SUPPORT OF WATER SUPPLY AND SANITATION – ISBN 978-92-64-05172-0 – © OECD/WWC 2008

advocacy. SDC promotes measures which aim for cleaner industrial production, as well as decentralised alternative energy projects. SDC supports the WEF (World Economic Forum) Water Initiative to increase stewardship and responsiveness of the private sector in water management in partnerships with the public sector and civil society.

SDC will continue to finance bilateral and multilateral water sector projects at least at its current level. SDC's multilateral funds will be opened up for water activities. Additional funds will be made available in emergencies and for disaster relief.

GEOGRAPHICAL AND SUB-SECTORAL FOCUS

In terms of presence on the field, SDC is currently supporting approximately 200 water projects. The geographical focus of SDC's support to the water sector remains in Asia and Africa, followed by Latin America. A significant increase is observed in East Europe. The portfolio is dominated by support in the area "Water for people" and "water for food" but investment in "water for nature" has significantly increased these last years. Concerning overall ODA, SDC decided to further concentrate its portfolio. By 2010, the number of priority countries will be narrowed down to 12.

RULES AND PRACTICES

Water is a common good, and access to it is a human right. Recognising the fundamental and inalienable nature of these two values, SDC supports a rights-based approach: making authorities responsible for respecting, protecting and fulfilling the right to access to sufficient, safe and affordable water for all people; empowering people to exercise their rights and responsibilities.

SDC's programmes are guided by six interdependent fields to ensure equitable, efficient and sustainable management of water resources:

- Social: SDC promotes a decentralised, bottom-up approach, in which participation by all segments of the society is essential. SDC promotes legal systems that acknowledge indigenous rights and rights of minority groups without discrimination and marginalisation. The recognition of indigenous knowledge is also advocated in national water strategies. SDC applies the Demand Responsive Approach (DRA), where consumers are empowered to make informed choices and where they have control over investment and operational decisions. Women are involved with equal rights as individual users and as partners for institutional development. SDC especially promotes access to water and sanitation in schools which is considered crucial for the attendance of girls.

- Environmental: SDC encourages the application of the precautionary principle, the "polluter pays" principle, and the implementation of compensation schemes for environmental services. In all activities affecting water and land use, SDC applies the approach of integrated risk and disaster reduction.

- Economic: SDC favours an approach where the consumer covers the costs of operation, maintenance and (at least partially) replacement. SDC supports mechanisms facilitating access to bank credits for the local public, private and civil sectors especially for poor communities. SDC explores the potential of solidarity fund-mobilisation by civil societies in the North, e.g. with an additional solidarity cent on the ordinary water rates or on bottled water.

- Institutional: SDC supports institutional development in accordance with good water governance, in particular institutions that permit an appropriate and locally adjusted sharing of roles and responsibilities between government authorities, civil society and the private sector. SDC encourages reform processes that promote decentralisation and participation. SDC works to resolve international cross-border conflicts of interests and promotes regional organisations and commissions for the joint management of shared water resources between neighbouring states.

- Technological: SDC supports socially acceptable, efficient, affordable, cost effective and environmentally friendly technologies, as well as local businesses that produce such technologies. Technology development in partnership between universities of the North and the South, national governmental organisations and NGOs are encouraged.

- Knowledge: SDC fosters and supports national and international networks, centres of knowledge and communities of practice as drivers to generate and transfer of knowledge. SDC actively seeks co-operation and information exchange with other federal agencies, international institutions and organisations as well as with the private sector and NGOs.

Switzerland states that planning, decision-making and management of resources (including financial) must be deferred to the lowest possible level, according to the subsidiarity principle. While governments must remain the owners of the resource and carry the ultimate responsibility for all legal and regulatory aspects of water management, they can delegate or transfer the right to use and/or the task of distribution to private operators under transparent and fair conditions. To avoid abuse of monopolistic power it is strongly recommended that the public sector retains the main ownership of public water supply and sanitation assets.

COORDINATION WITH OTHER ACTORS

SDC has the responsibility to coordinate the Swiss Interdepartmental group working in water and development.

SDC participates in policy dialogue at bilateral and multilateral level, and advocates a high profile for water on the international political agenda. Switzerland is highly committed to several multilateral institutions related to the water sector, in particular with the Water Supply and Sanitation Collaborative Council (WSSCC) for sanitation advocacy and increase coverage, the Water and Sanitation Program (WSP) for national reforms, large scaling-up of relevant approaches, the International Union for Conservation of Nature (IUCN) for the aspects of environmental and ecosystem services and the Global Water Partnership (GWP) for all the approach related with IWRM.

SDC supports the development of an enabling environment for private sector participation through multi-stakeholder dialogues, and advocates for an adequate system of rules and regulations, particularly where it enhances the role and potential of small-scale providers.

SDC, the State Secretariat for Economic Affairs (SECO) and the main multinational reinsurance company Swiss_RE have developed policy principles and guidelines for public-private partnerships (PPP) between the public authorities and the local, national or international private sector: it aims to establish good governance structures, and to increase involvement of civil society. Emphasis is also on improving the sustainability and poverty focus of PPPs. The key factors of such partnerships are: a sound political and legal framework, local ownership and leadership, effective regulatory mechanisms, adequate framework for investment, coordination with PRSPs, and smoothly functioning supply chains.

SDC in partnership with the World Economic Forum (WEF) facilitates the development of PPP between enterprises using water in their processes and Public sector and Civil society. Various important PPPs have been launched in India and South Africa.

SDC in partnership with Swiss municipal authorities and municipal water services has developed the "Solidarité EAU" platform to support a North-South decentralized co-operation for water and sanitation services. In this frame, SDC also encourages public-public-partnerships (PUPs) in water supply services.

Web References:

SDC Website http://www.deza.ch/en/Home/Themes/Water

UNITED KINGDOM
Aid at a glance - Water supply and sanitation

Summary statistics

Unless otherwise stated, figures refer to 2005-2006 annual average commitments expressed in 2006 constant prices.

Total aid to water supply and sanitation			
	USD million	Aid to water by United Kingdom as a share of total aid by United Kingdom	Aid to water by United Kingdom as a share of total DAC members' aid to water
United Kingdom	112.2	3%	2%
For reference, total DAC	4974.0	8%	100%

Top ten recipients of aid to water supply and sanitation			
		Aid to water by United Kingdom to that recipient as a share of	
	USD million	total aid by United Kingdom to that recipient*	total DAC members' aid to water to that recipient
Bangladesh	33.2	10%	19%
Ghana	7.1	4%	19%
Congo, Dem. Rep.	6.3	10%	19%
Sierra Leone	4.6	11%	79%
Sudan	3.9	3%	27%
Nepal	2.7	3%	14%
India	2.4	0%	0%
Guyana	2.0	80%	40%
China	1.5	2%	0%
Montserrat	1.5	20%	100%

Aid to all water-related sectors	
	USD million
Water supply and sanitation	112.2
Water transport	1.3
Hydro-electric power plants	0.0
Agricultural water resources	0.0
Total water-related aid	*113.5*

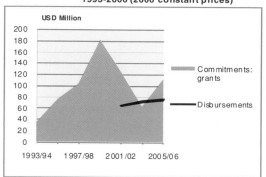

Annual aid to water supply and sanitation, 1993-2006 (2006 constant prices)

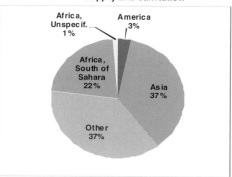

Regional distribution of aid to water supply and sanitation

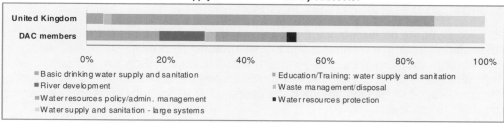

Water supply and sanitation aid by subsector

* % of sector allocable aid

The United Kingdom commissioned a study to further assess DFID spending in the water sector which suggested a rather higher level of spending, £ 128.8 million in financial year 2005-2006 (approximately USD 238 million). The study tried to identify all activities related to some extent to water. The search was based on water-related sector codes including when aid for water was extended as part of humanitarian aid, water-related objectives (activities reported under non-water-related sectors but marked as contributing to water-related objectives), and keywords. For each identified activity, the proportion of amount to take into account as "aid for water" was estimated.

UNITED KINGDOM
Development Co-operation Policy in the water supply and sanitation sector

This policy brief was produced prior to the launch of DFID's new water policy in October 2008.

POLICY STATEMENT

The United Kingdom's strategy is defined by a Water Action Plan, produced in 2004. DFID deals with water issues as an integral part of other projects in health and education. In its target strategy paper on water, DFID states that national governments must lead the way and its efforts must contribute to and be guided by their Poverty Reduction Strategies Papers (PRSP). Thus, Poverty Reduction Budget Support (PRBS) is one of DFID's most important bilateral funding mechanisms, at the core of its development assistance policy.

DFID's policy paper on water and sanitation focuses on 3 main areas: managing water resources, sanitation and good governance. DFID's primary aim is to move the issue of sanitation further up the political agenda. Concerning good governance, as DFID regards the concept as central to the change that is needed, DFID's commitments on increased aid in 2005, in particular relating to Africa, were made in return for a commitment to better governance. In its programmes, DFID focuses on state capability, responsiveness and accountability, strives to improve financial management, and works in particular with citizens, civil society groups, parliamentarians and the media. The United Kingdom also intends to adopt a new "quality of governance" assessment to guide the way in which aid is provided, and to launch a new £100 million Governance and Transparency Fund.

DFID especially helps local service providers (public and private, formal and informal) and local governments, which have to cope with huge population growth in towns and cities. Its actions aim at providing water and sanitation to the poorest, expanding coverage, improving services, and making sure that systems are well managed and maintained. In this regard, it is particularly important to involve all sections of a community in planning and implementation.

The 2006 White Paper sets out DFID's general commitments on water and sanitation. DFID is committed to increase spending on the essential public services, including water and sanitation, to at least half of the United Kingdom's direct support to developing countries. In particular, DFID will double its support to water and sanitation in sub-Saharan Africa to £95 million a year by 2007-2008, and more than double funding again to £200 million by 2010-2011 (all DFID's commitments are based on the definition of the "water-related sector" agreed at the Rio Summit, 1992 – Agenda 21).

The United Kingdom also committed to increasing its funding for innovative technologies for water treatment, purification and sanitation.

GEOGRAPHICAL AND SUB-SECTORAL FOCUS

DFID tries to concentrate its assistance on countries which have the largest numbers of poor people and the greatest regional and global influence on development. 16 partner countries, mainly located in sub-Saharan Africa and southern Asia, have signed with the United Kingdom a Public Service Agreement (PSA) which aims at achieving the MDGs, notably through a commitment for improved governance. In these PSA countries, DFID helps strengthen the national PRSP formulation processes, ensuring that water is given appropriate emphasis.

DFID has currently a target of allocating at least 90% of its bilateral funding for Low Income Countries (LIC) in 2005-2008. Within this allocation guideline, it is planned to scale-up aid to Africa and southern Asia, while de-emphasizing aid to Latin America. This will lead in particular to renewed intensity of support to regions most at danger of failing to meet the MDG targets, especially sub-Saharan Africa (DRC, Sudan, Ethiopia and Nigeria). The United Kingdom has proposed to move further into more complex and difficult environments for aid delivery, such as fragile states, especially those vulnerable to conflict.

 CRS AID ACTIVITIES IN SUPPORT OF WATER SUPPLY AND SANITATION – ISBN 978-92-64-05172-0 – © OECD/WWC 2008

AID MODALITIES

DFID provides predictable and flexible assistance using, as appropriate, either direct budget support, basic service grants (which earmark resources to one or more sector), or working through civil society, faith-based or other organisations. DFID gives a particular emphasis to the need for better aid predictability. DFID is able to set rolling triennial budget at the division and department levels and frequently enters into three-year arrangements with its partner countries. DFID is also considering longer-term arrangement with countries committed to poverty reduction and good governance, and has signed ten-year partnership arrangements with Sierra Leone, Rwanda and Afghanistan.

Since 2000, the United Kingdom has actively promoted new funding mechanisms, such as the International Finance Facility (IFF) proposed in 2003.

RULES AND PRACTICES

The United Kingdom is active in capacity-building initiatives, such as the Research-inspired Policy and Practice Learning (RIPPLE) in Ethiopia and the Nile Basin, a five-year programme led by the Overseas Development Institute (ODI), which aims to advance evidence-based learning on water supply and sanitation, financing and sustainability. The project, that trains several masters-level students, helps to improve data collection and water mapping and to strengthen local capacity to deliver water to the area. DFID also supports Water and Sanitation for the Urban Poor (WSUP). This partnership between the private, public and civil society sectors builds capacity of local service authorities and communities alike. It is currently operating in Mozambique, Kenya, Madagascar and India, and will reach 35,000 people by 2008, rising to half a million by 2009.

DFID has launched an initiative of south-south decentralised co-operation, based on knowledge sharing. In 2007, the agency funded workshops in Asia and Africa to explore whether regional partnerships between public operators were wanted. The workshops helped bring utility managers together and share lessons learned about reform and joint working, so that public utilities can replicate successes across different parts of the globe. DFID also funded a report by Loughborough University's Water Engineering Development Centre (WEDC) on water operators' partnerships and will fund the WSP to co-ordinate a regional water operators' partnership in Africa.

COORDINATION WITH OTHER ACTORS

In the water sector, DFID channels about one third of its assistance (37% in 2005-2006) through multilateral organisations such as the World Bank and United Nations organisations. In order to improve the way international system works, DFID has decided to focus its support on a smaller number of key international partnerships and networks.

In particular, DFID strongly supports the EU Water Initiative and is working to reform and empower it to meet its objectives on improving the effectiveness of EU development assistance on the ground. DFID also provides significant support to 3 other key programmes and institutions: 27% (almost £1m a year) of the core funding for the Water Supply and Sanitation Collaborative Council (WSSCC); over £53 million for the Water and Sanitation Programme (WSP) over 10 years; 18% of the total core funding of the Global Water Partnership (GWP) to improve coordination of all of the international action that concerns water.

DFID also support several regional programmes, and funding programmes such as the Public-Private Infrastructure Advisory Facility (PPIAF). DFID has pledged £6 million (2008 USD 11.8 million) in technical assistance to support the Africa Development Bank's Rural Water Supply and Sanitation Initiative.

In order to advocate greater effort and co-operation of international stakeholders, DFID published in 2006 the Global Call to Action on water and sanitation, known as the Five Ones: one global annual report on water and sanitation to monitor progress towards reaching the MDG targets, one global annual high-level meeting to agree on priority action, and within each country, one water and sanitation plan, one group to co-ordinate it and one lead UN body for the sector at the national level.

To improve co-ordination among its partners, the United Kingdom held its External Water Forum with over 120 stakeholders from NGOs, trade unions, academia and other government organisations. DFID works with civil society organisations in all regions to help them demand better access to water and sanitation. For example, the United Kingdom supports the work of WaterAid, that is a leading advocate for increasing water and sanitation access and initiated innovative ways of delivering to poor people, such as the Community Led Total Sanitation programme and community based urban service delivery. Finally, DFID and the World Bank Water and Sanitation Programme launched a domestic private sector participation initiative in 2005.

Web References:

Financial Support to the Water Sector, 2004-2006
http://www.dfid.gov.uk/pubs/files/water-sector-finance06.pdf

DFID's Water Action Plan 2004 http://www.dfid.gov.uk/pubs/files/wateractionplan.pdf

UNITED STATES
Aid at a glance - Water supply and sanitation

Summary statistics

Unless otherwise stated, figures refer to 2005-2006 annual average commitments expressed in 2006 constant prices.

* % of sector allocable aid

Total aid to water supply and sanitation

	USD million	Aid to water by United States as a share of total aid by United States	Aid to water by United States as a share of total DAC members' aid to water
United States	**902.5**	**5%**	**18%**
For reference, total DAC	4974.0	8%	100%

Top ten recipients of aid to water supply and sanitation

Aid to water by United States to that recipient as a share of

	USD million	total aid by United States to that recipient*	total DAC members' aid to water to that recipient
Iraq	742.6	16%	99%
Palestinian Adm. Areas	59.6	30%	69%
Jordan	53.0	31%	49%
India	16.3	13%	3%
Afghanistan	9.0	1%	41%
Sudan	5.3	3%	36%
Yemen	3.4	18%	7%
Kenya	2.8	1%	4%
Egypt	0.7	0%	1%
China	0.4	1%	0%

Aid to all water-related sectors **

	USD million
Water supply and sanitation	902.5
Water transport	37.5
Hydro-electric power plants	1.1
Agricultural water resources	146.7
Total water-related aid	1087.8

Annual aid to water supply and sanitation, 1993-2006 (2006 constant prices)

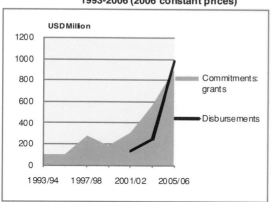

Regional distribution of aid to water supply and sanitation

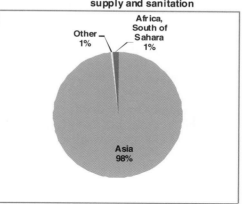

Water supply and sanitation aid by subsector

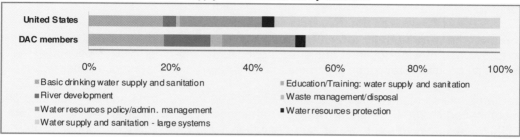

** The United States assess their total water-related aid to USD 1217.5 million. This figure takes into account water components of projects recorded in the environment sector (USD 20.6 million) and water aid associated with humanitarian aid (USD 119.1 million).

UNITED STATES
Development Co-operation Policy in the water supply and sanitation sector

POLICY STATEMENT

Senator Paul Simon Water for the Poor Act of 2005 (the WfP Act) makes the provision of safe water and sanitation services in developing countries a component of U.S. foreign assistance. The U.S. water policy is enumerated in ten action points in Section 3 of the WfPAct with the mandate "to provide affordable and equitable access to safe water and sanitation in developing countries" within the context of sound water resources management.

The WfP Act requires the Secretary of State, in consultation with the U.S. Agency for International Development (USAID) and other U.S. Government agencies, to develop and implement a strategy. Seen as a global challenge, the U.S. strategy draws on the growing body of internationally-endorsed principles and practices in water supply, management and productivity to achieve two internationally-agreed goals on drinking water and sanitation – to halve by the year 2015, the proportion of people unable to reach or afford safe drinking water [Millennium Declaration] and the proportion of people without access to basic sanitation [Johannesburg Plan of Implementation].

In the U.S. response to this challenge, aid for water and sanitation is part of a larger, integrated foreign assistance strategy where the specific approach depends strongly on the country context. The principles guiding resource allocations vary according to objective and depend on those factors that are most important for aid effectiveness and development results. While donors' resources are important, water and sanitation issues are fundamentally local challenges that require commitment, leadership, investments, and action by local, national, and regional governments and civil society.

U.S. Goal: In a water-secure world, individuals and countries would have reliable and sustainable access to an acceptable quantity and quality of water to meet human, livelihood, ecosystem, and production needs while reducing the risks of extreme hydrological events to people, the environment, and economies. Helping countries to achieve water security is fundamental to creating a safer and more prosperous world.

U.S. objectives to achieve this goal are:

- **Increase access to, and effective use of, drinking water and sanitation to improve human health**. This includes both short- and long-term sustainable access to drinking water and adequate sanitation, as well as activities to improve hygiene.

- **Improve water resources management**. This includes optimizing the benefits of drinking water among competing uses while ensuring human needs are met and environmental resources are protected. It also involves supporting regional efforts in managing shared waters (both surface and ground) and managing and/or adapting to hydrological variability and the risks of floods and droughts.

- **Increase the productivity of water resources.** This includes maximizing the efficient and productive use of water used in industrial, agricultural, and other consumptive sectors, as well as supporting pollution prevention programs and other programs that reduce water losses.

The United States has committed nearly $2 billion in 2005-06 for water aid and an estimated $900 million have been committed in 2007. In 2008, Congress enacted legislation that appropriated "...not less than $300 000 000 ... for safe drinking water and sanitation supply projects..." distributed over eight fund accounts. This is first time that funds have been appropriated specifically to implement the WfP Act in addition to other funds for water and sanitation. All funds support five approaches to achieve the water and sanitation objectives: capacity building, institutional strengthening and policy/regulatory reform;

diplomatic engagement; direct investment; investments in science and technology; and working through partnerships, especially in the engagement of nongovernmental and private partners to bring additional resources to address the issues.

GEOGRAPHICAL AND SUB-SECTORAL FOCUS

While over 80 countries receive water and sanitation aid, the United States has identified 36 countries where water will be a priority for U.S. foreign assistance based on country need and where U.S. assistance can make the most meaningful impact: Sub-Saharan Africa (16), Asia and the Pacific (11), Middle East (5), Europe/Eurasia (3), and Latin America/Caribbean (1). Both urban and rural areas are targeted.

Safe water access, basic sanitation, water/sanitation policy and governance, and host country strategic information capacity are the key sub-sector foci of U.S. aid for water. Additionally, the United States recognizes the vital link between emergency and development assistance - better design of emergency assistance support long-term development needs and provide a building block for sustainable access to water and sanitation services. According, over 10% of U.S. aid for water is channeled through disaster assistance programs.

AID MODALITIES

Grant assistance through programs/projects is the primary aid modality. The United States has also developed and expanded upon a number of new mechanisms – such as the Global Development Alliance and the Development Credit Authority – to creatively engage nongovernmental partners, including the commercial private sector and private philanthropic organizations, to mobilize additional resources to address key development challenges.

RULES AND PRACTICES

The Department of State and USAID have begun work on a joint strategic framework on water— "Addressing Water Challenges in the Developing World: A Framework for Action," whose purpose is to provide embassies and USAID missions with guidelines for developing activities within their host countries to achieve U.S. objectives on water. It also serves to inform country-level counterparts and other members of the international water community about the U.S. Government approach to the water sector, facilitating improved collaboration, communication, and shared learning. The recommendations of the Framework for Action are based on an assessment of U.S. investments in the sector and the limitations and comparative advantages of the U.S. Government and its partners.

PERFORMANCE EVALUATION

Assessment of country needs is a key factor in deciding whether and how water should be a component in a specific country plan. As an integral part of the Foreign Assistance Framework, there are standard indicators to measure both what is being accomplished with U.S. Government foreign assistance funds and the collective impact of foreign and host-government efforts to advance country development. Assessments of indicators are incorporated in annual Foreign Assistance Performance Reports and Performance Plans. Some key indicators include: total burden of water-related diseases, water-related disease outbreaks, renewable water available per capita, human/economic vulnerability to variable water flow and precipitation, stored water available per capita per day, and dependence on shared surface water or groundwater.

COORDINATION WITH OTHER ACTORS

More than 15 U.S. Federal agencies are involved in international water issues. Of these, three receive direct appropriations related to water in developing countries: USAID, the Millennium Challenge Corporation (MCC), and the Department of Defense. A Water Technical Earmark Group (TEG), comprised of functional and regional bureau staff, has been organized to assist Posts to meet the statutory requirements of the WfP Act in country assistance plans. To achieve foreign assistance goals in a specific country, U.S. investments in the water sector are made as part of a comprehensive country plan where inputs from a wide array of stakeholders in the partner country and the international water community are solicited.

The United States contributes to the general budgets of a number of international organizations that support freshwater projects around the world, as well as water and sanitation services in the context of emergency relief. The United States is a member of, makes financial contributions to, and exercises leadership in seven multilateral development banks – notably the World Bank Group and the Global Environment Facility – that support freshwater projects around the world. The United States does not fund water programs directly through its core contributions to these international organizations, although a percentage of this contribution is spent on water-related programs.

The United States actively solicits input from the NGO community, civil society entities, and the private sector on water/sanitation issues and contracts with them as executive agents to implement the U.S. water and sanitation assistance activities. Through new mechanisms, like the Global Development Alliance, the United States partners with the private sector to bring more resources to address water and sanitation needs.

Web References:

Senator Paul Simon Water for the Poor Act of 2005, H.R. 1973 [109th]
http://www.govtrack.us/congress/billtext.xpd?bill=h109-1973

U.S. Department of State
http://www.state.gov/g/oes/water/ for links to reports/documents/releases associated with the WfP Act.
http://www.state.gov/documents/organization/105643.pdf , Framework for Action on Water.
http://www.state.gov/f/ - State/Director of U.S. Foreign Assistance and Administrator of USAID
HTTP://WWW.STATE.GOV/G/OES/ - STATE/ BUREAU OF OCEANS AND INTERNATIONAL ENVIRONMENTAL AND SCIENTIFIC AFFAIRS
HTTP://WWW.STATE.GOV/P/NEA/RLS/RPT/2207/ - SECTION 2207 REPORT TO CONGRESS ON THE USE OF IRAQ RELIEF AND RECONSTRUCTION FUNDS

U.S. Agency for International Development (USAID)
http://www.usaid.gov/our_work/environment/water/
http://www.usaid.gov/our_work/environment/water/disaster.html - Disaster Assistance
http://www.usaid.gov/our_work/environment/water/gda.html - Global Development Alliance
http://www.usaid.gov/our_work/environment/water/dca.html - Development Credit Authority
http://www.usaid.gov/our_work/environment/water/congress_reports/2006_water_report_to_congress.pdf
http://www.usaid.gov/our_work/environment/water/congress_reports/water_congr_rpt_final05.pdf
http://pdf.usaid.gov/pdf_docs/PDACA138.pdf
http://www.usaid.gov/our_work/environment/water/congress_reports/drinkingwater_summary.pdf

Millennium Challenge Corporation http://www.mcc.gov/

U.S. Department of Defense http://www.sigir.mil/reports/quarterlyreports/Default.aspx

Foreign Assistance Framework and Performance Evaluation
http://www.state.gov/f/c23053.htm - Framework
http://www.state.gov/documents/organization/93447.pdf - Framework Structure and Definitions
http://www.state.gov/f/indicators/ - Framework Indicators
http://www.state.gov/documents/organization/101761.pdf - Master list of Common Indicators
http://www.state.gov/documents/organization/101764.pdf - Investing in People Indicator and Definitions
http://www.state.gov/documents/organization/101765.pdf - Economic Growth Indicators and Definitions
http://www.state.gov/documents/organization/101766.pdf - Humanitarian Assistance Indicators and Definitions
http://www.state.gov/documents/organization/101446.pdf - Fiscal Year 2007 Foreign Assistance Performance Report and Fiscal Year 2009 Performance Plan

PART 3

List of Aid Activities

2006

www.oecd.org/dac/stats/idsonline

AID ACTIVITIES IN SUPPORT OF WATER
2006 commitments, USD thousand

Source: Creditor Reporting System (CRS)

RECIPIENT / Donor	Agency	Sector	Amount USD thousand	Project description	CRS ID Number
Afghanistan					
AsDF		14010	63707	WESTERN BASINS WATER RESOURCES MANAGEMENT PROJECT	060031
Germany	GTZ	14010	1004	SECTOR REFORM WATER MANAGEMENT	2006006065
Germany	GTZ	14010	1255	DEVELOPMENT OF DECENTRALIZED WATER SUPPLY AND WASTE WATER MANAGEMENT	2006006051
Germany	BMZ	14010	126	INTEGRATED EXPERTS	2006002409
Germany	KFW	14020	1883	EXTENSION WATER SUPPLY, KABUL	2006000020
Germany	KFW	14020	14435	EXTENSION WATER SUPPLY II, KABUL	2006000019
Germany	BMZ	14030	126	BASIC DRINKING WATER SUPPLY AND BASIC SANITATION	2006005408
IDA		14010	31200	AFGHANISTAN: EMERGENCY NATIONAL SOLIDARITY PROJECT II	060091a
IDA		14010	29200	AFGHANISTAN URBAN WATER SECTOR PROJECT	060054a
IDA		14010	16000	AFGHANISTAN: EMERGENCY NATIONAL SOLIDARITY PROJECT (II SUPPLEMENTAL)	060046a
Ireland	DFA	14030	500	BASIC DRINKING WATER AND SANIT	060237
Japan	JICA	14020	49	WATER SUPPLY & SANIT. - LARGE SYST.	065590C
Japan	JICA	14040	0	RIVER DEVELOPMENT	065591C
Japan	JICA	14040	404	STUDY ON GROUNDWATER RESOURCES POTENTIAL IN KABUL BASIN IN ISLAMIC REPUB	060478C
Japan	JICA	14040	2837	STUDY ON GROUNDWATER RESOURCES POTENTIAL IN KABUL BASIN IN ISLAMIC REP.	060051C
Korea	KOICA	14010	19	TRANNIG PROGRAM	2006861572
Korea	KOICA	14030	18	DRINKING WATER PROJECT IN KABUL CITY IN AFGHANISTAN	2006862190
Netherlands	MFA	14030	120	BASIC DRINKING WATER SUPPLY AND BASIC SANITATION	2006001151
Norway	MFA	14030	220	SAFE WATER SUPPLY SYSTEM IN GUIZWAN DISTRICT - FARYAB	2006003914
Norway	MFA	14081	109	NCA WOMEN, WATER AND HEALTH	2006000223
UNICEF		14010	59	WATER RESOURCES POLICY/ADMIN. MGMT	060172
UNICEF		14030	161	SECTOR PLANS FOR HYGIENE, SANITATION AND WATER	060171
UNICEF		14030	975	HYGIENE, SANITATION AND WATER SUPPLY PROGRAMMES	060170
UNICEF		14030	614	COMMUNITY HYGIENE AND WATER SAFETY	060169
United States	DOD	14010	4478	WATER AND SANITATION	2006002859

Total aid for water in 2006 for   **Afghanistan**   **USD thousand 169499**

And as a share of aid to all recipient countries   **2.60%**

RECIPIENT / Donor	Agency	Sector	Amount USD thousand	Project description	CRS ID Number
Africa, regional					
AfDF		14010	13798	SADC: SHARED WATERCOURSES SUPPORT PROJECT	060059
Austria	ADA	14010	160	WATER RESOURCES POLICY/ADMIN. MGMT	2005000203
Canada	CIDA	14010	43	GATINEAU, QUÉBEC / 2007.05.24-25	060809
EC	CEC	14010	41558	ACP EU WATER FACILITY (2ND COMMITMENT) FOR GRANT CONTRACTS	2006100325
EC	CEC	14010	155768	ACP - EU WATER FACILITY	2006100234
IDA		14010	29700	SENEGAL RIVER BASIN MULTI-PURPOSE WATER RESOURCES DEVELOPMENT PROJECT	060062a

RECIPIENT Donor	Agency	Sector	Amount USD thousand	Project description	CRS ID Number
IDA		14020	4422	AFRICA STOCKPILES PROGRAMME - PROJECT 1 (MALI, MAROCCO AND TANZANIA)	060120b
Sweden	Sida	14010	88	EXCHANGE LVBC-HELCOM	2006000276
Sweden	Sida	14010	1831	PARTNERSHIP FUND 2006-08	2006000252
Sweden	Sida	14020	144	REVIEW WUP	2006002380
Sweden	Sida	14081	1383	244 ITWRM LAKE VICTORIA	2006008261
Sweden	Sida	14081	47	AFDB REVIEW	2006003375
UNICEF		14010	6	WATER RESOURCES POLICY/ADMIN. MGMT	063196
United Kingdom	DFID	14010	92	AFRICAN DEVELOPMENT BANK - RWSSI	060703
United Kingdom	DFID	14010	648	AFDB: AFRICAN WATER FACILITY	060702

Total aid for water in 2006 for **Africa, regional** USD thousand 249688
And as a share of aid to all recipient countries 3.83%

Albania

RECIPIENT Donor	Agency	Sector	Amount USD thousand	Project description	CRS ID Number
Austria	ADA	14020	34	WATER SUPPLY & SANIT. - LARGE SYST.	2002009037
Austria	Reg	14030	16	BASIC DRINKING WATER SUPPLY AND BASIC SANITATION	2006008300
Austria	BMF	14050	1506	WASTE MANAGEMENT: INTERCOASTAL ZONE MANAGEMENT AND CLEAN-UP	2006008065
EC	EDF	14050	6276	TREATMENT OF HAZARDOUS WASTE	2006201283
Greece	YPEHO DE	14010	8	HIGHER EFFICIENCY OF WATER-RELATED DEVELOPMENT	2006000310
Greece	YPEJ	14015	45	DRAWING UP OF A STUDY ON THE IMPACT BETWEEN MIKRI PRESPA LAKE & RIVER	2006000319
Greece	ALLOI	14015	22	DEV. OF METHODOLOGY FOR BRACKISH KARST SPRING'S EXPLOITATION	2006000317
Greece	ALLOI	14015	15	DEV. AND APPLICATION OF ANALYTICAL METHODOLOGIES AND REMEDIATION TECH.	2006000316
Italy	DGCS	14010	216	WATER RESOURCES POLICY/ADMIN. MGMT	060041
Japan	JICA	14020	84	WATER SUPPLY & SANIT. - LARGE SYST.	063578C
Japan	JICA	14020	1240	TC IN SEWERAGE	060116C
Luxembourg	MFA	14020	1697	WATER SUPPLY & SANIT. - LARGE SYST.	061065
Norway	MFA	14020	2206	REHABILITATION PROJECT - WATER SUPPLY	2006003756

Total aid for water in 2006 for **Albania** USD thousand 13365
And as a share of aid to all recipient countries 0.21%

Algeria

RECIPIENT Donor	Agency	Sector	Amount USD thousand	Project description	CRS ID Number
Belgium	DGCD	14010	313	WATER RESOURCES POLICY/ADMIN. MGMT	2005002328
Belgium	DGCD	14081	173	EDUC./TRNG:WATER SUPPLY & SANITATION	2005000317
Germany	BMZ	14010	99	REFORM OF THE WATER SECTOR IN THE MENA REGION	2006003342
Germany	Fed Min	14030	942	DECENTRALIZED WATER SUPPLIES AND SEWAGE DISPOSAL	2006011161
Germany	BMZ	14081	2	WATER EFFICIENT MGMT. OF WASTEWATER, TREATMENT & REUSE IN MEDITERRANEAN	2006003382
Greece	YPEHO DE	14010	8	HIGHER EFFICIENCY OF WATER-RELATED DEVELOPMENT	2006000301
Italy	LA	14015	3	WATER RESOURCES PROTECTION	061649
Japan	JICA	14020	33	WATER SUPPLY & SANIT. - LARGE SYST.	063627C
Japan	JICA	14040	1	RIVER DEVELOPMENT	063628C
Japan	JICA	14040	1679	TC IN WATER RESOURCES DEVELOPMENT	060084C

RECIPIENT Donor	Agency	Sector	Amount USD thousand	Project description	CRS ID Number
Korea	KOICA	14010	7	TRANINIG PROGRAM	2006861172
Spain	MFA	14010	24	AT IN OTC WATER PROJECT MONITORING	064223
Spain	AG	14010	53	TIRSINE RESTAURATION AND REHABILITATION	063195
Switzerland	SDC	14050	9	WASTE MANAGEMENT/DISPOSAL	2001001759

Total aid for water in 2006 for   **Algeria**   **USD thousand 3343**
And as a share of aid to all recipient countries   **0.05%**

America, regional

RECIPIENT Donor	Agency	Sector	Amount USD thousand	Project description	CRS ID Number
Canada	IDRC	14010	20	WATER RESOURCES POLICY/ADMIN. MGMT	060998b
Canada	IDRC	14050	13	WASTE MANAGEMENT/DISPOSAL	061230b
Japan	JICA	14020	26	WATER SUPPLY & SANIT. - LARGE SYST.	065259C

Total aid for water in 2006 for   **America, regional**   **USD thousand 59**
And as a share of aid to all recipient countries   **0.00%**

Angola

RECIPIENT Donor	Agency	Sector	Amount USD thousand	Project description	CRS ID Number
EC	CEC	14010	4699	ACP EU WATER FACILITY : UNICEF REDUC CHILD MORT & INCREAS SCHOOL ATT	2006100297
EC	CEC	14030	2510	AUGMENTATION ET PROLOGATION CF	2006100006
Germany	BMZ	14010	15	TRANSNET - TRANSBOUNDARY MANAGEMENT OF NATURAL RESOURCES IN SADC REGION	2006003315
Japan	MOFA	14030	3711	EMERGENCY RURAL WATER SUPPLY IN NEIGHBORING PROVINCES OF LUANDA	060136
Japan	JICA	14040	6	RIVER DEVELOPMENT	063759C
Japan	JICA	14040	127	GRANT AID CO-OPERATION IN WATER RESOURCES DEVELOPMENT	061868C
Spain	MUNIC	14030	4	IMPROVEMENT OF HYGIENIC CONDITIONS IN KWANZA SUL ANGOLA	069829
UNICEF		14010	66	WATER RESOURCES POLICY/ADMIN. MGMT	060088
UNICEF		14030	185	HYGIENE, SANITATION AND WATER SUPPLY PROGRAMMES	060089
United Kingdom	DFID	14010	460	UNICEF WATSAN	050963

Total aid for water in 2006 for   **Angola**   **USD thousand 11785**
And as a share of aid to all recipient countries   **0.18%**

Argentina

RECIPIENT Donor	Agency	Sector	Amount USD thousand	Project description	CRS ID Number
Germany	BMZ	14010	2	STRENGHTENING THE MANAGEMENT CAPACITIES OF ECUADORIAN WATER UTILITIES	2006003363
Japan	JICA	14020	39	WATER SUPPLY & SANIT. - LARGE SYST.	064913C
Japan	JICA	14050	54	WASTE MANAGEMENT/DISPOSAL	064914C
Spain	Misc	14010	0	EFFICIENT MANAGEMENT OF LOCAL RESOURCES BY USING MODELS OF SIMULATION	0610655
Spain	Misc	14010	6	INTEGRAL SYSTEMA OF DATA AND REPORTS ON WATER RESOURCES	0610375
Spain	Misc	14010	11	GENERAL GUIDELINES FOR A SUSTAINABLE MANAGEMENT OF URBAN WATER	0610352
Spain	MUNIC	14030	4	WATER SUPPLY FOR ABRA DE SAN ANTONIO	068323
Spain	MUNIC	14030	5	WATER FOR EL CHURCAL	068285
Spain	AG	14030	35	WATER IS LIFE	062235
Spain	AG	14081	40	TRAINING AND TECHNOLOGY TRANSFER CENTRE FOR SUSTAINABLE WATER ACCESS	062255

RECIPIENT Donor	Agency	Sector	Amount USD thousand	Project description	CRS ID Number

| | | | Total aid for water in 2006 for | Argentina | USD thousand 196 |
| | | | And as a share of aid to all recipient countries | | 0.00% |

Armenia

Greece	YPESD DA	14050	88	PROVISION OF GARBAGE COLLECTION TRUCK AND BINS & OF A TRUCK FOR ROAD	2006000327
IDA		14010	10500	SOCIAL INVESTMENT FUND III	060065a
Japan	JICA	14020	0	WATER SUPPLY & SANIT. - LARGE SYST.	065476C
Japan	JICA	14040	9	RIVER DEVELOPMENT	065477C
Japan	JICA	14040	194	RURAL WATER SUPPLY AND SEWERAGE SYSTEMS IN REPUBLIC OF ARMENI	061155C
Norway	MFA	14015	137	HYDROMETEOROLOGY IN ARMENIA	2006003366
Norway	MFA	14030	122	WATER SANITATION SIX COM. ARMENIA	2006001803

| | | | Total aid for water in 2006 for | Armenia | USD thousand 11050 |
| | | | And as a share of aid to all recipient countries | | 0.17% |

Asia, regional

Australia	AusAID	14010	280	MRC APP HYDROLOGICAL NETWORK IMPROVEMENT	2006000796
Australia	AusAID	14040	88	MRC BASIN DEVELOPMENT PLANNING	2006001299
Belgium	DGCD	14081	47	EDUC./TRNG:WATER SUPPLY & SANITATION	2001009888
Denmark	MFA	14040	1262	SUPPORT TO THE MRC ENVIRONMENT PROGRAMME	061195
Denmark	MFA	14040	7572	SUPPORT TO MRC BASIN DEVELOPMENT PLAN	061194
Finland	MFA	14040	4895	DECISION SUPPORT AND INFORMATION MANAGEMENT PROGRAMME	050105
Japan	JICA	14050	115	WASTE MANAGEMENT/DISPOSAL	066301C
Japan	JICA	14050	291	OVERSEAS DEVELOPMENT STUDY IN URBAN SANITATION	060742C
Netherlands	MFA	14010	1255	WATER RESOURCES POLICY/ADMIN. MGMT	2006000627
Norway	MFA	14010	4	TRAVEL SUPPORT WORKING GROUP ON WATER AND HEALTH	2006001177
Norway	MFA	14030	166	FACILITATOR TO THE PROTOCOL ON WATER AND HEALTH	2006001381
United Kingdom	DFID	14050	431	COMMUNITY LED TOTAL SANITATION	060614

| | | | Total aid for water in 2006 for | Asia, regional | USD thousand 16406 |
| | | | And as a share of aid to all recipient countries | | 0.25% |

Azerbaijan

Germany	KFW	14030	32446	COMMUNAL INFRASTRUCTURE II	2006000091
Japan	JICA	14020	0	WATER SUPPLY & SANIT. - LARGE SYST.	065487C
Norway	MFA	14050	193	WASTE MANAGEMENT	2006002884
UNICEF		14010	104	WATER RESOURCES POLICY/ADMIN. MGMT	060145

| | | | Total aid for water in 2006 for | Azerbaijan | USD thousand 32743 |
| | | | And as a share of aid to all recipient countries | | 0.50% |

CRS AID ACTIVITIES IN SUPPORT OF WATER SUPPLY AND SANITATION – ISBN 978-92-64-05172-0 – © OECD/WWC 2008

RECIPIENT Donor	Agency	Sector	Amount USD thousand	Project description	CRS ID Number

Bangladesh

RECIPIENT	Agency	Sector	Amount	Project description	CRS ID
AsDF		14010	41503	SECONDARY TOWNS WATER SUPPLY AND SANITATION PROJECT	060049
AsDF		14010	20855	SOUTHWEST AREA INTEGRATED WATER RESOURCES PLANNING & MANAGEMEN	060012
IDA		14010	4000	INVESTMENT PROMOTION AND FINANCING FACILITY	060136d
Italy	DGCS	14010	9	WATER RESOURCES POLICY/ADMIN. MGMT	060366
Italy	Art.	14020	16530	MODUNAGHAT WATER SUPPLY PROJECT	060006
Japan	JICA	14020	104	WATER SUPPLY & SANIT. - LARGE SYST.	065855C
Japan	JBIC	14020	105017	KARNAPHULI WATER SUPPLY PROJECT	063065
Japan	JICA	14020	102	ARSENIC MITIGATION ADVISOR TO DPHE	062421C
Japan	JICA	14040	148	RIVER DEVELOPMENT	065856C
Japan	JICA	14040	89	STORM WATER DRAINAGE SYSTEM IN DHAKA CITY (PHASE	062826C
Japan	JICA	14050	105	WASTE MANAGEMENT/DISPOSAL	065857C
Japan	JICA	14050	210	GRANT AID CO-OPERATION IN URBAN SANITATION	061046C
Korea	KOICA	14010	28	DISPATCH OF EXPERT	2006862252
Korea	KOICA	14010	7	TRANINIG PROGRAM	2006861680
Netherlands	MFA	14010	13231	SOUTH WEST WATER PROJECT: INTEGRATED PLANNING AT DECENTRALISED LEVEL	2006000091
Netherlands	MFA	14010	68512	WATER RESOURCES POLICY/ADMIN. MGMT	2003002314
Sweden	Sida	14020	353	BGD SAID WTP CONS STUDY	2006002308
Switzerland	SDC	14015	559	TANGUAR HAOR COMMUNITY	2006000509
UNICEF		14010	27	WATER RESOURCES POLICY/ADMIN. MGMT	060246
UNICEF		14030	689	HYGIENE, SANITATION AND WATER SUPPLY PROGRAMMES	060245
UNICEF		14030	671	COMMUNITY HYGIENE AND WATER SAFETY	060244
United Kingdom	DFID	14010	66237	SANITATION, HYGIENE EDUCATION AND WATER SUPPLY IN BANGLADESH (SHEWA-B)	060857
United Kingdom	DFID	14010	92	WATER RESOURCES POLICY/ADMIN. MANAGEMENT	060282

Total aid for water in 2006 for　　　　**Bangladesh**　　　　**USD thousand 339078**
And as a share of aid to all recipient countries　　　　**5.21%**

Barbados

RECIPIENT	Agency	Sector	Amount	Project description	CRS ID
Japan	JICA	14040	6	RIVER DEVELOPMENT	064530C
Japan	JICA	14040	94	TC IN RIVERS/SAND ARRESTATION	062658C

Total aid for water in 2006 for　　　　**Barbados**　　　　**USD thousand 100**
And as a share of aid to all recipient countries　　　　**0.00%**

Belarus

RECIPIENT	Agency	Sector	Amount	Project description	CRS ID
Germany	BMZ	14050	65	INTEGRATED EXPERTS	2006002441
Germany	L G	14081	3	WORKSHOP "INTEGRATED WATER AND RESOURCES MANAGEMENT"	2006012238
IDA		14050	500	PERSISTANT ORGANIC POLLUTANTS (POPS) ENVIRL ASSESSMENT TA	060372

RECIPIENT Donor	Agency	Sector	Amount USD thousand	Project description	CRS ID Number
				Total aid for water in 2006 for Belarus	**USD thousand 568**
				And as a share of aid to all recipient countries	**0.01%**

Benin

RECIPIENT Donor	Agency	Sector	Amount USD thousand	Project description	CRS ID Number
Belgium	MPRF	14030	20	BASIC DRINKING WATER SUPPLY AND BASIC SANITATION	2006003573
Belgium	MPRW	14030	21	BASIC DRINKING WATER SUPPLY AND BASIC SANITATION	2005001811
Belgium	DGCD	14030	114	BASIC DRINKING WATER SUPPLY AND BASIC SANITATION	2005000401
Belgium	DGCD	14030	529	HYDRAULIQUE ET ASSAINISSEMENT EN APPUI AU DEV. INSTITUTIONNEL	2004003712
Belgium	MPRW	14030	454	DEVELOPPEMENT AGRICOLE, EAU, ASSAINISSEMENT, SANTE, ARTISANAT	2002003807
Belgium	DGCD	14030	609	PADEAR -ATACORA DONGA EAU POTABLE ET ASSAINISSEMENT (PADEAR)	2002000328
Belgium	DGCD	14030	547	HYDRAULIQUE ET ASSAINISSEMENT EN APPUI AU DEV. INTÉGRÉ	2002000271
Canada	CIDA	14030	176	BENIN WATER WELLS PROJECT	060596
EC	CEC	14010	1259	ACP EU WATER FACILITY	2006100300
Germany	BMZ	14010	336	WATER RESOURCES POLICY AND ADMINISTRATIVE MANAGEMENT	2006005399
Germany	GTZ	14030	6276	BASIC DRINKING WATER SUPPLY AND BASIC SANITATION	2006006116
Germany	KFW	14030	5648	ADDUCTION D'EAU	2006000138
Ireland	DFA	14030	68	BASIC DRINKING WATER AND SANIT	060705
Italy	LA	14020	3	WATER SUPPLY & SANIT. - LARGE SYST.	062027
Netherlands	MFA	14010	32945	WATER RESOURCES POLICY/ADMIN. MGMT	2006000225
Netherlands	MFA	14020	23447	WATER SUPPLY & SANIT. - LARGE SYST.	2003012889
Netherlands	MFA	14030	2405	CONSTRUCTION DE 8 AEV DANS 4 DÉPARTEMENTS DU SUD DU BÉNIN	2006000464
Spain	MUNIC	14030	2	IMPROVEMENT IN THE PUMPING AND WATER EXTRACTION IN THE CATHOLIC MISSION OF BENIN	068255
Spain	AG	14040	628	COLLABORATION AGREEMENT BETWEEN THE GENERALITAT AND PEACE MESSENGER ASSOCIATION TO CARRY OUT THE PRO	062632
UNICEF		14010	207	WATER RESOURCES POLICY/ADMIN. MGMT	060856
				Total aid for water in 2006 for Benin	**USD thousand 75693**
				And as a share of aid to all recipient countries	**1.16%**

Bhutan

RECIPIENT Donor	Agency	Sector	Amount USD thousand	Project description	CRS ID Number
Austria	ADA	14030	92	BASIC DRINKING WATER SUPPLY AND BASIC SANITATION	2005000056
Japan	JICA	14020	32	WATER SUPPLY & SANIT. - LARGE SYST.	065618C
UNICEF		14010	45	WATER RESOURCES POLICY/ADMIN. MGMT	060320
				Total aid for water in 2006 for Bhutan	**USD thousand 169**
				And as a share of aid to all recipient countries	**0.00%**

Bilateral, unspecified

RECIPIENT Donor	Agency	Sector	Amount USD thousand	Project description	CRS ID Number
Australia	AusAID	14010	5	WATER QUALITY INITIATIVE	2006002631
Australia	AusAID	14030	1117	WATER QUALITY INITIATIVE	2006000434
Australia	AusAID	14081	28	WATER QUALITY INITIATIVE	2006001983
Austria	ADA	14010	6	WATER RESOURCES POLICY/ADMIN. MGMT	2005000024

CRS AID ACTIVITIES IN SUPPORT OF WATER SUPPLY AND SANITATION – ISBN 978-92-64-05172-0 – © OECD/WWC 2008

RECIPIENT Donor	Agency	Sector	Amount USD thousand	Project description	CRS ID Number
Belgium	DGCD	14081	121	EDUC./TRNG:WATER SUPPLY & SANITATION	2001007936
Canada	IDRC	14020	41	FOCUS CITIES : CAPACITY BUILDING FOR ECONOMIC ANALYSIS	061308b
Canada	IDRC	14030	88	FOGQUEST: IV INTERNATIONAL CONFERENCE ON FOG, FOG COLLECTION & DEW	061235b
Denmark	MFA	14010	1683	COMPETING FOR WATER: UNDERSTANDING CONFLICT & CO. IN LOCAL WATER GOVER	061199
Denmark	MFA	14030	597	KNOWLEDGE NETWORK ON WATER AND DEVELOPMENT	031012
EC	CEC	14010	1171	ACP EU WATER FACILITY : IMPROV MUNIC WASTEWATER MANAG ACP COAST CITIES	2006100323
EC	CEC	14010	3119	ACP EU WATER FACILITY	2006100322
EC	CEC	14010	1795	ACP EU WATER FACILITY : ENCODAGE DONNEES FINANC RELATIVES A EI PROG REG	2006100321
EC	CEC	14010	390	ACP EU WATER FACILITY : SIEGE : BPD WATER AND SANITATION -	2006100318
EC	CEC	14010	2117	ACP EU WATER FACILITY : INVOLV CIV SOC. IN MONITOR OF DEV IN WATER/SANIT	2006100308
EC	CEC	14010	10041	5TH GLOBAL (N+1) CONTRIBUTION AGREEMENT WITH UNICEF MANAGED	2006100173
EC	CEC	14010	12552	ACP - EU WATER FACILITY: SUPPORT TO AFRICA-EU	2006100171
EC	CEC	14010	25104	ACP EU WATER FACILITY RIDER 1 (SUPPORT TO AWF)	2006100169
EC	CEC	14010	6401	ACP - EU WATER FACILITY	2006100166
Finland	MFA	14010	251	MEDITERRANEAN ENVIRONMENTAL TECHNICAL ASSISTANCE PROGRAMME (METAP)	050104
Finland	MFA	14010	251	SUPPORT TO GLOBAL WATER PARTNERSHIP ORGANIZATION (GWPO)	043036
Finland	MFA	14030	444	FRAME AGREEMENT WITH NGO (FINNISH RED CROSS)	060481
France	MAE	14010	6088	CONTRIBUTION À LA BAD POUR LA FACILITÉ AFRICAINE DE L'EAU	2006900015
France	AFD	14015	1255	APPUI A LA COMMISSION DU FLEUVE MEKONG	2006155700
Germany	BMZ	14010	502	GWPO GLOBAL WATER PARTNERSHIP	2006010037
Germany	KFW	14020	1130	EDUCATION AND TRAINING	2006002028
Germany	BMZ	14030	251	WHO: JOINT MONITORING PROGRAMME FOR WATER SUPPLY AND SANITATION	2006010096
Germany	GTZ	14050	26	WASTE MANAGEMENT/DISPOSAL	2006006121
Germany	BMZ	14081	3	PREPARATORY ACTIVITIES FOR NEW PROJECTS	2006003393
Germany	BMZ	14081	2	WATER EFFICIENT MGMT OF WASTEWATER, TREATMENT & REUSE IN MEDITERRANEAN	2006003379
Germany	BMZ	14081	2	WATER EFFICIENT MGMT. OF WASTEWATER, TREATMENT AND REUSE	2006003378
Germany	BMZ	14081	11	WATER EFFICIENT MANAGEMENT OF WASTEWATER, TREATMENT AND REUSE	2006003377
Germany	KFW	14081	2121	EDUCATION AND TRAINING	2006002030
Greece	ALLOI	14081	2	SUPPORTING A CAMPAIGN FOR THE DAY OF WATER	2006000329
Ireland	DFA	14010	639	WATER POLICY /ADMINISTRATION	060202
Japan	JICA	14020	26	WATER SUPPLY & SANIT. - LARGE SYST.	066578C
Japan	JICA	14040	27	RIVER DEVELOPMENT	066579C
Netherlands	MFA	14010	3012	WATER: CAPNET 2ND PHASE	2006000561
Netherlands	MFA	14010	955	WATER RESOURCES POLICY/ADMIN. MGMT	2003003525
Netherlands	MFA	14020	477	WATER SUPPLY & SANIT. - LARGE SYST.	2003015513
Netherlands	MFA	14030	6588	BASIC DRINKING WATER SUPPLY AND BASIC SANITATION	2003010155
Netherlands	MFA	14050	8717	WASTE MANAGEMENT/DISPOSAL	2003015505
Netherlands	MFA	14081	11391	EDUC./TRNG:WATER SUPPLY & SANITATION	2003014321
Norway	MFA	14010	30	GUIDELINS FOR NATIONAL IWRM PLANS	2006004586
Norway	NORAD	14010	22	SERVICE DELIVERY IN FRAGILE STATES	2006003924
Norway	MFA	14010	624	UNDP TRUST FUND ON ENERGY AND ENVIRONMENT	2006003064
Norway	NORAD	14010	1871	GLOBAL WATER PARTNERSHIP 2006-2008	2006002507

RECIPIENT Donor	Agency	Sector	Amount USD thousand	Project description	CRS ID Number
Norway	MFA	14010	49	EXPO WORLD WATER FORUM 4	2006002119
Norway	NORAD	14010	85	FICTION - FACTS - STRATEGY FOR ENGAGEMENT IN WATER AND SANITATION	2006002109
Norway	NORAD	14010	30	GLOBAL WATER AND SANITATION EVALUATION	2006000994
Norway	NORAD	14030	156	WSSCC - WATER SUPPLY AND SANITATION COLL. COUNCIL	2006003916
Norway	MFA	14081	80	APPROPRIATE SANITATION IN THE DEVELOPING WORLD	2006002624
Portugal	ICP	14015	3	EU WATER INICIATIVE	060709
Sweden	Sida	14010	1330	ZAMBEZI IWRM	2006008262
Sweden	Sida	14010	2034	SCHOOL SANITATION	2006003378
Sweden	Sida	14010	11	APPRAISAL GREEN BLUE WATE	2006003358
Sweden	Sida	14010	2102	AGREEMENT SIWI	2006003273
Sweden	Sida	14010	5018	GWPO	2006003271
Sweden	Sida	14030	1356	SUPPORT ECOSANRES PILOT.. SUPPORT ECOSANRES PILOT	2006003362
Sweden	Sida	14030	69	STUDY NEW WATERINITIATIVE	2006003335
Sweden	Sida	14030	10172	ECOSANRES 2	2006003269
Sweden	Sida	14050	610	WASTE BURNING	2006002348
Sweden	Sida	14081	4069	CAP-NET CAPACITY BUILING	2006003360
Switzerland	SDC	14010	112	WATER RESOURCES POLICY/ADMIN. MGMT	1997000877
Switzerland	SDC	14010	762	GWP GLOBAL WATER PARTNERSHIP	1997000746
Switzerland	SDC	14015	16	WATER RESOURCES PROTECTION	1992000355
Switzerland	SDC	14030	184	BASIC DRINKING WATER SUPPLY AND BASIC SANITATION	1996001322
UNICEF		14010	70	WATER RESOURCES POLICY/ADMIN. MGMT	063018
UNICEF		14030	194	COMMUNITY HYGIENE AND WATER SAFETY	063017
United Kingdom	DFID	14010	957	EU WATER INITIATIVE & ACP - EU WATER FACILITY	050432
United Kingdom	DFID	14010	992	ENGINEERING RESEARCH PROGRAMME	041022
United Kingdom	DFID	14030	4416	WATER & SANITATION: WELL RESOURCE CENTRE	020120
United Kingdom	DFID	14050	431	COMMUNITY LED TOTAL SANITATION	060659
United States	AID	14010	1472	WATER RESOURCES POLICY/ADMIN. MGMT	2006002869
United States	AID	14010	1860	WATER	2006002864
United States	Misc	14030	15	BASIC DRINKING WATER SUPPLY AND BASIC SANITATION	2006003031
United States	TDA	14050	45	WASTE MANAGEMENT/DISPOSAL	2006003065

Total aid for water in 2006 for   **Bilateral, unspecified**   **USD thousand 151793**
And as a share of aid to all recipient countries   **2.33%**

Bolivia

RECIPIENT Donor	Agency	Sector	Amount USD thousand	Project description	CRS ID Number
Canada	CIDA	14030	2997	WATER AND SANITATION - PACSAS: SUPPORT TO UNICEF	060308
EC	EDF	14020	10167	PROGRAMME D'APPUI SECTORIEL EN APPROVISIONNEMENT D'EAU ET ASSAINISS.	2006200489
EC	EDF	14020	6025	PROGRAMME D'APPUI SECTORIEL EN APPROVISIONNEMENT D'EAU ET ASSAINISSEMENT	2006200488
Finland	MFA	14020	213	NGO SUPPORT / WATER PROJECT IN SACACA	013653
Germany	BMZ	14010	3	STRENGHTENING THE MANAGEMENT CAPACITIES OF ECUADORIAN WATER UTILITIES	2006003362
Germany	BMZ	14010	101	INTEGRATED EXPERTS	2006002410
Germany	KFW	14015	4074	WATER RESOURCES PROTECTION	2006000171

RECIPIENT Donor	Agency	Sector	Amount USD thousand	Project description	CRS ID Number
Germany	BMZ	14030	126	CAPACITY BUILDING FOR YOUTH IN WATER SUPPLY SYSTEMS IN BOLIVIA	2006008149
IDA		14010	7200	BOLIVIA URBAN INFRASTRUCTURE PROJECT	060039b
IDA		14020	12900	BOLIVIA URBAN INFRASTRUCTURE PROJECT	060039a
Japan	JICA	14020	83	WATER SUPPLY & SANIT. - LARGE SYST.	064948C
Japan	JICA	14020	87	PROGRAM EVALUATION 'PROGRAM FOR WATER SUPPLY IN THE POVERTY AREA'	062900C
Japan	JICA	14020	171	PROYECTO DE MEHORAMIENTO DEL SISTEMA DE AGUA POTABLE EN LA CIUDAD DE COC	061341C
Japan	JICA	14040	26	RIVER DEVELOPMENT	064949C
Japan	JICA	14040	94	WATER IS HEALTH AND LIFE (AGUA ES SALUD Y VIDA)	062681C
Japan	JICA	14050	24	WASTE MANAGEMENT/DISPOSAL	064950C
Netherlands	MFA	14010	12992	WATER: LAP WATERSHED PROGRAMME	2006000788
Netherlands	MFA	14030	3766	BASIC DRINKING WATER SUPPLY AND BASIC SANITATION	2006000389
Spain	MUNIC	14030	30	DRINKING WATER SUPPLY AND BASIC SANITATION FOR THE PEASANT COMMUNITIES	068701
Spain	MUNIC	14030	4	LETRINES' AND SHOWER INSTALLATION AT EL ALTO IN BOLIVIA	068470
Spain	MUNIC	14030	53	ACCESS TO A DRINKING WATER SYSTEM MANAGED BY THE PEASANT FAMILIES	067670
Spain	MUNIC	14030	8	LETRINES AND SHOWERS INSTALATION	067469
Spain	MUNIC	14030	17	CONTINUATION OF THE CLEAN-UP REFUSE MATERIALS	066406
Spain	AG	14030	95	MANOS UNIDAS (UNITED HANDS) CAMPAIGN AGAINST HUNGER	063874
Spain	AG	14030	113	ACCESS TO POTABLE WATER SYSTEMS WITH LIME FOR A DIGNIFIED LIFE BETANZOS AND WICHACA (BOLIVIA)	063337
Spain	AG	14030	75	HUMID ZONES CONSTRUCTED FOR WATER RESIDUALS PURIFICATIONS - SECOND PHASE	063281
Spain	AG	14030	197	IMPROVEMNET OF BASIC NEEDS IN WATER SUPPLY IN INDEGENOUS COMUNITIES	062551
Spain	AG	14030	60	POTABLE WATER SYSTEM- COMMUNITY OF LAGUNILLAS	061936
Spain	AG	14030	71	BASIC SANITATION IN TUMUPASA	061196
Spain	MUNIC	14040	19	IMPROVEMENT IN AGRICULTURAL PERFORMANCE AND FOOD SAFETY THROUGH DONATIONS OF SYSTEMS THAT MANAGE THE	068068
Spain	AG	14050	5	SOLID WASTE MANAGEMENT IN THE MUNIICPALITY OF TUPIZA	061484
Spain	AG	14081	44	IMPROVEMENT IN SALUBRITY OF 75 INHABITANTS OF ALTO SAN PEDRO	061646
Sweden	Sida	14030	4747	NEW PROGR WAT-SAN UNICEF	2006001857

Total aid for water in 2006 for   **Bolivia**   **USD thousand 66585**
And as a share of aid to all recipient countries   **1.02%**

Bosnia-Herzegovina

RECIPIENT Donor	Agency	Sector	Amount USD thousand	Project description	CRS ID Number
Austria	BMF	14020	222	WATER SUPPLY & SANIT. - LARGE SYST.	20060016gg
Austria	BMF	14020	1027	WATER SUPPLY FACILITY, ENGENEERING AND EQUIPMENT	200600016g
Austria	ADA	14030	15	BASIC DRINKING WATER SUPPLY AND BASIC SANITATION	20060000bb
Greece	YPEHO DE	14010	8	HIGHER EFFICIENCY OF WATER-RELATED DEVELOPMENT	2006000311
Italy	DGCS	14010	7	WATER RESOURCES POLICY/ADMIN. MGMT	060616
Italy	LA	14020	11	WATER SUPPLY & SANIT. - LARGE SYST.	061336
Norway	MFA	14010	218	GREEN BELT NATURAL HERITAGE	2006004400
Norway	MFA	14020	2310	WATER SUPPLY INVESTMENTS VRBAS	2006004595
Norway	MFA	14020	335	WATER SUPPLY PRJEDOR	2006002344
Norway	MFA	14050	249	WASTEWATER SYSTEM MANAGEMENT	2006003356
Spain	MFA	14040	220	SUPPORT FOR INTEGRAL MANAGEMENT OF THE NERETVA´S BASIN	064229

RECIPIENT Donor	Agency	Sector	Amount USD thousand	Project description	CRS ID Number
Switzerland	SDC	14010	130	WATER RESOURCES POLICY/ADMIN. MGMT	1996001305
Switzerland	SDC	14010	1776	WATER BIHAC	1996000441

Total aid for water in 2006 for   **Bosnia-Herzegovina**   **USD thousand 6528**
And as a share of aid to all recipient countries   **0.10%**

Botswana

RECIPIENT Donor	Agency	Sector	Amount USD thousand	Project description	CRS ID Number
Germany	BMZ	14010	15	TRANSNET - TRANSBOUNDARY MANAGEMENT OF NATURAL RESOUCES IN SADC REGION	2006003331
Japan	JICA	14020	67	WATER SUPPLY & SANIT. - LARGE SYST.	063776C
Norway	MFA	14030	52	PERSONNEL EXCHANGE	2006003601
United States	TDA	14020	500	WATER SUPPLY & SANIT. - LARGE SYST.	2006002965

Total aid for water in 2006 for   **Botswana**   **USD thousand 634**
And as a share of aid to all recipient countries   **0.01%**

Brazil

RECIPIENT Donor	Agency	Sector	Amount USD thousand	Project description	CRS ID Number
Belgium	MPRW	14050	2	WASTE MANAGEMENT/DISPOSAL	2005002102
Belgium	DGCD	14050	47	WASTE MANAGEMENT/DISPOSAL	2004005206
Canada	CIDA	14050	380	FS/REDONDA BIOGAS&WATER WASTE TREATMENT	060106
France	MINEFI	14050	522	VALORISATION DECHETS CURITIBA	2006003516
Germany	KFW	14030	3766	BASIC DRINKING WATER SUPPLY AND BASIC SANITATION	2006000204
Germany	L G	14050	11	WASTE RECYCLING IN SAO PAULO - PURCHASE OF A PICKUP-TRUCK	2006012149
Germany	BMZ	14050	111	WASTE MANAGEMENT/DISPOSAL	2006005413
Ireland	DFA	14030	44	BASIC DRINKING WATER AND SANIT	060828
Italy	LA	14020	13	WATER SUPPLY & SANIT. - LARGE SYST.	061349
Italy	LA	14040	221	INTEGRATED PROJECT FOR ENVIRONMENTALLY FRIENDLY WATER DEVELOPMENT	061960
Japan	JICA	14020	177	WATER SUPPLY & SANIT. - LARGE SYST.	064989C
Japan	JICA	14040	80	RIVER DEVELOPMENT	064990C
Japan	JICA	14040	509	STUDY ON INTERGRATED PLAN OF ENV. IMPROVEMENT IN CATCHMENT AREA OF	060368C
Japan	JICA	14040	1151	TC IN WATER RESOURCES DEVELOPMENT	060125C
Japan	JICA	14050	55	WASTE MANAGEMENT/DISPOSAL	064991C
Japan	JICA	14050	104	DEV. OF AN INTEGRATED SOLUTION RELATED TO INDUSTRIAL WASTE MANAGEMENT	062338C
Luxembourg	MFA	14030	20	BASIC DRINKING WATER SUPPLY AND BASIC SANITATION	061060
Spain	AG	14030	2	IMPROVE THE SOCIO-SANITARY AND ENVIRONMENTAL SITUATION	062197
Spain	AG	14030	25	WATER SUPPLY TO ASSURE NURISHMENT AND LIFE TO THE BAURU SETTLEMENT	061316
Spain	Misc	14050	2	ELECTRODIALYSIS ALICATION FOR THE TREATMENT OF WATER SOLUTION	0610658
United States	TDA	14020	170	WATER SUPPLY & SANIT. - LARGE SYST.	2006002974
United States	Misc	14050	44	WASTE MANAGEMENT/DISPOSAL	2006003066

Total aid for water in 2006 for   **Brazil**   **USD thousand 7455**
And as a share of aid to all recipient countries   **0.11%**

RECIPIENT	Agency	Sector	Amount USD thousand	Project description	CRS ID Number
Donor					

Burkina Faso

Belgium	DGCD	14020	21	WATER SUPPLY & SANIT. - LARGE SYST.	2004003736
Belgium	MPRF	14030	129	BASIC DRINKING WATER SUPPLY AND BASIC SANITATION	2006003414
Belgium	DGCD	14030	134	INTERVENTION EN HYDRAULIQUE VILLAGEOISE	2004005306
Belgium	DGCD	14030	73	BASIC DRINKING WATER SUPPLY AND BASIC SANITATION	2004003735
Belgium	MPRW	14050	2	WASTE MANAGEMENT/DISPOSAL	2005002103
Belgium	DGCD	14050	77	WASTE MANAGEMENT/DISPOSAL	2005000376
Germany	BMZ	14010	18	WATER RESOURCES POLICY AND ADMINISTRATIVE MANAGEMENT	2006005400
Germany	L G	14030	9	WATER FOR TORLA	2006012187
Germany	BMZ	14030	47	BORE WELL IN THE PIEA DEPARTMENT	2006004893
Japan	JICA	14040	212	D' APPROVISIONNEMENT EN EAU POTABLE DANS LES REGIONS DU PLATEAU CENTRAL	061033C
Spain	MFA	14010	865	WATER IN ÁFRICA: IMPROVEMENT IN MALÍ BURKINA FASSO SENEGAL Y GUINEA	064609
Spain	MUNIC	14030	13	WELL BORING AND EQUIPMENT OF A ACCEPTANCE FLAT FOR STREET CHILDREN	068819
Spain	MUNIC	14030	26	WELL PERFORATION AND FLOOR SHELTER CONSTRUCTION	066770
Spain	AG	14030	68	WATER SUPPLY THROUGH THE CONSTRUCTION OF FIVE WELLS AND HYDROGEOLOGICAL MODELIZATION IN FIVE COMMUNI	063248
Spain	AG	14030	34	TWELVE WATER WELLS	063113
Spain	AG	14030	18	CONSTRUCTION OF NEW WATER WELLS	062951
Spain	AG	14030	32	POTABLE WATER SUPPLY- BULSA	062178
Spain	AG	14030	285	IMPROVEMENTS IN THE SERVICE EXPLOITATION CONDITIONS	061977
Sweden	Sida	14050	34	WASTE MANAGEMENT BFA	2006000243
UNICEF		14010	96	WATER RESOURCES POLICY/ADMIN. MGMT	062972
UNICEF		14030	157	SECTOR PLANS FOR HYGIENE, SANITATION AND WATER	062974
UNICEF		14030	183	HYGIENE, SANITATION AND WATER SUPPLY PROGRAMMES	062973

Total aid for water in 2006 for   **Burkina Faso**   **USD thousand 2532**
And as a share of aid to all recipient countries   **0.04%**

Burundi

AfDF		14040	13239	PROJET AMENAGEMENT DES BASSINS VERSANTS	060064
Belgium	DGCD	14030	2510	PROGRAMME D'URGENCE HYDRAULIQUE	2006002742
Belgium	DGCD	14030	95	BASIC DRINKING WATER SUPPLY AND BASIC SANITATION	2004000326
Germany	KFW	14030	9995	PROGRAMME D'EAU	2006000252
IDA		14010	7650	PUBLIC WORKS AND EMPLOYMENT CREATION PROJECT-SUPPLEMENTAL	060070a
Italy	DGCS	14020	25	WATER SUPPLY & SANIT. - LARGE SYST.	060849
UNICEF		14010	21	WATER RESOURCES POLICY/ADMIN. MGMT	060468

Total aid for water in 2006 for   **Burundi**   **USD thousand 33536**
And as a share of aid to all recipient countries   **0.51%**

Cambodia

| France | AFD | 14020 | 13932 | INVESTIISEMENT CAPACITE USINE TRAITEMENT | 2006117800 |
| Germany | BMZ | 14040 | 48 | DEVELOPING POTENTIALS IN RURAL AREAS OF MEKONG RIPARIAN COUNTRIES | 2006003373 |

RECIPIENT Donor	Agency	Sector	Amount USD thousand	Project description	CRS ID Number
Germany	BMZ	14081	135	GIS FOR SUSTAINABLE 'WATER MANAGEMENT IN SIEM REAP	2006003392
Italy	LA	14020	1	WATER SUPPLY & SANIT. - LARGE SYST.	062059
Japan	JICA	14020	657	WATER SUPPLY & SANIT. - LARGE SYST.	065893C
Japan	JICA	14020	108	TC IN WATER SUPPLY	062253C
Japan	Oth. MIN	14030	33	BASIC DRINKING WATER SUPPLY AND BASIC SANITATION	066983C
Japan	MOFA	14030	3703	RURAL DRINKING WATER SUPPLY IN KAMPONG CHAM PROVINCE	050104
Japan	JICA	14040	105	RIVER DEVELOPMENT	065894C
Japan	JICA	14040	215	FLOOD PROTECTION AND DRAINAGE IMPROVEMENT PROJECT	061020C
Japan	JICA	14040	358	GRANT AID CO-OPERATION IN RIVERS/SAND ARRESTATION	060574C
Japan	JICA	14040	404	TC IN WATER RESOURCES DEVELOPMENT	060479C
Japan	JICA	14050	15	WASTE MANAGEMENT/DISPOSAL	065895C
Japan	JICA	14050	127	SOLID WASTE MANAGEMENT IMPROVEMENT PROJECT IN PHNOM PENH	061873C
Japan	JICA	14050	210	STRENGTHENING OF SOLID WASTE MANAGEMENT FOR MUNICIPALITY OF PHNOM PENH	061048C
Korea	KOICA	14010	15	TRANINIG PROGRAM	2006861706
Korea	KOICA	14020	12	DS OF WATERWORKS IN CAMBODIA	2006862202
UNICEF		14010	125	WATER RESOURCES POLICY/ADMIN. MGMT	060518
UNICEF		14030	293	SECTOR PLANS FOR HYGIENE, SANITATION AND WATER	060520

Total aid for water in 2006 for   **Cambodia**   USD thousand 20496

And as a share of aid to all recipient countries   0.31%

Cameroon

RECIPIENT Donor	Agency	Sector	Amount USD thousand	Project description	CRS ID Number
AfDF		14020	37658	PROJET D'ASSAINISSEMENT DE YAOUNDE (PADY)	060062
Belgium	DGCD	14030	434	ADDUCTION EAU MAROUA REHABILITATION EAU POTABLE	2001001133
Canada	IDRC	14030	313	MAÎTRISE DE L'ASSAINISSEMENT DANS UN ÉCOSYSTÈME URBAIN DE YAOUNDÉ ()	061010b
Germany	BMZ	14020	387	RURAL WATER DEVELOPMENT, DICESE YAGOUA	2006008115
Germany	BMZ	14030	514	BASIC DRINKING WATER SUPPLY AND BASIC SANITATION	2006005410
Germany	L G	14081	15	EDUC./TRNG:WATER SUPPLY & SANITATION	2006012375
Japan	JICA	14020	427	GRANT AID CO-OPERATION IN WATER SUPPLY	060455C
Japan	MOFA	14030	4424	LE PROJET D'HYDRAULIQUE RURALE	060120
Japan	JICA	14040	3	RIVER DEVELOPMENT	063803C
Spain	MUNIC	14030	8	REHABILITATION OF WATER SUPPLY FOR KIYAN KUMBO CAMEROON	068246
Spain	MUNIC	14030	5	PROJECT ON WATER SUPPLY REHABILITATION FOR KIYAN KUMBO CAMEROON	068214
Spain	MUNIC	14030	13	IMPROVEMENT OF WATER SUPPLY FOR A HEALTH CENTRE	067429
UNICEF		14010	36	WATER RESOURCES POLICY/ADMIN. MGMT	060550

Total aid for water in 2006 for   **Cameroon**   USD thousand 44238

And as a share of aid to all recipient countries   0.68%

Cape Verde

RECIPIENT Donor	Agency	Sector	Amount USD thousand	Project description	CRS ID Number
EC	CEC	14020	24150	PROGRAMME D'APPROVISIONNEMENT EN EAU POTABLE DES VILLES	2006100048
Japan	JICA	14020	5	WATER SUPPLY & SANIT. - LARGE SYST.	063817C

     CRS AID ACTIVITIES IN SUPPORT OF WATER SUPPLY AND SANITATION – ISBN 978-92-64-05172-0 – © OECD/WWC 2008

RECIPIENT Donor	Agency	Sector	Amount USD thousand	Project description	CRS ID Number
Japan	JICA	14040	5	RIVER DEVELOPMENT	063818C
Luxembourg	MFA	14010	79	WATER RESOURCES POLICY/ADMIN. MGMT	061085
Luxembourg	MFA	14020	36	WATER SUPPLY & SANIT. - LARGE SYST.	061075
Luxembourg	MFA	14050	1	WASTE MANAGEMENT/DISPOSAL	061088
Spain	AG	14030	238	REINFORCEMENT IN WATER SUPPLY IN THE ISLE OF MAIO CABO VERDE	061682

Total aid for water in 2006 for **Cape Verde** USD thousand **24513**
And as a share of aid to all recipient countries **0.38%**

Central African Rep.

EC	CEC	14010	1287	ACP EU WATER FACILITY : EAU POTABLE & ASSAINISSEMENT	2006100291
UNICEF		14010	158	WATER RESOURCES POLICY/ADMIN. MGMT	060584

Total aid for water in 2006 for **Central African Rep.** USD thousand **1444**
And as a share of aid to all recipient countries **0.02%**

Central Asia, regional

EC	EDF	14040	6276	TRANSBOUNDARY MANAGEMENT OF KURA RIVER BASIN	2006201222
Germany	Fed Min	14010	13	WATER RESOURCES POLICY/ADMIN. MGMT	2006011017
Switzerland	SDC	14010	287	CENTRAL ASIA: CANAL AUTOMATION FERGANA	2003001703
Switzerland	SDC	14010	1253	REGIONAL WATER: CENTRE FOR HYDROMET	2001000779
Switzerland	SDC	14010	763	IWRM REGIONAL WATER FERGANA	2001000663
Switzerland	SDC	14010	89	WATER RESOURCES POLICY/ADMIN. MGMT	2001000039

Total aid for water in 2006 for **Central Asia, regional** USD thousand **8681**
And as a share of aid to all recipient countries **0.13%**

Chad

AfDF		14030	19123	PROGRAMME D'ALIMENTATION EN EAU POTABLE	060026
Germany	BMZ	14010	2	TRANSNET - TRANSBOUNDARY MANAGEMENT OF NATURAL RESOURCES IN SADC REGION	2006003309
Italy	LA	14030	38	BASIC DRINKING WATER SUPPLY AND BASIC SANITATION	062244
Japan	JICA	14020	98	GRANT AID CO-OPERATION IN WATER SUPPLY	062523C
Japan	JICA	14020	241	LE PROJET D'APPROVISIONNEMENT EN POTABLE DES ZONES RURALES DE LA REPUBLI	060891C
Spain	MUNIC	14030	6	ACCESS TO DRINKING WATER IN EAST CHAD	068738
Spain	MUNIC	14030	21	"SOLAR PUMPING SYSTEM"	067103
Spain	MUNIC	14030	14	POTABLE WATER ACCESS FOR THE SUDANESE REFIUGEE POPULATION	066719
UNICEF		14010	194	WATER RESOURCES POLICY/ADMIN. MGMT	060633

Total aid for water in 2006 for **Chad** USD thousand **19737**
And as a share of aid to all recipient countries **0.30%**

Chile

RECIPIENT Donor	Agency	Sector	Amount USD thousand	Project description	CRS ID Number
Germany	BMZ	14010	3	STRENGHTENING THE MANAGEMENT CAPACITIES OF ECUADORIAN WATER UTILITIES	2006003361
Japan	JICA	14020	14	WATER SUPPLY & SANIT. - LARGE SYST.	065025C
Japan	JICA	14050	66	WASTE MANAGEMENT/DISPOSAL	065026C
Luxembourg	MFA	14030	6	BASIC DRINKING WATER SUPPLY AND BASIC SANITATION	061619

Total aid for water in 2006 for **Chile** **USD thousand 89**
And as a share of aid to all recipient countries **0.00%**

China

RECIPIENT Donor	Agency	Sector	Amount USD thousand	Project description	CRS ID Number
Australia	AusAID	14010	4	ACEDP PREPARATION	2006002683
Australia	AusAID	14010	57	REGIONAL IMPACTS OF REVEGETATION ON WATER RESOURCES OF THE LOESS PLATEA	2006001491
Australia	AusAID	14010	675	CHINA WATER ENTITLEMENTS AND TRADING ACTIVITY	2006000484
Austria	BMF	14020	1619	DELIVERY OF PUMPS FOR A PUMP STATION	200600006g
Austria	BMF	14020	2192	POTABLE WATER PLANT: ELECTRO MECHANISM EQUIPMENT	200600002g
Austria	BMF	14050	791	SOLID WASTE TREATMENT FACILITY	200600003g
Canada	CIDA	14020	121	VS/USINE DE POMPES À EAUX -CHINE - VS/USINE DE POMPES À EAUX -CHINE	060089
France	MINEFI	14010	941	ASSIST. TECH. ENVIRON. URBAIN DU YUNNAN	2006003513
France	MINEFI	14020	377	QUALITE EAUX RIVIERE DES PERLES	2006003508
Germany	Fed Min	14015	499	SUSTAINABLE USE OF THE RESOURCE WATER	2006011112
Germany	BMZ	14015	90	INTEGRATED EXPERTS	2006002418
Germany	L G	14020	1461	SEMI-CENTRALIZED SUPPLY AND DISPOSAL SYSTEMS FOR URBAN AREAS IN CHINA	2006012022
Germany	L G	14020	500	COMPARISON OF DESINFECTION PROCESSES IN EFFLUENT OF PUBLIC WASTEWATER	2006012021
Germany	Fed Min	14020	1727	WATER SUPPLY	2006011120
Germany	BMZ	14030	77	IMPROVING ACCESS TO WATER TO RURAL PEOPLE IN PROVINCE OF SHANDONG	2006008143
Germany	BMZ	14030	734	REDUCING POVERTY THROUGH DISSEMINATION OF APPROPRIATE ENV. TECHNOLOGIES	2006004449
Germany	BMZ	14040	7	DEVELOPING POTENTIALS IN RURAL AREAS OF MEKONG RIPARIAN COUNTRIES	2006003374
Germany	L G	14050	2	WASTE MANAGEMENT/DISPOSAL	2006012402
Germany	L G	14050	75	ASIA-LINK PROJECT (EU) / PROF. DR. MICHAEL NELLES	2006011529
Germany	Fed Min	14050	18	GERMAN-VIETNAMESE SEMINAR "SOIL PROTECTION & REMEDIATION OF CONTAM.'	2006011028
Germany	BMZ	14050	203	INTEGRATED EXPERTS	2006002442
IDA		14020	5000	GEF-NINGBO WATER AND ENVIRONMENT PROJECT	060373
Japan	PRF	14015	23	WATER RESOURCES PROTECTION	067353C
Japan	JICA	14020	80	WATER SUPPLY & SANIT. - LARGE SYST.	065935C
Japan	JBIC	14020	53969	GUANGXI YULIN CITY ENVIRONMENT IMPROVEMENT PROJECT	063061
Japan	JBIC	14020	63557	HARBIN CITY WATER ENVIRONMENT IMPROVEMENT PROJECT	063060
Japan	JBIC	14020	109107	KUNMING CITY WATER ENVIRONMENT IMPROVEMENT PROJECT	063051
Japan	JICA	14020	91	TC IN SEWERAGE	062752C
Japan	JICA	14040	315	RIVER DEVELOPMENT	065936C
Japan	JICA	14040	441	HUMAN RESOURCE DEVELOPMENT PROJECT FOR WATER RESOURCES P.R.C	061279C
Japan	JICA	14040	279	THE STUDY ON THE IMPROVEMENT OF THE WATER RIGHTS SYSTEMS	060771C
Japan	JICA	14040	793	TC IN RIVERS/SAND ARRESTATION	060199C
Japan	JICA	14040	1390	TC IN WATER RESOURCES DEVELOPMENT	060098C

CRS AID ACTIVITIES IN SUPPORT OF WATER SUPPLY AND SANITATION – ISBN 978-92-64-05172-0 – © OECD/WWC 2008

RECIPIENT Donor	Agency	Sector	Amount USD thousand	Project description	CRS ID Number
Japan	PRF	14050	22	WASTE MANAGEMENT/DISPOSAL	067354C
Japan	JICA	14050	51	WASTE MANAGEMENT/DISPOSAL	065937C
Japan	ODC	14081	8	EDUC./TRNG:WATER SUPPLY & SANITATION	067421C
Korea	KOICA	14010	4	TRANINIG PROGRAM	2006861734
Luxembourg	MFA	14030	25	BASIC DRINKING WATER SUPPLY AND BASIC SANITATION	060737
Norway	NORAD	14020	234	YANJI WASTEWATER TREATMENT PLANT	2006002463
Spain	MUNIC	14030	18	DRINKING WATER SUPPLY FOR 7 FAR-OFF SETTLEMENTS IN TIBET	066325
Sweden	Sida	14050	50	WASTE MANAGEMENT/DISPOSAL	2006002531
Sweden	Sida	14050	610	CHN CLEANER PRODUCTION	2006002471
Switzerland	seco	14020	122	WATER SUPPLY & SANIT. - LARGE SYST.	2006002833
Switzerland	SDC	14030	48	BASIC DRINKING WATER SUPPLY AND BASIC SANITATION	1986000276
Switzerland	SDC	14050	29	WASTE MANAGEMENT/DISPOSAL	1986000269
UNICEF		14010	84	WATER RESOURCES POLICY/ADMIN. MGMT	060683
UNICEF		14030	179	HYGIENE, SANITATION AND WATER SUPPLY PROGRAMMES	060684
United Kingdom	DFID	14010	46	RURAL WATER SUPPLY & SANITATION REFORM PROGRAMME	050618
United States	TDA	14015	1	WATER RESOURCES PROTECTION	2006002932
United States	TDA	14020	541	WATER SUPPLY & SANIT. - LARGE SYST.	2006002962

Total aid for water in 2006 for   China   **USD thousand 249285**
And as a share of aid to all recipient countries   3.83%

Colombia

France	MINEFI	14040	549	AMENAG./GESTION BASSIN VERSANT	2006003505
Germany	BMZ	14010	3	STRENGTHENING THE MANAGEMENT CAPACITIES OF ECUATORIAN WATER UTILITIES	2006003358
Italy	LA	14040	100	RIVER DEVELOPMENT	061690
Japan	JICA	14020	72	WATER SUPPLY & SANIT. - LARGE SYST.	065058C
Japan	JICA	14040	11	RIVER DEVELOPMENT	065059C
Japan	JICA	14040	621	STUDY ON MONITORING AND EARLY WARNING SYSTEM FOR LANDSLIDES AND FLOODS	060274C
Norway	MFA	14050	15	PERSONNEL EXCHANGE	2006004505
Spain	MUNIC	14030	106	SEWAGE MASTER PLAN: CONSTRUCTION OF RESIUDUAL WATER COLLECTORS	066724
Spain	AG	14030	118	IMPROVEMNET IN SANITATION ENVIRONMANTAL AND HABITABILITY CONDITIONS	062556
Spain	MUNIC	14050	1	CLOSURE OF THE TRASH DUMP FROM SARAVENA	069112
Spain	Misc	14081	3	COURSE IN HYDROLOGIC AND HYDROCHEMICAL PROCESS IN WATER-BEARINGS	0610239

Total aid for water in 2006 for   Colombia   **USD thousand 1599**
And as a share of aid to all recipient countries   0.02%

Comoros

IDA		14010	1000	ADDITIONAL FINANCING FOR THE SERVICES SUPPORT CREDIT	060085c
UNICEF		14010	30	WATER RESOURCES POLICY/ADMIN. MGMT	060725

RECIPIENT Donor	Agency	Sector	Amount USD thousand	Project description	CRS ID Number

<table>
<tr><td colspan="2">Total aid for water in 2006 for</td><td colspan="2">Comoros</td><td>USD thousand 1030</td></tr>
<tr><td colspan="2">And as a share of aid to all recipient countries</td><td colspan="3">0.02%</td></tr>
</table>

Congo, Dem. Rep.

Donor	Agency	Sector	Amount	Project description	CRS ID Number
Belgium	MPRF	14030	13	BASIC DRINKING WATER SUPPLY AND BASIC SANITATION	2006003459
Belgium	DGCD	14030	127	BASIC DRINKING WATER SUPPLY AND BASIC SANITATION	2004000327
Belgium	DGCD	14050	113	WASTE MANAGEMENT/DISPOSAL	2005000887
EC	CEC	14010	3138	ACP EU WATER FACILITY : ALIM. EAU POTABLE COMMUNES PERIPH KINSHASA-EST	2006100292
EC	CEC	14010	4393	ACP - EU WATER FACILITY	2006100288
Germany	GTZ	14010	1255	WATER SECTOR REFORM	2006006060
Germany	BMZ	14010	15	TRANSNET - TRANSBOUNDARY MANAGEMENT OF NATURAL RESOURCES IN SADC REGION	2006003314
Germany	KFW	14020	35145	URBAN WATER SUPPLY SECONDARY TOWNS	2006000841
Ireland	DFA	14030	96	BASIC DRINKING WATER AND SANIT	060609
Spain	MUNIC	14030	34	BUILDING OF A WELL TO SUPPLY WATER TO THE UNDERPRIVILEGED POPULATION	068907
Spain	MUNIC	14030	11	SET UP OF A POTABLE WATER WELL IN THE COMMUNITY OF KIMWENZA	068017
Spain	MFA	14030	15	REPAIRMENT AND CONTITIONING OF THE POTABLE WATER SUPPLY	064054
Spain	MFA	14030	70	POTABLE WATER WELL	063929
Spain	AG	14030	53	POTABLE WATER SUPPLY FOR 13 VILLAGES IN CONGO.	062176
Spain	AG	14030	94	POTABLE WATER SUPPLY IN MUKILA	062022
Spain	AG	14030	45	SOCIO-SANITARY PROJECTS OF POTABLE WATER SUPPLY BY CONDUCTION	061605
Switzerland	SDC	14030	22	BASIC DRINKING WATER SUPPLY AND BASIC SANITATION	1980000885
UNICEF		14010	36	WATER RESOURCES POLICY/ADMIN. MGMT	060780
UNICEF		14030	543	HYGIENE, SANITATION AND WATER SUPPLY PROGRAMMES	060781
United Kingdom	DFID	14010	4600	WATER SUPPLY TO MBUJI MAYI	060551
United Kingdom	DFID	14020	4574	WATER AND SANITATION PROGRAMME	050232

<table>
<tr><td colspan="2">Total aid for water in 2006 for</td><td colspan="2">Congo, Dem. Rep.</td><td>USD thousand 54391</td></tr>
<tr><td colspan="2">And as a share of aid to all recipient countries</td><td colspan="3">0.84%</td></tr>
</table>

Congo, Rep.

Donor	Agency	Sector	Amount	Project description	CRS ID Number
AfDF		14030	1471	ETUDE DU PROGRAMME AEPA	060053
UNICEF		14010	0	WATER RESOURCES POLICY/ADMIN. MGMT	060748

<table>
<tr><td colspan="2">Total aid for water in 2006 for</td><td colspan="2">Congo, Rep.</td><td>USD thousand 1471</td></tr>
<tr><td colspan="2">And as a share of aid to all recipient countries</td><td colspan="3">0.02%</td></tr>
</table>

Cook Islands

Donor	Agency	Sector	Amount	Project description	CRS ID Number
Australia	AusAID	14010	222	WATER SUPPLY INVESTIGATIONS 2003	2003000690
Japan	JICA	14020	30	WATER SUPPLY & SANIT. - LARGE SYST.	066310C
New Zealand	NZAid	14030	402	NIWA - BUILDING CAPACITY IN WATER QUALITY MEASUREMENT IN COOK ISLANDS	060613

Total aid for water in 2006 for				Cook Islands	USD thousand 654	
And as a share of aid to all recipient countries					0.01%	

Costa Rica

Germany	BMZ	14020	42	LOCAL AGENDA 21 IN CENTRAL AMERICA	2006003369
Germany	GTZ	14050	3766	COMPETITIVENESS AND ENVIRONMENT PROGRAMME	2006006142
IDA		14010	3000	MAINSTREAMING MARKET-BASED INSTRUMENTS FOR ENVIR. MANAGEMENT PROJECT	060109c
Japan	JICA	14020	87	WATER SUPPLY & SANIT. - LARGE SYST.	064537C
Japan	JBIC	14020	128875	METROPOLITAN SAN JOSE ENVIRONMENT IMPROVEMENT PROJECT	063017
Japan	JICA	14050	45	WASTE MANAGEMENT/DISPOSAL	064538C
Spain	MFA	14040	502	INTEGRAL PROJECT OF REGIONAL IMPORTANCE BASIN OF THE FRIO RIVER	065104
Spain	Misc	14050	4	TECHNICAL-ECONOMIC VIABILITY FOR THE IMPLEMENTATION OF PLANTS OF SEWAGE TREATMENT	0610038
UNICEF		14010	4	WATER RESOURCES POLICY/ADMIN. MGMT	060823

Total aid for water in 2006 for				Costa Rica	USD thousand 136324	
And as a share of aid to all recipient countries					2.09%	

Cote d'Ivoire

Germany	BMZ	14010	3	TRANSNET - TRANSBOUNDARY MANAGEMENT OF NATURAL RESOURCES IN SADC REGION	2006003311
UNICEF		14010	76	WATER RESOURCES POLICY/ADMIN. MGMT	061403
UNICEF		14030	199	HYGIENE, SANITATION AND WATER SUPPLY PROGRAMMES	061404

Total aid for water in 2006 for				Cote d'Ivoire	USD thousand 278	
And as a share of aid to all recipient countries					0.00%	

Croatia

EC	EDF	14020	12803	KARLOVAC WATER AND WASTEWATER PROGRAMME	2006200076
EC	EDF	14050	7531	BIKARAC REGIONAL WASTE MANAGEMENT CENTRE	2006200081
Germany	L G	14020	36	FEASABILITY STUDYEXTENSION OF DUCT SYSTEM	2006012239
Greece	YPEHO DE	14010	8	HIGHER EFFICIENCY OF WATER-RELATED DEVELOPMENT	2006000312
Norway	MFA	14050	461	ESTABLISHMENT OF REG. WASTE MANAGEMENT	2006001176

Total aid for water in 2006 for				Croatia	USD thousand 20839	
And as a share of aid to all recipient countries					0.32%	

Cuba

Belgium	MPRW	14050	254	FORMATION PROFESSIONNELLE, GESTION DES DECHETS, ENVIRONNEMENT	2001009767
Germany	BMZ	14030	16	IMPROVEMENT OF WATER SUPPLY IN THE DISTRICT OF BOYEROS, HAVANNA	2006004889
Japan	JICA	14020	143	WATER SUPPLY & SANIT. - LARGE SYST.	064563C
Japan	JICA	14040	15	RIVER DEVELOPMENT	064564C
Japan	JICA	14050	19	WASTE MANAGEMENT/DISPOSAL	064565C

RECIPIENT Donor	Agency	Sector	Amount USD thousand	Project description	CRS ID Number
Japan	JICA	14050	381	TC IN URBAN SANITATION	060528C
Spain	MUNIC	14015	17	PDHL 2005: BOCA DE DOS RÍOS GUAMÁ	069412
Spain	MUNIC	14015	18	SUPORT TO THE RESTORATION OF THE BOCA DE DOS RÍOS AQUEDUCT	066517
Spain	MUNIC	14030	8	REPLACEMENT OF THE CINDUCTOR VALVES AND URAN ZONES NETS IN "SAN JOSÉ"	069663
Spain	MUNIC	14030	21	CHLORINE PILLS PRODUCTION FOR WATER TREATMENT AND DESINFECTION	069494
Spain	MUNIC	14030	14	PDHL 2005: WATER - SONGO LA MAYA	069409
Spain	MUNIC	14030	1	IMPROVEMENT IN THE WATER SUPPLY FOR THE POPULAR COUNCIL VILLA 1	069036
Spain	MUNIC	14030	13	"SUPPORT FOR COMMUNAL HYGIENE SERVICES IN THE CITY LA HABANA"	068179
Spain	MUNIC	14030	15	WATER SUPPLY FOR THE ESTABLISHMENT OF MANAGUACO	068077
Spain	MUNIC	14030	31	CREATION OF A WELL FIELD POTABLE WATER SUPPLY.	067879
Spain	MUNIC	14030	18	NETWORK EXPANSION ACQUEDUCTS AND SEWAGE	067335
Spain	MUNIC	14030	15	SUPPORT TO INTEGRAL DEVELOPMENT OF A RURAL SETTLEMENT	066515
Spain	AG	14030	114	SUPPLY AND REUTILIZATION OF PURE WATER	063527
Spain	AG	14030	22	AQCUISITION OF PUMPING EQUIPMENT FOR POTABLE WATER	062294
Spain	MFA	14040	75	ENVIRONMENTAL EVALUATION AND DIAGNOSIS OF HYDRIC RESOURCES	065152
Spain	MUNIC	14050	6	PURCHASE OF NEW CONTAINERS	068891
Spain	MUNIC	14050	15	ACQUISITION OF NEW TRASH CONTAINERS FOR THE CITY LA HABANA	067884
Spain	MUNIC	14050	6	ACQUISITION OF TRASH CONTAINERS FOR THE CITY OF LA HABANA	067867
Spain	MUNIC	14050	147	IMPROVEMENT IN HABITABILITY AND ENVIRONMENTAL HYGIENE CONDITIONS	067284
Spain	AG	14050	107	SUPPORT FOR BASIC HYGENE SERVICES IN THE CITY LA HABANA"	061330
Spain	Misc	14081	18	HAVANA PROJECT: INTEGRAL MANAGEMENT OF WATER MASTER' S DEGREE	0610558

Total aid for water in 2006 for   Cuba   USD thousand 1507
And as a share of aid to all recipient countries   0.02%

Djibouti

RECIPIENT Donor	Agency	Sector	Amount USD thousand	Project description	CRS ID Number
EC	CEC	14020	13556	APPUI INSTITUTIONNEL POLITIQUE SECTORIELLE EAU ET REHABILITATION	2006100050
UNICEF		14010	18	WATER RESOURCES POLICY/ADMIN. MGMT	060885
United States	TDA	14020	1	WATER SUPPLY & SANIT. - LARGE SYST.	2006002990

Total aid for water in 2006 for   Djibouti   USD thousand 13575
And as a share of aid to all recipient countries   0.21%

Dominican Republic

RECIPIENT Donor	Agency	Sector	Amount USD thousand	Project description	CRS ID Number
EC	CEC	14010	4590	ACP EU WATER FACILITY IN 63 POOR RURAL COMMUNITIES (BORDER)	2006100312
EC	CEC	14010	1883	WATER RESOURCES POLICY/ADMIN. MGMT	2006100051
Japan	JICA	14020	40	WATER SUPPLY & SANIT. - LARGE SYST.	064586C
Japan	JICA	14050	145	WASTE MANAGEMENT/DISPOSAL	064587C
Japan	JICA	14050	507	STUDY FOR INTEGRATED MANAGEMENT OF URBAN SOLID WASTE IN SANTODOMINGO NAT	060370C
Japan	JICA	14050	819	TC IN URBAN SANITATION	060191C
Spain	MUNIC	14015	53	BUILDING OF A RURAL AQUEDUCT IN SAN FRANCISCO COMMUNITY	066536
Spain	MUNIC	14030	22	IMPROVEMENT OF SANITATION CONDITIONS IN BATEY EL FAO, GANDULES & SAN JOSÉ	068622

CRS AID ACTIVITIES IN SUPPORT OF WATER SUPPLY AND SANITATION – ISBN 978-92-64-05172-0 – © OECD/WWC 2008

RECIPIENT Donor	Agency	Sector	Amount USD thousand	Project description	CRS ID Number
Spain	MUNIC	14030	55	DRINKING WATER SUPPLY HEALTH INSTALLATIONS	067668
Spain	MUNIC	14030	33	IMPROVEMENT IN THE HYGIENIC SANITARY CONDITIONS	067430
Spain	MUNIC	14030	7	CONSTRUCTION OF A WATER DEPOSIT IN AN ACQUEDUCT	067177
Spain	MUNIC	14030	10	IMPROVEMENT IN HYGIENIC SANITARY AND ENVIRONMENTAL CONDITIONS	066854
Spain	MUNIC	14030	12	WATER SUPPLY REFORESTATION AND ENVIRONMENTAL EDUCATION	066851
Spain	MUNIC	14030	54	CONSTRUCTION OF A RURAL AQUEDUCT SYSTEM WITH SOLAR PANNELS IN THE ZONES OF SITIO NUEVO AND MONGOTE	066786
Spain	AG	14030	98	BASIC AND NATURAL ENVIRONMENT SANITATION	065895
Spain	MFA	14030	560	POTABLE WATER PROGRAM AND SANITATION IN THE RURAL ZONES 2ª PHASE (2006)	065164
Spain	AG	14030	88	BASIC SANITATION IN EL VISO	062571
Spain	AG	14030	44	IMPROVED SANITATION CONDITIONS	062122
Spain	AG	14030	73	POTABLE WATER ACCESS AND IMPROVEMENT OF THE SANITARY HYGIENIC CONDITIONS PROMOTING ORGANIZATIONAL D	062105

Total aid for water in 2006 for   **Dominican Republic**   **USD thousand 9092**

And as a share of aid to all recipient countries   **0.14%**

Ecuador

RECIPIENT Donor	Agency	Sector	Amount USD thousand	Project description	CRS ID Number
Belgium	DGCD	14020	219	PROVISION D'EAU POTABLE ET ASSAINISSEMENT À CAÑAR	2004000321
Belgium	MPRF	14030	88	BASIC DRINKING WATER SUPPLY AND BASIC SANITATION	2005001640
Belgium	DGCD	14030	102	AMORCE D'UN SYSTÈME D'EAU POTABLE DANS LA SIERRA	2004000324
Belgium	DGCD	14030	153	SYSTÈMES D'IRRIGATION INCORPORÉS DANS UNE GESTION DURABLE DU BASSIN	2004000322
Belgium	DGCD	14030	181	BASIC DRINKING WATER SUPPLY AND BASIC SANITATION	2004000320
Belgium	DGCD	14030	528	EAU POTABLE POUR LA SIERRA NORTE	2002000258
Canada	CIDA	14081	88	MUSQUEAM/TORTORAS COMMUNITY PARTNERSHIP	060007
Germany	BMZ	14010	43	STRENGHTENING THE MANAGEMENT CAPACITIES OF ECUADORIAN WATER UTILITIES	2006003357
Japan	JICA	14020	69	WATER SUPPLY & SANIT. - LARGE SYST.	065094C
Japan	JICA	14020	159	GRANT AID CO-OPERATION IN WATER SUPPLY	061454C
Japan	MOFA	14020	3196	IMPROV. OF THE WATER SYSTEM OF IBARRA	060124
Japan	MOFA	14020	17259	IMPROV. OF THE WATER SYSTEM OF HUAQUILLAS AND ARENILLAS	060083
Japan	MOFA	14020	421	IMPROV. OF WATER SYSTEM OF HUAQUILLAS CITY AND ARENILLAS CITY	060017
Japan	JICA	14040	108	FOLLOW-UP CO-OPERATION FOR GROUNDWATER DEVELOPMENT PROJECT F.Y.2006	062251C
Korea	Kexim	14030	43630	THE POTABLE WATER SUPPLY EXPANSION PROJECT	2006861149
Spain	AG	14010	283	COLLABORATION AGREEMENT(GENERALITAT AND THE VALENCIA FUNDS)	062625
Spain	MUNIC	14030	14	ENVIRONMENTAL SANITATION	069145
Spain	MUNIC	14030	5	CHUMABI PROYECT. WATER WELL PERFORATION IN RURAL COMMUNITIES	068407
Spain	MUNIC	14030	19	CHUMAVÍ PROJECT. WATER WELL PERFORATION	068192
Spain	MUNIC	14030	63	CONSTRUCTION OF A POTABLE WATER SYSTEM AND LETRINIZATION	068015
Spain	MUNIC	14030	56	IRRIGATION SYSTEM FOR THE INTEGRAL FARM SAN CARLOS	067692
Spain	MUNIC	14030	9	POTABLE WATER	067494
Spain	MUNIC	14030	7	POTABLE WATER FOR COMMUNITIES OF SAN NICOLÁS DE CAR.	067170
Spain	AG	14030	166	POTABLE WATER IN RURAL COMMUNITIES OF SIERRA CENTRAL AND ORIENTE.	062444
Spain	AG	14030	130	AUTOCONSTRUCTION OF WATER AND LETRINE NETWORKS IN 5 MARGINAL COMMUNITIES OF MILAGRO	061919

RECIPIENT / Donor	Agency	Sector	Amount USD thousand	Project description	CRS ID Number
Spain	AG	14030	111	CONSTRUCTION OF FOUR INTEGRAL WATER SYSTEMS FOR HUMAN CONSUMPTION	061621
Spain	AG	14030	13	RIOBAMBA PROJECT. WELL PERFORATIO IN SAN ANTONIO DE PADUA	061174
Spain	MFA	14040	803	ORDER MANAGEMENT AND DEVELOPMENT OF THE HYDROGRAPHIC BASIN OF CATAMAYO	065248
Sweden	Sida	14010	8	PROCUR ADV CONAMU CONCOPE PROCUR ADV ECUADOR	2006002296
UNICEF		14010	11	WATER RESOURCES POLICY/ADMIN. MGMT	060918
United States	Misc	14020	58	WATER SUPPLY & SANIT. - LARGE SYST.	2006002979

Total aid for water in 2006 for   **Ecuador**   **USD thousand 68000**
And as a share of aid to all recipient countries   **1.04%**

Egypt

RECIPIENT / Donor	Agency	Sector	Amount USD thousand	Project description	CRS ID Number
Germany	GTZ	14010	5021	WATER AND SANITATION MANAGEMENT	2006006072
Germany	BMZ	14010	224	REFORM OF THE WATER SECTOR IN THE MENA REGION	2006003344
Germany	BMZ	14010	2	TRANSNET - TRANSBOUNDARY MANAGEMENT OF NATURAL RESOURCES IN SADC REGION	2006003306
Germany	L G	14050	2	WASTE MANAGEMENT/DISPOSAL	2006012400
Germany	BMZ	14050	112	INTEGRATED EXPERTS	2006002447
Germany	BMZ	14081	4	EFFICIENT MANAGEMENT OF WASTEWATER, TREATMENT AND REUSE	2006003384
Greece	YPEHO DE	14010	8	HIGHER EFFICIENCY OF WATER-RELATED DEVELOPMENT	2006000300
Japan	JICA	14020	142	WATER SUPPLY & SANIT. - LARGE SYST.	063707C
Japan	JICA	14020	197	GRANT AID CO-OPERATION IN WATER SUPPLY	061124C
Japan	JICA	14020	306	MANAGEMENT CAPACITY OF OPERATION AND MAINTENANCE	060698C
Japan	MOFA	14020	20816	UPGRADING OF EL MAHALA EL KOBRA WATER TREATMENT PLANT	060122
Japan	JICA	14040	108	RIVER DEVELOPMENT	063708C
Netherlands	MFA	14010	5546	SUPPORT COORDINATION IN NWRP IMPLEMENTATION	2006000063
Netherlands	MFA	14020	21892	WATER SUPPLY & SANIT. - LARGE SYST.	2003011249
UNICEF		14010	28	WATER RESOURCES POLICY/ADMIN. MGMT	062943
United States	TDA	14050	4	WASTE MANAGEMENT/DISPOSAL	2006003069

Total aid for water in 2006 for   **Egypt**   **USD thousand 54412**
And as a share of aid to all recipient countries   **0.84%**

El Salvador

RECIPIENT / Donor	Agency	Sector	Amount USD thousand	Project description	CRS ID Number
Germany	BMZ	14020	17	LOCAL AGENDA 21 IN CENTRAL AMERICA	2006003367
Japan	JICA	14020	35	WATER SUPPLY & SANIT. - LARGE SYST.	064618C
Japan	JICA	14050	22	WASTE MANAGEMENT/DISPOSAL	064619C
Luxembourg	MFA	14010	52	WATER RESOURCES POLICY/ADMIN. MGMT	061093
Luxembourg	MFA	14020	989	WATER SUPPLY & SANIT. - LARGE SYST.	061091
Spain	MUNIC	14030	8	DRINKING WATER SUPPLY FOR RURAL COMMUNITIES OF LA LIBERTAD-SURPHASE III	069924
Spain	MUNIC	14030	55	SECURE DRINKING WATER SUPPLY IN RURAL COMMUNITIES OF LA LIBERTAD-SOUTH	069916
Spain	MUNIC	14030	5	SECURITY IN DRINKING WATER SUPPLY	069907
Spain	MUNIC	14030	262	WATER SUPPLY FOR SEGUNDO MONTES COMMUNITY	069359
Spain	MUNIC	14030	8	PROVISION OF RURAL POTABLE WATER	069185

CRS AID ACTIVITIES IN SUPPORT OF WATER SUPPLY AND SANITATION – ISBN 978-92-64-05172-0 – © OECD/WWC 2008

RECIPIENT Donor	Agency	Sector	Amount USD thousand	Project description	CRS ID Number
Spain	MUNIC	14030	8	OPERATION OF A COMMUNITY SHOP AND INSTALLATION OF A WATER PUMP	068579
Spain	MUNIC	14030	11	REPLACEMENT OF THE WATER PUMP PLANT	067828
Spain	MUNIC	14030	151	INTRODUCTION OF A SEWAGE SYSTEM FOR BLACK WATER	067728
Spain	MUNIC	14030	19	INTEGRAL MANAGEMENT OF THE HYDIC RESOURCE LOS MILAGROS RIVER	067427
Spain	MUNIC	14030	87	CONSTRUCTION OF LETRINES AND ENVIRONMENTAL SANITATION	067324
Spain	MUNIC	14030	26	RESCUE PRODUCTIVE HERITAGE FROM THE DISTRICT OF AGUA ZARCA AND R	067204
Spain	MUNIC	14030	55	STRENGTHENING OF ENVIRONMENTAL UNITS	067028
Spain	MUNIC	14030	31	PROGRAM ON INTEGRAL SANITATION	066964
Spain	MUNIC	14030	13	PIPAGE REPAIR AND CONSTRUCTION OF A DISTRIBUTION TANK	066739
Spain	MUNIC	14030	57	SECURITY FOR DRINKING WATER SUPPLY	066409
Spain	MUNIC	14030	88	DRINKING WATER SUPPLY. EL SALVADOR	066361
Spain	MUNIC	14030	113	DRINLING WATER SUPLY FOR THE CANTÓN TUTULTEPEQUE. EL SALVADOR	066302
Spain	AG	14030	37	IMPROVEMENT OF BASIC HEALTH AND SANITATION FOR 180 FAMILIES IN LLOBASCO MUNICIPALITY. EL SALVADOR	065856
Spain	MFA	14030	251	STRENGTHENING OF ENVIRONMENTAL MANAGEMENT	064957
Spain	MFA	14030	188	POTABLE WATER SUPPLY IN THE CANTONS OF LOMITAS Y EL NIÑO	064884
Spain	AG	14030	73	IMPROVEMENT IN POTABLE WATER SUPPLY	063860
Spain	AG	14030	26	PEACE AND SOLIDARITY FOUNDATION OF EUSKADI	063846
Spain	AG	14030	237	CONSTRUCTION OF A POTABLE WATER SYSTEM AND BASIC SANITATION	063471
Spain	AG	14030	181	PROGRAM OF SAFETY IN SANITATION SERVICES AND POTABLA WATER SUPPLY	063384
Spain	AG	14030	50	DIRECTING PLAN OF SANITATION INFRASTRUCTURE AND POTABLE WATER SUPPLY	063193
Spain	AG	14030	176	SAFETY IN SANITATION AND WATER SUPPLY SERVICES	062678
Spain	AG	14030	103	SPOTABLE WATER SERVICE IN THE COMMUNITY BUENA VISTA	062281
Spain	AG	14030	154	POTABLE WATER SUPPLY FOR THE POPULATION OF NUEVA GRANADA	062195
Spain	AG	14030	423	BUILDING CLEAN AND HEALTHY MUNICIPALITIES	062014
Spain	AG	14030	56	WATER SANITATION AND ENVIRONMENTAL EDUCATION	061866
Spain	AG	14030	34	IMPROVEMNET IN SANITARY CONDITIONS	061804
Spain	AG	14030	118	IMPLEMENTATION OF AN INTEGRAL SYSTEM TO TREAT BLACK WATERS	061609
Spain	AG	14030	67	PROMOTING HEALTHY MUNICIPALITIES	061432
Spain	AG	14030	113	POTABLE WATER INTRODUCTION IN THE CASERÍO LA ARENERA AND TRINIDAD CANTÓN	061393
Spain	AG	14030	90	SANITATION AND NUTRITION IN RURAL COMMUNITIES	061272
Spain	AG	14030	340	INSTALATION OF THE SYSTEM OF HYDAULICS NETWORK IN THE COMMUNITIES OF SUCHINANGO EL OLIMPO ANDLAS DE	061036
Spain	Misc	14030	2	DRINKING WATER SUPLY SECURITY IN THE RURALES COMMUNITIES OF LA LIBERTAD (3 F)	0610250
Spain	MFA	14040	138	PROJECT TO SUPPORT CREATION OF ORGANISMS IN THE BASIN	065064
Spain	MUNIC	14050	63	INTEGRAL MANAGEMENT OF SOLID WASTE IN SAN SALVADOR	069455
Spain	MFA	14050	151	SUPPORT INTEGRAL MANAG. OF THE COMMUNITY´S WASTE AND WATER RESIDUAL	065096
Spain	MFA	14050	251	SUSTAINABLE MANAGEMENT OF SOLID WASTE	064915
Spain	AG	14050	107	IMPROVEMNET IN HEALTH CONDITIONS	063486
Spain	AG	14050	5	WASTES A ROAD FOR SOCIOLABORAL INSERTION	061489
UNICEF		14010	0	WATER RESOURCES POLICY/ADMIN. MGMT	060931

Total aid for water in 2006 for				**El Salvador**	**USD thousand 5543**
And as a share of aid to all recipient countries					**0.09%**

Equatorial Guinea

Spain	MUNIC	14030	11	BUILDING OF A DRINKING WATER WELL IN EL BUEN PASTOR SCHOOL. MALABO	069844
Spain	MUNIC	14030	6	ENVIRONMENTAL SANITATION. BATA. ECUATORIAL GUINEA	068031
Spain	MFA	14030	57	REHABILITATION OF THE POTABLE WATER SYSTEM MUNICIPALITY OF NKUE	064163
Spain	MFA	14030	16	REPAIRMENT OF POTABLE WATER INSTALLATION AND CONSTRUCTION OF A TOWN WATER PUMP IN BASILÉ	064068
Spain	MFA	14030	39	REPAIRMENT REFORM AND NEW OPPENING OF WATER SUPPLY IN BALOEN	064024

Total aid for water in 2006 for				**Equatorial Guinea**	**USD thousand 130**
And as a share of aid to all recipient countries					**0.00%**

Eritrea

Germany	BMZ	14030	100	PRESERVATION OF WATER AND MEASURES TO PROTECT THE ENVIRONMENT IN JEGAR	2006004898
Germany	L G	14040	13	CONSTRUCTION OF GABIONS	2006012157
Italy	DGCS	14010	330	WATER RESOURCES POLICY/ADMIN. MGMT	060742
Italy	LA	14020	84	WATER SUPPLY & SANIT. - LARGE SYST.	061673
Japan	JICA	14020	90	THE PROJECT FOR URBAN WATER SUPPLY IN DEBUB REGION	062785C
Japan	JICA	14020	680	GRANT AID CO-OPERATION IN WATER SUPPLY	060246C
Japan	MOFA	14030	438	URBAN WATER SUPPLY IN DEBUB REGION	060182
Netherlands	MFA	14010	630	WATER RESOURCES POLICY/ADMIN. MGMT	2006000712
Netherlands	MFA	14030	2155	BASIC DRINKING WATER SUPPLY AND BASIC SANITATION	2006000395
Norway	MFA	14030	624	WATER TO ANSEBA	2006004324
Norway	MFA	14040	5	CONSULTANCY SERVICE FOR ASSESSING THE STATUS OF THE CONSTRUCTION OF SHIKETI DAM	2006001644
Spain	MUNIC	14030	34	DRINKING WATER SUPPLY FOR DASE POPULATION (ERITREA)	069992
UNICEF		14010	80	WATER RESOURCES POLICY/ADMIN. MGMT	061006
UNICEF		14030	988	COMMUNITY HYGIENE AND WATER SAFETY	061005

Total aid for water in 2006 for				**Eritrea**	**USD thousand 6250**
And as a share of aid to all recipient countries					**0.10%**

Ethiopia

AfDF		14020	64151	RURAL WATER SUPPLY & SANITATION PROGRAM	060069
Austria	Reg	14030	3	BASIC DRINKING WATER SUPPLY AND BASIC SANITATION	2006008633
Belgium	DGCD	14050	403	PREVENTION ET TRAITEMENT DECHETS PESTICIDES OBSOLETES	2004001148
EC	CEC	14010	377	ACP EU WATER FACILITY : RURAL WATER SUPPLY AND SANITATION PROJECT	2006100315
EC	CEC	14010	935	ACP EU WATER FACILITY:ACCESS WATER AND SANITATION SERVICES	2006100314
EC	CEC	14010	1667	ACP EU WATER FACILITY : WATER DEV, SANIT, HYG PROMOTI ON AND ENV PROJ	2006100313
EC	CEC	14010	1706	ACP EU WATER FACILITY : INTEGRATED COMMUNITY DEV PROG WATER EXTENS PROJ	2006100307
EC	CEC	14010	3625	ACP - EU WATER FACILITY	2006100304

RECIPIENT Donor	Agency	Sector	Amount USD thousand	Project description	CRS ID Number
EC	CEC	14010	13553	ACP EU WATER FACILITY : OFFICE FOR PROGRESS OF MDGS ON WATER & SANIT.	2006100298
EC	CEC	14010	20710	SMALL TOWN WATER & SANITATION PROJECT - ETHIOPIA	2006100148
EC	EIB	14030	20710	SMALL TOWN WATER & SANITATION PROGRAM	2006300022
Finland	MFA	14030	218	FRAME AGREEMENT WITH NGO (SAVE THE CHILDREN)	060331
Finland	MFA	14030	717	NGO SUPPORT / WATER AND HYGIENE PROJECT (WHPJ) IN JIMMA AREA, ETHIOPIA	060129
Finland	MFA	14030	11297	RURAL WATER SUPPLY AND ENVIRONMENTAL PROJECT IN ETHIOPIA	060123
Germany	BMZ	14010	90	INTEGRATED EXPERTS	2006002411
Germany	L G	14050	25	ORGANIC WASTE	2006012189
Greece	YPEJ	14040	64	CONSTRUCTION OF TWO WATER RESERVOIRS IN DAMOT GALE	2006000326
IDA		14010	21500	PROTECTION OF BASIC SERVICES	060013c
Ireland	DFA	14030	657	BASIC DRINKING WATER AND SANIT.	060181
Italy	DGCS	14010	282	WATER RESOURCES POLICY/ADMIN. MGMT	061057
Italy	LA	14020	16	WATER SUPPLY & SANIT. - LARGE SYST.	061703
Italy	LA	14050	126	WASTE MANAGEMENT/DISPOSAL	061910
Japan	JICA	14020	159	WATER SUPPLY & SANIT. - LARGE SYST.	063870C
Japan	MOFA	14030	198	WATER SUPPLY IN AFAR REGION	060188
Japan	MOFA	14030	4674	WATER SUPPLY IN SOUTHERN NATIONS, NATIONALITIES & PEOPLES'REGIONAL STATE	060128
Japan	JICA	14040	191	RIVER DEVELOPMENT	063871C
Japan	JICA	14040	132	TC IN WATER RESOURCES DEVELOPMENT	061782C
Japan	JICA	14040	137	THE PROJECT FOR RURAL WATER DEVELOPMENT IN THE AFAR STATE	061727C
Japan	JICA	14040	434	THE PROJECT FOR RURAL WATER SUPPLY AND REHABILITATION IN TIGRAY REGION	060791C
Japan	JICA	14040	520	GRANT AID CO-OPERATION IN WATER RESOURCES DEVELOPMENT	060355C
Japan	JICA	14040	587	GROUNDWATER DEV. AND WATER SUPPLY TRAINING CENTER PHASE 2(ETHIOPIA WATER	060304C
Korea	KOICA	14030	250	THE DEVELOPMENT OF GROUND WATER IN BORENA PROJECT	2006862146
Norway	NORAD	14030	245	WATER MANAGEMENT, HEALTH AND HYGIENE EDUCATION	2006002470
Norway	MFA	14030	687	RURAL WATER SUPPLY	2006002338
Norway	MFA	14030	302	WATER AND SANITATION PROJECT	2006000902
Spain	AG	14010	377	WATER PROGRAM IN ETHIOPIA.	063441
Spain	AG	14010	14	WATER IS LIFE	062916
Spain	MFA	14015	74	DEVELOPMENT OF AQUIFER RESOURCES IN THE ANGAR VALLEY	063931
Spain	MUNIC	14030	5	WELL PERFORATION ON THE ANGAR VALLEY	069223
Spain	MUNIC	14030	4	3 WATER PUMPS FOR ETHIOPÍA	068879
Spain	MUNIC	14030	34	DEVELOPMENT OF WATER-BEARING INFRASTRUCTURES TO IMPROVE THE AGRICULTURAL PRODUCTION IN VALLE DE ANGA	068788
Spain	MUNIC	14030	4	BUILDING OF A RESERVOIR. VALLE DE ANGAR GUTEN - ETHYOPIA	068771
Spain	MUNIC	14030	16	BUILDING OF 2 DAMS TO STORAGE WATER	068494
Spain	MUNIC	14030	19	WELL PERFORATION IN THE ANGAR GUTEN VALLEY. ETHIOPIA	068039
Spain	MUNIC	14030	5	CONSTRUCTION OF A DAM TO STORE WATER IN THE VALLEY OF ANGAR GUTEN " (ETHIOPIA)	067317
Spain	MFA	14030	236	MORTALITY REDUCATION AND MOBILITY OF THE ETHIOPIAN RURAL POPULATION GIVING SPECIAL ATTENTION TO WOM	064753
Spain	AG	14030	54	ENGINEERING ASSOCIATION FOR CO-OPERATION	063696
Spain	AG	14030	188	ETHIOPÍAN WATER. STABLISHMENT OF A SUPPORT BANK FOR THE COMMUNAL MANAGEMENT PROJECTS FOR POTABLE WAT	061323
Switzerland	SDC	14030	11	BASIC DRINKING WATER SUPPLY AND BASIC SANITATION	1980000901

RECIPIENT Donor	Agency	Sector	Amount USD thousand	Project description	CRS ID Number
UNICEF		14010	55	WATER RESOURCES POLICY/ADMIN. MGMT	060965
UNICEF		14030	2016	HYGIENE, SANITATION AND WATER SUPPLY PROGRAMMES	060964
UNICEF		14030	240	COMMUNITY HYGIENE AND WATER SAFETY	060963
United Kingdom	DFID	14010	856	SUPPORT TO POLICY DEVELOPMENT IN THE WATER SECTOR	060152

Total aid for water in 2006 for   **Ethiopia**   **USD thousand 175554**
And as a share of aid to all recipient countries   **2.70%**

Europe, regional

RECIPIENT Donor	Agency	Sector	Amount USD thousand	Project description	CRS ID Number
Greece	YPEPU	14015	126	DEVELOPMENT OF A SYSTEM TO ADMINISTER FLOODING FLOWS OF THE EVROS RIVER	2006000320
Norway	MFA	14010	98	WATER RESEARCH: NORSK INSTITUTT	2006001605
Switzerland	SDC	14015	38	WATER RESOURCES PROTECTION	2006000591
Switzerland	SDC	14015	185	SAVA RIVER	2006000225
United States	Misc	14050	72	WASTE MANAGEMENT/DISPOSAL	2006003064

Total aid for water in 2006 for   **Europe, regional**   **USD thousand 519**
And as a share of aid to all recipient countries   **0.01%**

Fiji

RECIPIENT Donor	Agency	Sector	Amount USD thousand	Project description	CRS ID Number
Japan	JICA	14020	58	WATER SUPPLY & SANIT. - LARGE SYST.	066323C
Japan	JICA	14040	29	RIVER DEVELOPMENT	066324C
Japan	JICA	14050	22	WASTE MANAGEMENT/DISPOSAL	066325C
Korea	KOICA	14010	12	TRANINIG PROGRAM	2006861924
New Zealand	NZAid	14050	746	"NIWA - SUSTAINABLE WASTE TREATMENT SYSTEMS FOR FIJIAN VILLAGES"	060684
New Zealand	NZAid	14081	5	"WATER TREATMENT TRAINING"	060034

Total aid for water in 2006 for   **Fiji**   **USD thousand 872**
And as a share of aid to all recipient countries   **0.01%**

Gambia

RECIPIENT Donor	Agency	Sector	Amount USD thousand	Project description	CRS ID Number
Belgium	MPRF	14030	37	BASIC DRINKING WATER SUPPLY AND BASIC SANITATION	2005001639
IDA		14010	600	COMMUNITY-DRIVEN DEVELOPMENT PROJECT	060032e
Japan	MOFA	14030	2543	RURAL WATER SUPPLY	060143
Spain	MUNIC	14030	4	WATER PUMP INSTALLATION IN A COMMUNITY. GAMBIA	066549
UNICEF		14010	1	WATER RESOURCES POLICY/ADMIN. MGMT	061062

Total aid for water in 2006 for   **Gambia**   **USD thousand 3185**
And as a share of aid to all recipient countries   **0.05%**

Georgia

RECIPIENT Donor	Agency	Sector	Amount USD thousand	Project description	CRS ID Number
EC	EDF	14010	5	WATER GOVERNANCE REFORM IN GEORGIA	2006200047
Finland	MFA	14010	854	ENVIRONMENTAL MONITORING SYSTEM IN GEORGIA	060260

    CRS AID ACTIVITIES IN SUPPORT OF WATER SUPPLY AND SANITATION – ISBN 978-92-64-05172-0 – © OECD/WWC 2008

RECIPIENT Donor	Agency	Sector	Amount USD thousand	Project description	CRS ID Number
Germany	KFW	14020	1325	MUNICIPAL INFRASTRUCTURE I	2006000435
Germany	KFW	14020	21437	MUNICIPAL INFRASTRUCTURE I	2006000432
Japan	JICA	14020	0	WATER SUPPLY & SANIT. - LARGE SYST.	065500C
Sweden	Sida	14030	3608	KUTAISI WATER EBRD	2006008158
Sweden	Sida	14030	1933	POTI WATER EBRD	2006008157
Switzerland	SDC	14030	48	BASIC DRINKING WATER SUPPLY AND BASIC SANITATION	2004001494
Switzerland	SDC	14030	223	GEO WATER SUPPLY SIGNAGHI / LAGODEKHI	2004001493
UNICEF		14010	88	WATER RESOURCES POLICY/ADMIN. MGMT	061096

Total aid for water in 2006 for   **Georgia**   **USD thousand 29520**
And as a share of aid to all recipient countries   **0.45%**

Ghana

RECIPIENT Donor	Agency	Sector	Amount USD thousand	Project description	CRS ID Number
AfDF		14020	67667	ACCRA SEWERAGE IMPROVEMENT PROJECT (ASIP)	060056
Belgium	DGCD	14020	8057	PRÊT D'ÉTAT À ÉTAT - EXPANSION D'UN SYSTÈME D'APPROVISIONNEMENT EN EAU	2006003006
Belgium	DGCD	14030	3998	BONIFICATION DES INTÉRÊTS	2006003298
Belgium	DGCD	14050	7923	PRET D'ETAT A ETAT - ASSAINISSEMENT DE LA BAIE D'ELMINA	2006003001
Germany	BMZ	14010	6	REFORM OF THE WATER SECTOR IN THE MENA REGION	2006003346
Germany	BMZ	14010	1	TRANSNET - TRANSBOUNDARY MANAGEMENT OF NATURAL RESOURCES IN SADC REGION	2006003312
Netherlands	MFA	14010	780	WATER RESOURCES POLICY/ADMIN. MGMT	2006001059
Spain	MUNIC	14030	21	CONSTRUCTION OF SEVEN TOILETS AND A WATER DEPOSIT FOR A PREPSCHOOL	066727
Spain	AG	14030	25	CONSTRUCTION OF TWO WATER WELLS IN LIBERATION BARRACKS AND RECCE QUARTER GUARDS	062301
Spain	AG	14030	13	CONSTRUCTION OF A WATER WELL IN KOTOKROM	062300
UNICEF		14010	1	WATER RESOURCES POLICY/ADMIN. MGMT	061112
UNICEF		14030	242	SECTOR PLANS FOR HYGIENE, SANITATION AND WATER	061111
UNICEF		14030	286	HYGIENE, SANITATION AND WATER SUPPLY PROGRAMMES	061110
United Kingdom	DFID	14010	14167	DFID SUPPORT TO THE GHANA WATER SECTOR	060172

Total aid for water in 2006 for   **Ghana**   **USD thousand 103189**
And as a share of aid to all recipient countries   **1.58%**

Grenada

RECIPIENT Donor	Agency	Sector	Amount USD thousand	Project description	CRS ID Number
EC	CEC	14020	8410	SOUTHERN GRENADA WATER SUPPLY IMPROVEMENT PROJECT	2006100076

Total aid for water in 2006 for   **Grenada**   **USD thousand 8410**
And as a share of aid to all recipient countries   **0.13%**

Guatemala

RECIPIENT Donor	Agency	Sector	Amount USD thousand	Project description	CRS ID Number
Austria	Reg	14030	20	BASIC DRINKING WATER SUPPLY AND BASIC SANITATION	2006008608
Germany	BMZ	14020	17	LOCAL AGENDA 21 IN CENTRAL AMERICA	2006003365
Germany	BMZ	14030	282	BASIC DRINKING WATER SUPPLY FOR RURAL COMMUNITIES	2006008150
Japan	JICA	14020	92	WATER SUPPLY & SANIT. - LARGE SYST.	064648C

RECIPIENT Donor	Agency	Sector	Amount USD thousand	Project description	CRS ID Number
Japan	JICA	14020	254	GRANT AID CO-OPERATION IN WATER SUPPLY	060830C
Japan	MOFA	14020	5670	REHABILITATION AND COSTRUCTION OF THE WATER TREATMENT PLANT	060114
Japan	JICA	14050	28	WASTE MANAGEMENT/DISPOSAL	064649C
Japan	JICA	14050	527	WATER ENV. IMPROVEMENT IN METROPOLITAN AREA	060349C
Netherlands	MFA	14015	5387	WATER MANAGEMENT	2006000371
Netherlands	MFA	14015	2605	WATER RESOURCES PROTECTION	2003015471
Spain	MFA	14010	803	STRENGTHENING PROJECT OF THE GROUP OF MUNICIPALITIES IN RIO NARANJO	065140
Spain	MFA	14010	201	STRENGTHENING OF THE GROUP OF COMMUNITIES KAQCHILELES DEL LAGO ATITLAN	065123
Spain	MFA	14010	351	SUPPORT FOR THE COMMUNITY OF TSOLOJYA IN THE BASIN OF THE ATITLAN LAKE FOR MUNICIPAL STRENGTHENING	065111
Spain	AG	14010	239	SUBSIDY FOR PLANIFICATION AND WATER MANAGEMENT AND SANITATION IN SOLOLÁ GUATEMALA	063880
Spain	MUNIC	14030	31	INDIVIDUAL TANKS FOR RAIN WATER COLLECTION	069599
Spain	MUNIC	14030	23	IMPROVEMENT OF THE BASIC INFRASTRUCTURES FOR DRIKING WATER SUPPLY	069569
Spain	MUNIC	14030	3	WATER SUPPLY FOR SANTA ISABEL COMMUNITY	068869
Spain	MUNIC	14030	9	RURAL INTEGRATION FOR THE ORGANIZATIONAL STRENGTHENING BY MEANS OF WATER AND BASIC SANITATION SYSTEM	068859
Spain	MUNIC	14030	4	HEALTH. WATER SUPPLY	068809
Spain	MUNIC	14030	19	"WATER SUPPLY AND HOUSING CONDITIONING IN SANTA DELMI"	067903
Spain	MUNIC	14030	25	FEASIBILITY AND PROJECT FORMULATION STUDY. PROVISION OF WATER SERVICES IN THE MAMSOHUE	067802
Spain	MUNIC	14030	125	WATER AND SANITATION IN 18 COMMUNITIES OF THE HIGH BASIN OF THE ATITLÁN LAKE	067716
Spain	MUNIC	14030	130	INTRODUCTION OF POTABLE WATER IN THE HEAD MUNICIPALITY OF CHAMPERICO	067715
Spain	MUNIC	14030	13	WATER SUPPLY BY MANUAL PUMPING	067637
Spain	MUNIC	14030	73	FAMILY PROGRAM FOR WATER SUPPLY THROUGH FAMILY CISTERNS	067406
Spain	MUNIC	14030	111	IMPROVING POTABLE WATER SYSTEMS FOR 220 FAMILIES	067344
Spain	MUNIC	14030	23	POTABLA WATER INTRODUCTION FOR 200 FAMILIES	067191
Spain	MUNIC	14030	68	INTRODUCTION OF POTABLE WATER IN THE NEIGHBORHOODS OF CALVARIO & LA JOYA	067112
Spain	MUNIC	14030	34	PROVIDE A WATER TANK TO THE MUNICIPALITY OF CHAMPERICO	067063
Spain	MUNIC	14030	17	POTABLE WATER INTRODUCTION IN THE HEADLINE OF CHAMBERICO	066884
Spain	MUNIC	14030	31	RURAL ORGANIZATIONAL STRENGTHENING THROUGH WATER SYSTEM & SANITATION	066783
Spain	MUNIC	14030	45	HEALTH AND WATER SOCIAL MANAGEMENT	066762
Spain	AG	14030	63	PROJECT FOR THE BASIC INFRASTRUCTURES IMPROVEMENT (WATER SUPPLY)	065800
Spain	MFA	14030	521	PROJECT SUPPORT TO IMPROVE POTABLE WATER SUPPLY SYSTEM	065136
Spain	MFA	14030	251	ACCESS TO POTABLE WATER	064393
Spain	AG	14030	232	NAZIOARTEKO ELKARTASUNA SOLIDARITY	063887
Spain	AG	14030	253	WATER ACCESS AND AGRICULTURAL PRODUCTION WITH A FOOD SOVERIGNTY APPROACH	063462
Spain	AG	14030	292	IMPROVEMENT IN POTABLE WATER AND SANITATION SERVICES	062537
Spain	AG	14030	253	SOCIAL MANGEMENT OF HEALTH AND WATER (GESACRUZ B-9)	062382
Spain	AG	14030	262	UNIQUE WATER PROJECT: SUSTAINABLE AND COMMUNAL PROGRAM FOR WATER SUPPLY AND CONSERVATION OG HYDIC RE	061989
Spain	AG	14030	141	HEALTH IN COMMUNITIES EL AGUACATE AND TUICHPECH	061906
Spain	AG	14030	24	POST EMERGENCY ATTENTION FOR FAMILIES VICTIMS OF THE TROPICAL STORM STAN IN THE MUNICIPALITY OF SOLO	061805
Spain	AG	14030	53	BASIC SANITATION PLAN: POTABLE WATER SUPPLY FOR TWO VILLAGES IN QUICHÉ	061619
Spain	AG	14030	34	INTEGRAL PROJECT OF WATER AND BASIC SANITATION FOR RURAL DEVELOPMENT IN THE COMMUNITY OF SABALPOP C	061271

 CRS AID ACTIVITIES IN SUPPORT OF WATER SUPPLY AND SANITATION – ISBN 978-92-64-05172-0 – © OECD/WWC 2008

RECIPIENT Donor	Agency	Sector	Amount USD thousand	Project description	CRS ID Number
Spain	AG	14030	55	INTEGRAL PROJECT OF WATER AND BASIC SANITATION FOR RURAL DEVELOPMENT IN THE COMMUNITY OF PACHALÍ CH	061270
Spain	AG	14030	50	IMPLEMENTATION OF PLUVIAL IN THE EL CÓNDOR NEIGHBORHOOD JUTIAPA	061158
Spain	AG	14030	132	HYDRIC SOLUTIONS PROGRAM FOR RURAL FAMILIES OF ALTA VERAPAZ GUATEMALA	061140
Spain	AG	14030	47	INTEGRAL PROJECT OF WATER AND BASIC SANITATION FOR RURAL DEVELOPMENT	061137
Spain	Misc	14030	6	DRINKING WATER SUPPLY FOR SAN ANTONIO SINACHÉ HAMLET QUICHÉ GUATEMALA. PROJECT FINANCED BY THE NOT	0610518
Spain	MFA	14040	92	PROTECTION OF MICROBASINS WATER PRODUCERS	065161
Spain	Misc	14040	1	MODEL OF URBAN INTERACTION OF THE LA RIBERA AREA. USE OF A WATER REGIME IN THE SEMIDRY BASIN OF VALL	0610653
Spain	AG	14050	628	COLLABORATION AGREEMENT	062637

Total aid for water in 2006 for   **Guatemala**   **USD thousand 20650**
And as a share of aid to all recipient countries   **0.32%**

Guinea

EC	CEC	14010	2342	ACP EU WATER FACILITY : EAU POTABLE ET ASSAINISSEMENT / REP DE GUINEE	2006100319
IDA		14010	1750	VILLAGE COMMUNITY SUPPORT PROGRAM - PHASE I - ADDITIONAL FINANCING GRANT	060114b
Japan	JICA	14020	263	GRANT AID CO-OPERATION IN WATER SUPPLY	060806C
Japan	MOFA	14020	5799	LE PROJET D'ACCROISSEMENT DE LA PRODUCTION D'EAU POTABLE A CONAKRY	060140
Japan	JICA	14040	25	RIVER DEVELOPMENT	063957C
Spain	MFA	14010	865	WATER IN ÁFRICA: IMPROVEMENT IN MALÍ BURKINA FASSO SENEGAL Y GUINEA	064610
UNICEF		14010	122	WATER RESOURCES POLICY/ADMIN. MGMT	061156

Total aid for water in 2006 for   **Guinea**   **USD thousand 11167**
And as a share of aid to all recipient countries   **0.17%**

Guinea-Bissau

EC	CEC	14010	511	ACP EU WATER FACILITY : WATER & SANITATION ACCESS IN RURAL AREAS	2006100306
IDA		14010	4050	MULTI-SECTOR INFRASTRUCTURE REHABILITATION PROJECT	060102b
Portugal	ICP	14050	125	SANITATION SYSTEM AND WASTE COLLECTING IN BISSAU	060303
Spain	MUNIC	14030	12	WATER AND SANITATION SUPPLY FOR SCHOOLS AND SMALL HAMLETS	069967
Spain	MFA	14030	562	WATER AND HEALTH AGREEMENT IN SCHOOLS	064734
Spain	AG	14030	338	WATER SUPPLY	061903

Total aid for water in 2006 for   **Guinea-Bissau**   **USD thousand 5598**
And as a share of aid to all recipient countries   **0.09%**

Guyana

Japan	JICA	14020	8	WATER SUPPLY & SANIT. - LARGE SYST.	065118C
Japan	JICA	14020	258	GRANT AID CO-OPERATION IN WATER SUPPLY	060826C
Japan	MOFA	14020	5593	WATER SUPPLY IN CORRIVERTON	060152
UNICEF		14010	20	WATER RESOURCES POLICY/ADMIN. MGMT	060381
United Kingdom	DFID	14020	4048	GUYANA WATER AND SANITATION SECTOR PROJECT	001172

RECIPIENT Donor	Agency	Sector	Amount USD thousand	Project description	CRS ID Number
				Total aid for water in 2006 for Guyana	**USD thousand 9927**
				And as a share of aid to all recipient countries	**0.15%**

Haiti

RECIPIENT Donor	Agency	Sector	Amount USD thousand	Project description	CRS ID Number
Belgium	DGCD	14010	26	WATER RESOURCES POLICY/ADMIN. MGMT	2004000346
Belgium	DGCD	14030	118	VOLET ASSAINISSEMENT DANS LE CADRE D'UN DÉVELOPPEMNT INTÉGRÉ	2004000314
Belgium	DGCD	14030	111	SOUTIEN À UNE GESTION DES EAUX JUSTE, PARTICIPATIVE ET DURABLE	2004000313
Belgium	DGCD	14030	345	SOUTIEN AUX ACTIONS LOCALES ET AUX COMMUNAUTÉS DANS LA GESTION INTÉGRÉE	2004000312
Belgium	DGCD	14030	186	BASIC DRINKING WATER SUPPLY AND BASIC SANITATION	2004000311
Belgium	MPRW	14050	310	FORMATION PROFESSIONNELLE, ENVIRONNEMENT, GESTION DES DECHETS	2001007907
EC	CEC	14010	1874	ACP EU WATER FACILITY	2006100316
EC	CEC	14010	566	ACP EU WATER FACILITY:REHAB & RENF SYST ALIM. EAU POTABLE ANSE ROUGE	2006100311
EC	CEC	14010	861	ACP EU WATER FACILITY : APPUI RESEAUX COTIERS	2006100296
EC	CEC	14010	1548	ACP EU WATER FACILITY : EAU & ASSAIN COMMUNES D'AQUIN ET ST LOUIS DU SUD	2006100287
EC	CEC	14010	3766	ACP EU WATER FACILITY : APPUI SOC CIVILE RENF EAU & ASSAIN CAP HAITIEN	2006100286
IDB Sp.Fund		14030	15000	RURAL WATER AND SANITATION PROGRAM	060082
Ireland	DFA	14030	63	BASIC DRINKING WATER AND SANIT	060735
Luxembourg	MFA	14030	100	BASIC DRINKING WATER SUPPLY AND BASIC SANITATION	060332
Norway	MFA	14010	434	WATER RESOURCES POLICY/ADMIN. MGMT	2006000707
Spain	MFA	14015	188	PROTECTION OF HYDRIC RESOURCES OF THE EASTERN REGION IN THE SOUTHEASTERN HYDROGEOGRAPHIC UNIT	065135
Spain	AG	14015	168	CONSOLIDATION OF THE PERDRENALE'S RIVER CHANNEL	061962
Spain	MFA	14030	377	PROGRAME OF POTABLE WATER AND SANITATION	065130
				Total aid for water in 2006 for Haiti	**USD thousand 26040**
				And as a share of aid to all recipient countries	**0.40%**

Honduras

RECIPIENT Donor	Agency	Sector	Amount USD thousand	Project description	CRS ID Number
Canada	CIDA	14020	485	FLOOD PREVENTION IN HONDURAS	060551
Canada	CIDA	14030	4408	SAFE WATER AND SANITATION SERVICES - PASOS III	060315
Germany	BMZ	14020	17	LOCAL AGENDA 21 IN CENTRAL AMERICA	2006003366
Italy	DGCS	14010	451	WATER RESOURCES POLICY/ADMIN. MGMT	060355
Italy	DGCS	14020	1194	WATER SUPPLY & SANIT. - LARGE SYST.	060354
Japan	JICA	14020	64	WATER SUPPLY & SANIT. - LARGE SYST.	064692C
Japan	JICA	14020	244	GRANT AID CO-OPERATION IN WATER SUPPLY	060874C
Japan	JICA	14020	262	EL PROYECTO DE ABASTECIMIENTO DE AGUA PARA EL AREA URBANA DE TEGUCIGALPA	060813C
Japan	Oth. MIN	14030	33	BASIC DRINKING WATER SUPPLY AND BASIC SANITATION	066780C
Japan	JICA	14040	42	RIVER DEVELOPMENT	064693C
Korea	KOICA	14010	8	TRANINIG PROGRAM	2006861391
Spain	MFA	14010	251	PROJECT TO STRENGTHEN CAPACITIES FOR WATER MANAGEMENT	065080
Spain	MFA	14010	314	PROJECT TO STRENGTHEN MUNICIPAL MANAGEMENT OF WATER SERVICES	065075
Spain	AG	14010	377	IMPROVEMENT OF DRINKABLE WATER ACCESS MANAGEMENT	061010

CRS AID ACTIVITIES IN SUPPORT OF WATER SUPPLY AND SANITATION – ISBN 978-92-64-05172-0 – © OECD/WWC 2008

RECIPIENT Donor	Agency	Sector	Amount USD thousand	Project description	CRS ID Number
Spain	MUNIC	14030	19	"IMPROVEMNET IN POTABLE WATER SUPPLY FOR 316 FAMILIES"	068166
Spain	MUNIC	14030	55	CONSTRUCTION OF POTABALE AND BLACK WATER NETWORKS	068156
Spain	MUNIC	14030	44	SANITARY SEWAGE AND TREATMENT PLANT FOR THE MUNICIPALITY OF SAN IGNACIO. DEPARTMENT OF FRANCISCO MOR	068030
Spain	MUNIC	14030	15	SANITARY SEWAGE IN THE MUNICIPALITY OF PROGRESO-YORO	067418
Spain	MUNIC	14030	8	SANITARY SEWAGE; COLONY 19 JULIO I AND II STAGE AND COLONY 1 MARZO	067299
Spain	MFA	14030	314	PROJECT ON MUNICIPALMANAGEMENT OF POTABLE WATER SYSTEM IN TELA	065079
Spain	MFA	14030	402	PROJECT ACCESS TO POTABLE WATER IN SANTA ROSA OF COPAN PHASE III	065065
Spain	MFA	14030	173	EFFICIENT ACCESS TO WATER QUALITY FOR INHABITANTS IN 16 RURAL COMMUNITIES	064887
Spain	AG	14030	89	CONSTRUCTION OF SANITARY MODULES	062203
Spain	AG	14030	351	CONSTRIUCTION OF A WATER SUPPLY SYSTEM IN ADEQUATE QUANTITY AND QUALITY FOR THE POPULATION OF THE MU	061901
Spain	AG	14030	377	EFFICIENT ACCESS TO QUALITY WATER FOR THE PEOPLE OF 16 RURAL COMMUNITIES OF THE MUNICIPALITY OF FLOR	061892
Spain	AG	14030	148	SEWAGE SYSTEM	061865
Spain	AG	14030	38	POTABLE WATER PUBLIC SUPPLY FOR THE COMMUNITIES BUENA ESPERANZA I AND BUENA ESPERANZA II IN THE MUNI	061317
Spain	AG	14030	144	CONSTRUCTION OF SEWAGE SYSTEM AND A TREATMENT PLANT IN ADEQUATE QUANTITY AND QUALITY FOR THE MUNICI	061204
Spain	AG	14030	101	BASIC SANITATION OF THE PLAN GRANDE COMMUNITY AND THE REINO UNICO COLONY THROUGH THE CONSTRUCTION IF	061201
Spain	AG	14030	223	CONSTRUCTION OF A WATER SUPPLY SYSTEM	061035
Spain	AG	14030	371	CONSTRUCTION OF SEWAGE SYSTEM AND A TREATMENT PLANT	061029
Spain	AG	14030	329	IMPROVEMENT IN HEALTH CONDITIONS	061020
Sweden	Sida	14030	3255	PILOT LOCAL DEVELOPMT W&S	2006001992
UNICEF		14010	73	WATER RESOURCES POLICY/ADMIN. MGMT	061257

Total aid for water in 2006 for   **Honduras**   **USD thousand 14676**

And as a share of aid to all recipient countries   **0.23%**

India

Australia	AusAID	14030	71	INDEPENDENT COMPLETION REPORTS	2006001392
Canada	CIDA	14050	471	WASTE MANAGEMENT - MUNICIPAL IMPROVEMENTS-INDIA	060086
EC	EDF	14030	93511	STATE PARTNERSHIP PROG. - RAJASTHAN	2006201336
Germany	BMZ	14010	251	UNEP DAMS AND DEVELOPMENT PROJECT (PHASE 2)	2006010072
Germany	BMZ	14010	18	CONGRESS PARTICIPATIONS	2006003364
Germany	BMZ	14015	209	EFFECTIVE DEVELOPMENT THROUGH PROTECTION OF RESOURCES	2006004880
Germany	BMZ	14015	78	INTEGRATED EXPERTS	2006002420
Germany	KFW	14020	1255	RURAL WATER SUPPLY	2006000571
Germany	KFW	14020	3766	RURAL WATER SUPPLY	2006000524
Germany	L G	14030	36	PLANNING AND IMPLEMENTATION OF DECENTRALIZED WASTEWATER TREATMENT SYSTEM	2006011587
Germany	BMZ	14030	65	IRRIGATION AND TRAINING PROGRAMME	2006008145
Germany	BMZ	14030	531	IMPROVING LIVELIHOODS	2006004450
Germany	BMZ	14030	94	INTEGRATED EXPERTS	2006002434
Germany	BMZ	14050	86	UNEP E-WASTE RECYCLING SYSTEM	2006010073
Germany	BMZ	14081	9	WATER EFFICIENT MANAGEMENT OF WASTEWATER, TREATMENT AND REUSE	2006003389

RECIPIENT Donor	Agency	Sector	Amount USD thousand	Project description	CRS ID Number
IDA		14010	133980	PUNJAB RURAL WATER SUPPLY AND SANITATION	060142a
IDA		14010	90	TECHNICAL ASSISTANCE FOR INFRASTRUCTURE PPPS	060093c
IDA		14010	112800	UTTARANCHAL RURAL WATER SUPPLY AND SANITATION PROJECT	060036a
IDA		14020	12320	PUNJAB RURAL WATER SUPPLY AND SANITATION	060142d
IDA		14020	1200	UTTARANCHAL RURAL WATER SUPPLY AND SANITATION PROJECT	060036e
Ireland	DFA	14030	393	BASIC DRINKING WATER AND SANIT	060281
Italy	LA	14020	44	WATER SUPPLY & SANIT. - LARGE SYST.	061742
Italy	DGCS	14020	18	WATER SUPPLY & SANIT. - LARGE SYST.	060414
Italy	LA	14040	17	RIVER DEVELOPMENT	061765
Japan	JBIC	14015	66400	HUSSAIN SAGAR LAKE AND CATCHMENT AREA IMPROVEMENT PROJECT	063026
Japan	JICA	14020	151	WATER SUPPLY & SANIT. - LARGE SYST.	065731C
Japan	JICA	14020	1116	AUGMENTATION OF WATER SUPPLY AND SANITATION FOR GOA STATE	060130C
Japan	JICA	14020	1891	TC IN SEWERAGE	060070C
Japan	JBIC	14020	243625	BANGALORE WATER SUPPLY AND SEWERAGE PROJECT	053029
Japan	MOFA	14030	5180	DEVELOPMENT OF GROUNDWATER IN THE STATE OF UTTAR PRADESH	060001
Japan	JICA	14040	62	RIVER DEVELOPMENT	065732C
Japan	JICA	14040	88	COUNTRY/SECTOR PROGRAM FORMUL COST(INDIA)	062839C
Japan	JICA	14040	94	TECHNOLOGY TRANSFER ON CONTROL AND ABATEMENT OF RIVER POLLUTION	062663C
Japan	JICA	14040	127	GRANT AID CO-OPERATION IN WATER RESOURCES DEVELOPMENT	061865C
Japan	JICA	14050	10	WASTE MANAGEMENT/DISPOSAL	065733C
Japan	JBIC	14050	30790	KOLKATA SOLID WASTE MANAGEMENT IMPROVEMENT PROJECT	063027
Japan	JICA	14050	95	STRENGTHENING CAPACITY ON RESTORATION AND MANAGEMENT OF HUSSAINS	062616C
Luxembourg	MFA	14030	24	BASIC DRINKING WATER SUPPLY AND BASIC SANITATION	061622
Spain	MUNIC	14015	54	PROGRAM TO RECOVER NATURAL RESOURCES OF 3 HYDROGEOGRAPHIC BASINS	066963
Spain	MUNIC	14030	4	CONSTRUCTION OF A WELL IN THE "CALASANZ MIDDLE SCHOOL" OF KAMDA	068231
Spain	AG	14030	287	ALBOAN FOUNDATION.	063852
Spain	AG	14030	373	PHASE 2 PREVENTIVE HEALTH WATER AND SANITATION IN 44 COMMUNITIES	061907
Spain	AG	14030	8	REINFORCEMENT OF THE SANITATION CONDITIONS IN A SHELTER FOR CHILDREN	061121
Spain	Misc	14030	56	ACCESS TO DRINKING WATER FOR THE UNDERPRIVILEGED FAMILIES OF VALIYAPARA NIRAPPEKADA VALIYAKANDOM	0610333
Spain	MUNIC	14040	52	DEVELOPMENT OF THE HYDROGRAPHIC BASINS IN KUTALAPALLI AND PEMMANAKUNTAPALLI	069470
Spain	MUNIC	14040	37	DEVELOPMENT OF THE HYDROGRAPHICAL BASINS OF CHOWTAKUNTAPALLI AND KONDAKINDA	069396
Switzerland	SDC	14010	103	WATER RESOURCES POLICY/ADMIN. MGMT	1995000355
Switzerland	SDC	14010	1094	ISPWD-K INDO-SWISS PARTICIPATIVE WATERSHED KARNATAKA	1995000091
Switzerland	SDC	14015	798	MSSRF SWAMINATHAN RESEARCH FOUNDATION	1998000911
Switzerland	SDC	14015	160	SMALL ACTIONS COORDINATION OFFICE INDIA	1997000757
UNICEF		14030	544	SECTOR PLANS FOR HYGIENE, SANITATION AND WATER	061287
UNICEF		14030	1545	HYGIENE, SANITATION AND WATER SUPPLY PROGRAMMES	061286
UNICEF		14030	1575	COMMUNITY HYGIENE AND WATER SAFETY	061285
United Kingdom	DFID	14010	4784	WATER AND SANITATION PROGRAMME	060296
United States	Misc	14010	50	WATER RESOURCES POLICY/ADMIN. MGMT	2006002887
United States	AGR	14010	427	WATER RESOURCES POLICY/ADMIN. MGMT	2006002879

RECIPIENT Donor	Agency	Sector	Amount USD thousand	Project description	CRS ID Number
United States	AID	14010	1155	WATER RESOURCES POLICY/ADMIN. MGMT	2006002870
United States	AID	14010	8655	IMPROVED ACCESS TO CLEAN ENERGY AND WATER IN SELECTED STATES	2006002861
United States	TDA	14050	4	WASTE MANAGEMENT/DISPOSAL	2006003068

Total aid for water in 2006 for   **India**   **USD thousand 732736**
And as a share of aid to all recipient countries   **11.25%**

Indonesia

RECIPIENT Donor	Agency	Sector	Amount	Project description	CRS ID
Australia	AusAID	14010	3	THE IMPACT OF CHANGING AGROFORESTRY MOSAICS ON CATCHMENT WATER YIELD	2005000844
Australia	AusAID	14010	996	WASPOLA	1998001402
Australia	AusAID	14010	54	WASPOLA (PHASE 1&2)	1998000402
Australia	AusAID	14030	2	WB WS&S FOR LOW INCOME COMMUNITES PH II	2006002843
Australia	AusAID	14030	415	PAMSIMAS DESIGN & WSS SCOPING MISSION	2006000622
Australia	AusAID	14030	614	TEMPORARY HOUSING ASSISTANCE	2006000514
Australia	AusAID	14030	1445	PERMANENT HOUSING ASSISTANCE	2006000274
Australia	AusAID	14030	1599	WB WS&S FOR LOW INCOME COMMUNITES	2006000251
Canada	IDRC	14020	1058	ECONOMIC INCENTIVES FOR IMPROVED WATER SANITATION & SOLID WASTE SERVICES	061169b
Canada	IDRC	14050	343	DECENTRALIZED URBAN SOLID WASTE MANAGEMENT IN INDONESIA	060642b
France	MINEFI	14010	1072	TRANSFERT EAUX WEST TARUM CANAL	2006003530
Germany	BMZ	14010	100	INTEGRATED EXPERTS	2006002412
Germany	BMZ	14015	100	INTEGRATED EXPERTS	2006002419
Germany	Fed Min	14030	47	DECENTRALIZED WATER SUPPLIES AND SEWAGE DISPOSAL	2006011182
Germany	BMZ	14030	1185	WATER SUPPLY AND CREDIT PROGRAMME IN SOUTH-EAST SULAWESI	2006004897
Germany	BMZ	14030	86	DEMAND-ORIENTED INTRODUCTION AND DISSEMINATION OF DECENTRALIZED SERVICE	2006004448
Germany	L G	14081	75	TRAINING SEMINAR: TSUNAMI TECHNOLOGIES AND PRIVATE/PUBLIC MANAGEMENT	2006012207
IDA		14010	21000	ID-ACEH-INFRA. RECONSTR ENABLING PROGRAM (IREP)	060080a
IDA		14010	59125	THIRD WATER SUPPLY AND SANITATION FOR LOW INCOME COMMUNITIES PROJECT	060043a
IDA		14020	45375	THIRD WATER SUPPLY AND SANITATION FOR LOW INCOME COMMUNITIES PROJECT	060043b
Japan	JICA	14020	261	WATER SUPPLY & SANIT. - LARGE SYST.	065980C
Japan	JICA	14020	108	STUDY ON REGIONAL WATER SUPPLY DEVELOPMENT PLAN FOR GREATER YOGYAKARATA	062254C
Japan	JICA	14020	118	TC IN WATER SUPPLY	062040C
Japan	JICA	14020	123	RURAL WATER SUPPLY IN PROVINCE OF NUSA TENGGARA BARAT AND NU	061949C
Japan	JICA	14020	155	GRANT AID CO-OPERATION IN WATER SUPPLY	061498C
Japan	Oth. MIN	14030	60	BASIC DRINKING WATER SUPPLY AND BASIC SANITATION	067034C
Japan	JICA	14040	534	RIVER DEVELOPMENT	065981C
Japan	JBIC	14040	140052	INTEGRATED WATE RRESOURCES AND FLOOD MANAGEMENT PROJECT FOR SEMARANG	063014
Japan	JICA	14040	100	SABO ENGINEERING	062472C
Japan	JICA	14040	105	WATER SUPPLY GUNUNGKIDUL REGENCY OF YOGYAKARTA SPECIAL TERRITORY	062302C
Japan	JICA	14040	174	STUDY ON CAPACITY DEVELOPMENT FOR JENEBERANG RIVER BASIN MANAGEMENT IN R	061299C
Japan	JICA	14040	515	THE STUDY ON COUNTERMEASURES FOR SEDIMENTATION IN WONOGIRI M	060361C
Japan	JICA	14040	2677	TC IN WATER RESOURCES DEVELOPMENT	060108C
Japan	JICA	14050	18	WASTE MANAGEMENT/DISPOSAL	065982C

RECIPIENT / Donor	Agency	Sector	Amount USD thousand	Project description	CRS ID Number
Japan	ODC	14081	4	EDUC./TRNG:WATER SUPPLY & SANITATION	067426C
Korea	KOICA	14010	4	TRANINIG PROGRAM	2006861753
Netherlands	MFA	14030	6276	WATER SECTOR: ACEH	2006000162
Sweden	Sida	14030	4408	WSP ENVION SANITATION IND ENVIRONMENTAL SANITATION	2006003277
Switzerland	SDC	14050	79	WASTE MANAGEMENT/DISPOSAL	1986000273
UNICEF		14010	126	WATER RESOURCES POLICY/ADMIN. MGMT	061336
United Kingdom	DFID	14020	57	WATER AND SANITATION TECHNICAL CO-OPERATION ASSIGNMENT	060275

Total aid for water in 2006 for   **Indonesia**   **USD thousand 290645**
And as a share of aid to all recipient countries   **4.46%**

Iran

RECIPIENT / Donor	Agency	Sector	Amount USD thousand	Project description	CRS ID Number
Greece	YPEJ	14030	38	WATER SUPPLY & SANITATION TO THE TOWN OF BAM	2006000325
Greece	YPEJ	14030	38	BASIC DRINKING WATER SUPPLY AND BASIC SANITATION FOR THE TOWN OF BAM	2006000323
Italy	DGCS	14010	545	WATER RESOURCES POLICY/ADMIN. MGMT	060026
Japan	JICA	14020	35	WATER SUPPLY & SANIT. - LARGE SYST.	065269C
Japan	JICA	14020	1821	TC IN WATER SUPPLY	060075C
Japan	JICA	14040	283	RIVER DEVELOPMENT	065270C
UNICEF		14010	45	WATER RESOURCES POLICY/ADMIN. MGMT	061367

Total aid for water in 2006 for   **Iran**   **USD thousand 2804**
And as a share of aid to all recipient countries   **0.04%**

Iraq

RECIPIENT / Donor	Agency	Sector	Amount USD thousand	Project description	CRS ID Number
Japan	MOFA	14010	959	SUPPORT FOR ENVIRONMENTAL MANAGEMENT OF IRAQI MARSHLANDS	062133
Japan	JICA	14020	220	WATER SUPPLY & SANIT. - LARGE SYST.	065298C
Japan	JICA	14020	584	TC IN WATER SUPPLY	060485C
Japan	JICA	14020	518	UNDEFINED	060359C
Japan	JICA	14020	684	THE FEASIBILITY STUDY ON BAGHDAD WATER SUPPLY SYSTEM IMPROVEMENT PROJECT	060243C
Japan	JICA	14040	6	RIVER DEVELOPMENT	065299C
Japan	JICA	14040	223	TC IN WATER RESOURCES DEVELOPMENT	060976C
Korea	KOICA	14010	149	TRANINIG PROGRAM	2006861532
Korea	KOICA	14020	527	ERBIL WATER SUPPLY AND SEWERAGE SYSTEM IMPROVEMENT PROJECT	2006862131
Norway	MFA	14010	715	WATER RESOURCES POLICY/ADMIN. MGMT	2006002244
Norway	MFA	14020	1808	WATER INTERVENTION IN SOUTH IRAK	2006000570
Norway	MFA	14030	814	WATER INTERVENTION IN CENTRAL IRAK	2006001389
UNICEF		14010	85	WATER RESOURCES POLICY/ADMIN. MGMT	061390
United States	DOD	14010	138251	WATER AND SANITATION	2006002857
United States	DOD	14015	574	WATER RESOURCES PROTECTION	2006002926
United States	DOD	14015	5284	WATER RESOURCES AND SANITATION: DAM REPAIR, REHAB. AND NEW CONSTRUCTION	2006002923
United States	DOD	14020	3404	WATER SUPPLY & SANIT. - LARGE SYST.	2006002963
United States	DOD	14020	2000	UMM QASR TO BASRA WATER PIPELINE AND TREATMENT PLANT	2006002959

CRS AID ACTIVITIES IN SUPPORT OF WATER SUPPLY AND SANITATION – ISBN 978-92-64-05172-0 – © OECD/WWC 2008

RECIPIENT Donor	Agency	Sector	Amount USD thousand	Project description	CRS ID Number
United States	DOD	14020	8129	WATER PIPELINE AND TREATMENT PLANT	2006002948
United States	DOD	14020	23375	WATER RESOURCES AND SANITATION/PUBLIC WORKS PROJECTS-- SEWERAGE	2006002947
United States	DOD	14020	37933	WATER RESOURCES AND SANITATION / PUBLIC WORKS PROJECTS - SEWERAGE	2006002945
United States	DOD	14020	60373	WATER RESOURCES AND SANITATION: PUMPING STATIONS AND GENERATORS	2006002944
United States	DOD	14020	140558	WATER RESOURCES AND SANITATION/PUBLIC WORKS PROJECTS-- POTABLE WATER	2006002943
United States	DOD	14020	127833	WATER RESOURCES AND SANITATION / PUBLIC WORKS PROJECTS - POTABLE WATER	2006002942
United States	DOD	14030	56	BASIC DRINKING WATER SUPPLY AND BASIC SANITATION	2006003029
United States	DOD	14030	20263	WATER RESOURCES AND SANITATION/PUBLIC WORKS PROJECTS-- POTABLE WATER	2006003021
United States	DOD	14030	42917	WATER RESOURCES AND SANITATION/PUBLIC WORKS PROJECTS - POTABLE WATER	2006003020
United States	DOD	14040	13282	WATER RESOURCES AND SANITATION: DAM REPAIR, REHAB, AND NEW CONSTRUCTION	2006003049
United States	DOD	14040	41455	WATER RESOURCES AND SANITATION. DAM. REPAIR, REHAB, AND NEW CONSTRUCTION	2006003048
United States	DOD	14050	2720	WATER RESOURCES AND SANITATION: OTHER SOLID WASTE MANAGEMENT	2006003060

Total aid for water in 2006 for Iraq **USD thousand 675698**
And as a share of aid to all recipient countries 10.37%

Jamaica

Japan	JICA	14020	46	WATER SUPPLY & SANIT. - LARGE SYST.	064738C
Japan	JICA	14020	143	CAPACITY BUILDING FOR THE MAINTENANCE OF WATER SUPPLY FACILITIES	061633C
Japan	JICA	14050	25	WASTE MANAGEMENT/DISPOSAL	064739C
Korea	KOICA	14010	8	TRANINIG PROGRAM	2006861397

Total aid for water in 2006 for Jamaica **USD thousand 221**
And as a share of aid to all recipient countries 0.00%

Jordan

Germany	Fed Min	14010	2664	INTEGRATED WATER RESOURCE MANAGEMENT	2006011237
Germany	GTZ	14010	8786	MANAGEMENT OF WATER RESOURCES PROGRAMME	2006006070
Germany	BMZ	14010	218	REFORM OF THE WATER SECTOR IN THE MENA REGION	2006003352
Germany	BMZ	14020	84	INTEGRATED EXPERTS	2006002422
Germany	KFW	14020	13892	WATER SUPPLY GREATER AMMANN III	2006000687
Germany	BMZ	14081	202	WATER EFFICIENT MANAGEMENT OF WASTEWATER, TREATMENT AND REUSE	2006003387
Greece	YPEHO DE	14010	8	HIGHER EFFICIENCY OF WATER-RELATED DEVELOPMENT	2006000302
Italy	DGCS	14010	8	WATER RESOURCES POLICY/ADMIN. MGMT	060035
Italy	Art.	14020	9288	GREATER AMMAN'S WATER SUPPLY SYSTEM	060015
Japan	JICA	14020	183	WATER SUPPLY & SANIT. - LARGE SYST.	065328C
Japan	JICA	14020	89	CAPACITY DEVELOPMENT PROJECT FOR NONREVENUE WATER REDUCTION IN JORDAN	062824C
Japan	JICA	14020	353	GRANT AID CO-OPERATION IN WATER SUPPLY	060583C
Japan	JICA	14020	789	TC IN WATER SUPPLY	060200C
Japan	MOFA	14020	4390	IMPROV. OF THE WATER SUPPLY FOR THE ZARQA DISTRICT	060144
Japan	JICA	14040	57	RIVER DEVELOPMENT	065329C
Japan	JICA	14050	49	WASTE MANAGEMENT/DISPOSAL	065330C

RECIPIENT Donor	Agency	Sector	Amount USD thousand	Project description	CRS ID Number
Spain	MFA	14030	561	DEVELOPMENT PROGRAME FOR FIGHT AGAINST POVERTY	064317
United States	AID	14010	33	WATER RESOURCES POLICY/ADMIN. MGMT	2006002888
United States	AID	14010	45095	ENHANCED INTEGRATED WATER RESOURCES MANAGEMENT	2006002858

Total aid for water in 2006 for **Jordan** USD thousand **86751**
And as a share of aid to all recipient countries **1.33%**

Kazakhstan

Norway	MFA	14010	247	NATIONAL INTEGRATED WATER RESOURCES	2006002775
Norway	MFA	14030	295	VILLAGE WATER SYSTEMS	2006003042

Total aid for water in 2006 for **Kazakhstan** USD thousand **542**
And as a share of aid to all recipient countries **0.01%**

Kenya

Austria	ADA	14030	59	BASIC DRINKING WATER SUPPLY AND BASIC SANITATION	20060022ch
Belgium	DGCD	14030	348	BARINGO WATER PROGRAMME	2004001171
Belgium	DGCD	14030	24	BASIC DRINKING WATER SUPPLY AND BASIC SANITATION	2001001047
EC	CEC	14010	4800	ACP EU WATER FACILITY	2006100294
EC	CEC	14010	628	ACP EU WATER FACILITY : SAGE WATER PROVISION & SUST. WATER MANAGMT	2006100293
Finland	MFA	14030	669	FRAME AGREEMENT WITH NGO (FINNISH RED CROSS)	060467
Finland	MFA	14030	218	FRAME AGREEMENT WITH NGO (SAVE THE CHILDREN)	060332
Finland	MFA	14030	19	NGO SUPPORT / SINKING OF BORE HOLES FOR NURSERY SCHOOLS	060140
Finland	MFA	14081	125	SAFEWATER AND SANITATION FOR CHILDREN IN PRIMARY SCHOOLS IN KENYA	060290
France	AFD	14020	37655	INFRASTRUCTURES D'EAU ET ASSAINISSEMENT	2006100400
France	AFD	14050	21338	AMELIORATION GESTION DES DECHETS SOLIDES	2006106400
Germany	GTZ	14010	2259	WATER SECTOR REFORM	2006006048
Germany	BMZ	14010	106	WATER RESOURCES POLICY AND ADMINISTRATIVE MANAGEMENT	2006005401
Germany	BMZ	14010	73	TRANSNET - TRANSBOUNDARY MANAGEMENT OF NATURAL RESOURCES IN SADC REGION	2006003317
Germany	BMZ	14020	232	WATER SUPPLY AND SANITATION - LARGE SYSTEMS	2006005403
Germany	BMZ	14030	32	CONSTRUCTION OF AN ELEVATED TANK FOR WATER, MBAGATHI	2006004895
Germany	KFW	14030	628	DEVELOPMENT OF THE WATER SECTOR II	2006000808
Germany	KFW	14030	11924	DEVELOPMENT OF THE WATER SECTOR, II	2006000807
Ireland	DFA	14030	606	BASIC DRINKING WATER AND SANIT	060239
Italy	DGCS	14010	9	WATER RESOURCES POLICY/ADMIN. MGMT	060570
Italy	LA	14015	63	WATER RESOURCES PROTECTION	061278
Italy	LA	14020	148	WATER SUPPLY & SANIT. - LARGE SYST.	061496
Italy	DGCS	14040	4	RIVER DEVELOPMENT	060513
Japan	JICA	14020	101	WATER SUPPLY & SANIT. - LARGE SYST.	063997C
Japan	JICA	14020	134	THE PROJECT FOR RURAL WATER SUPPLY IN THE REPUBLIC OF KENYA	061760C
Japan	JICA	14020	246	GRANT AID CO-OPERATION IN WATER SUPPLY	060864C
Japan	MOFA	14030	4270	RURAL WATER SUPPLY	060178

CRS AID ACTIVITIES IN SUPPORT OF WATER SUPPLY AND SANITATION – ISBN 978-92-64-05172-0 – © OECD/WWC 2008

RECIPIENT Donor	Agency	Sector	Amount USD thousand	Project description	CRS ID Number
Japan	JICA	14040	483	STUDY ON INTEGRATED FLOOD MANAGEMENT FOR NYANDO RIVER BASIN IN REPUBLIC	060389C
Korea	KOICA	14010	9	DRINKING WATER DEVELOPMENT PROJECT IN EAST AFRICA	2006862174
Korea	KOICA	14030	370	DEVELOPMENT & REHABILITATION OF GROUND WATER WELLS	2006862151
New Zealand	NZAid	14030	65	"SUSTAINABLE WATER RESOURCES: KENYA"	060318
Norway	NORAD	14015	18	WATER RIGHTS AND SANITATION ADVOCACY	2006003091
Spain	AG	14015	98	IMPROVEMENT OF WATER ACCESS FOR SEMI-NOMADIC INHABITANTS OF THE TURKANA DISTRICT PLATEUS	062121
Spain	MUNIC	14030	2	CONSTRUCTION OF A WELL TO IMPROVE WATER SUPPLY IN THE NORTH TURKANA	069095
Spain	MUNIC	14030	5	STREGNTHENING OF WATER-BEARING AND DEVELOPMENT INFRASTRUCTURE	068777
Spain	MUNIC	14030	10	ACCESS TO DRINKING WATER IN TURKANA (KENYA)	068734
Spain	MUNIC	14030	13	"STRENGTHENING OF WATER RESOURCES AND DEVELOPMENT INFRASTRUCTURES	068172
Spain	MUNIC	14030	3	CONSTRUCTION OF A WELL TO BETTER SUPPLY WATER TO NORTH TURKANA AREA	067653
Spain	MUNIC	14030	28	DEVELOPMENT OF WATER RESOURCES IN THE DISTRICT OF TURKANA KENIA	067354
Spain	MUNIC	14030	13	POTABLE WATER ACCESSIBILITY FOR INHABITANTS IN THE NORTHEAST ZONE OF THE TUKANA DISTRICT	067212
Spain	AG	14030	48	WATER ACCESSABILITY FOR INHABITANTS OF THE NORTHEASTER ZONE DISTRICT OF TURKANA	062456
Spain	AG	14030	53	WELL CONSTRUCTION TO IMPROVE WATER SUPPLY IN THE NORTHEASTERN ZONE OF TURKANA.	062454
Spain	AG	14030	42	POTABLE WATER ACCESIBILITY POR THE POPULATION OF TURKANA AND ITS CATTLE	061604
Spain	AG	14030	135	IMPROVEMENT IN THE WATER ACCESS FOR NOMADIC POPULATIONS IN THE MOUNTAINS OF TURKANAAND THEIR CATTLE	061053
UNICEF		14010	22	WATER RESOURCES POLICY/ADMIN. MGMT	061490
UNICEF		14030	146	HYGIENE, SANITATION AND WATER SUPPLY PROGRAMMES	061491
United States	AID	14010	1000	MILLENNIUM WATER ALLIANCE	2006002867
United States	AID	14010	3172	MILLENIUM WATER ALLIANCE	2006002862

Total aid for water in 2006 for   **Kenya**   **USD thousand 92450**
And as a share of aid to all recipient countries   **1.42%**

Kiribati

Australia	AusAID	14010	24	NURSE UPSKILLING STUDY	2006001974

Total aid for water in 2006 for   **Kiribati**   **USD thousand 24**
And as a share of aid to all recipient countries   **0.00%**

Korea, Dem. Rep.

Greece	ALLOI	14015	75	BOREHOLE-TO-SURFACE ELECTRICAL RESISTIVITY TOMOGRAPHY TECHNOLOGY	2006000318
UNICEF		14010	2	WATER RESOURCES POLICY/ADMIN. MGMT	061553
UNICEF		14030	123	HYGIENE, SANITATION AND WATER SUPPLY PROGRAMMES	061554

Total aid for water in 2006 for   **Korea, Dem. Rep.**   **USD thousand 200**
And as a share of aid to all recipient countries   **0.00%**

Kyrgyz Republic

Germany	Fed Min	14020	27	DECENTRALISED WASTE WATER MANAGEMENT	2006011014

RECIPIENT Donor	Agency	Sector	Amount USD thousand	Project description	CRS ID Number
IDA		14010	1200	SECOND VILLAGE INVESTMENT PROJECT	060112d
Japan	JICA	14020	0	WATER SUPPLY & SANIT. - LARGE SYST.	065526C
Japan	JICA	14020	165	PROJECT FOR IMPROVEMENT OF THE WATER ENV. IN CHOLPON-ATA CITY	061394C
Norway	MFA	14050	39	WASTE MANAGEMENT IN BISHKEK	2006003519
UNICEF		14010	18	WATER RESOURCES POLICY/ADMIN. MGMT	061523
United Kingdom	DFID	14010	110	DEVELOP A STRATEGY FOR RURAL WATER SUPPLY & SANITATION	060357

Total aid for water in 2006 for **Kyrgyz Republic** USD thousand **1560**
And as a share of aid to all recipient countries **0.02%**

Laos

RECIPIENT Donor	Agency	Sector	Amount USD thousand	Project description	CRS ID Number
Belgium	DGCD	14040	564	RENFORCEMENT CAPACITES LAO NATIONAL MEKONG COMMITTEE (LNMC)	2004001155
Germany	BMZ	14040	29	DEVELOPING POTENTIALS IN RURAL AREAS OF MEKONG RIPARIAN COUNTRIES	2006003371
Japan	JICA	14020	299	WATER SUPPLY & SANIT. - LARGE SYST.	066028C
Japan	JICA	14020	85	TC IN WATER SUPPLY	062955C
Japan	JICA	14020	186	CAPACITY DEVELOPMENT OF WATER SUPPLY SYSTEM	062695C
Japan	MOFA	14020	24699	VIENTIANE WATER SUPPLY DEVELOPMENT	060092
Japan	Oth. MIN	14030	60	BASIC DRINKING WATER SUPPLY AND BASIC SANITATION	067061C
Japan	MOFA	14030	1	BASIC DRINKING WATER SUPPLY AND BASIC SANITATION	063413C
Japan	MOFA	14030	361	VIENTIANE WATER SUPPLY DEVELOPMENT	060028
Japan	JICA	14040	162	RIVER DEVELOPMENT	066029C
Japan	JICA	14040	335	TC IN RIVERS/SAND ARRESTATION	060627C
Korea	KOICA	14010	9	TRANINIG PROGRAM	2006861785
Korea	KOICA	14040	800	FS ON RIVERBANK PROTECTION AND PARK ALONG THE MEKONG RIVER	2006862193
Norway	NORAD	14081	49	TRAINING: NORTHERN AND CENTRAL REGION WATER SUPPLY AND SANITATION PROJECT	2006004014
UNICEF		14010	88	WATER RESOURCES POLICY/ADMIN. MGMT	061582

Total aid for water in 2006 for **Laos** USD thousand **27727**
And as a share of aid to all recipient countries **0.43%**

Lebanon

RECIPIENT Donor	Agency	Sector	Amount USD thousand	Project description	CRS ID Number
Germany	BMZ	14010	14	REFORM OF THE WATER SECTOR IN THE MENA REGION	2006003350
Germany	GTZ	14030	3766	BASIC DRINKING WATER SUPPLY AND BASIC SANITATION	2006006115
Germany	KFW	14030	2510	REHABILITATION OF WATER AND WASTEWATER SYSTEMS	2006000873
Germany	BMZ	14081	143	WATER EFFICIENT MANAGEMENT OF WASTEWATER, TREATMENT AND REUSE	2006003385
Greece	YPEHODE	14010	8	HIGHER EFFICIENCY OF WATER-RELATED DEVELOPMENT	2006000303
IDA		14010	9000	LB-MUNICIPAL INFRASTRUCTURE	060124c
Italy	DGCS	14020	6	WATER SUPPLY & SANIT. - LARGE SYST.	060037
Japan	MOFA	14030	954	REESTABLISHMENT OF WATER SUPPLIES IN CONFLICT-AFFECTED AREAS OF LEBANON	060249
Japan	MOFA	14030	858	EARLY RECOVERY ASSISTANCE TO MUNICIPALITIES OF SOUTH LEBANON	060248
Spain	MFA	14030	7	WATER PUMP AND ELECTRIC TRANSFORMER. PALESTINE REFUGEE CAMP	064363
Sweden	Sida	14030	72	EMERG. RECONSTR. LIBANON	2006002628

CRS AID ACTIVITIES IN SUPPORT OF WATER SUPPLY AND SANITATION – ISBN 978-92-64-05172-0 – © OECD/WWC 2008

RECIPIENT Donor	Agency	Sector	Amount USD thousand	Project description	CRS ID Number
UNICEF		14010	40	WATER RESOURCES POLICY/ADMIN. MGMT	061614

Total aid for water in 2006 for   **Lebanon**   **USD thousand 17377**
And as a share of aid to all recipient countries   **0.27%**

Lesotho

RECIPIENT Donor	Agency	Sector	Amount USD thousand	Project description	CRS ID Number
Germany	BMZ	14010	15	TRANSNET - TRANSBOUNDARY MANAGEMENT OF NATURAL RESOURCES IN SADC REGION	2006003330
Germany	BMZ	14030	50	INTEGRATED EXPERTS	2006002436
Ireland	DFA	14030	2970	BASIC DRINKING WATER AND SANIT	060042
Japan	JICA	14020	31	WATER SUPPLY & SANIT. - LARGE SYST.	064031C
Norway	MFA	14020	45	PERSONNEL EXCHANGE	2006002865

Total aid for water in 2006 for   **Lesotho**   **USD thousand 3112**
And as a share of aid to all recipient countries   **0.05%**

Liberia

RECIPIENT Donor	Agency	Sector	Amount USD thousand	Project description	CRS ID Number
Germany	BMZ	14050	176	SANITATION AND WASTE DISPOSAL IN 4 PARTS IN MONROVIA TOWN	2006008152
IDA		14010	990	EIP SUPPLEMENTAL COMPONENT	060121e
IDA		14010	3000	EMERGENCY INFRASTRUCTURE PROJECT	060079c
IDA		14020	4125	EIP SUPPLEMENTAL COMPONENT	060121b
UNICEF		14010	57	WATER RESOURCES POLICY/ADMIN. MGMT	061636
UNICEF		14030	203	HYGIENE, SANITATION AND WATER SUPPLY PROGRAMMES	061635

Total aid for water in 2006 for   **Liberia**   **USD thousand 8551**
And as a share of aid to all recipient countries   **0.13%**

Libya

RECIPIENT Donor	Agency	Sector	Amount USD thousand	Project description	CRS ID Number
Greece	YPEHO DE	14010	8	HIGHER EFFICIENCY OF WATER-RELATED DEVELOPMENT	2006000304

Total aid for water in 2006 for   **Libya**   **USD thousand 8**
And as a share of aid to all recipient countries   **0.00%**

Macedonia (TFYR)

RECIPIENT Donor	Agency	Sector	Amount USD thousand	Project description	CRS ID Number
Austria	ADA	14010	334	WATER RESOURCES POLICY/ADMIN. MGMT	2006000194
Austria	ADA	14020	39	WATER SUPPLY & SANIT. - LARGE SYST.	2001009051
Germany	KFW	14010	628	IRRIGATION PROGRAMME IN SOUTHERN VARDAR VALLEY	2006001035
Greece	YPEHO DE	14010	8	HIGHER EFFICIENCY OF WATER-RELATED DEVELOPMENT	2006000313
Italy	DGCS	14015	286	WATER RESOURCES PROTECTION	060086
Japan	JICA	14050	36	WASTE MANAGEMENT/DISPOSAL	063566C
Norway	MFA	14030	4	EVALUATION WATER PROJECTS	2006002382
Norway	MFA	14030	12	QUALITY ARSURING OF REPORT	2006001746
Switzerland	seco	14010	771	REHABILITATION OF PUMPING STATIONS OHRID EAST	2006002726

RECIPIENT / Donor	Agency	Sector	Amount USD thousand	Project description	CRS ID Number
Switzerland	SDC	14015	195	RESTORATION GOLEMA RIVER	2005000040
Switzerland	SDC	14015	263	WATER RESOURCES PROTECTION	2000001545
Switzerland	SDC	14015	239	RIVER MONITORING SYSTEM	2000000040
Switzerland	seco	14020	5665	MK: BEROVO URBAN SANITATION PROJECT	2006002548

Total aid for water in 2006 for   **Macedonia (TFYR)**   **USD thousand 8480**

And as a share of aid to all recipient countries   **0.13%**

Madagascar

RECIPIENT / Donor	Agency	Sector	Amount USD thousand	Project description	CRS ID Number
AfDF		14030	75022	PROGRAMME D'ALIMENTATION EN EAU POTABLE	060067
Belgium	DGCD	14030	116	EAU POTABLE ET ASSAINISSEMENT L'ANDROMBA	2005001210
Belgium	DGCD	14030	180	BASIC DRINKING WATER SUPPLY AND BASIC SANITATION	2005001209
Germany	BMZ	14010	17	TRANSNET - TRANSBOUNDARY MANAGEMENT OF NATURAL RESOURCES IN SADC REGION	2006003324
Germany	BMZ	14030	324	WATERSHED DEVELOPMENT BY PROMOTION OF RURAL SELFHELP	2006008144
IDA		14010	3600	MG-COMMUNITY DEVELOP. FUND/ ADDITIONAL FINANCING (FID IV) - SUPPLEMENTAL	060095d
Japan	JICA	14040	122	RIVER DEVELOPMENT	064046C
Japan	JICA	14040	336	ETUDE SUR L'APPROVISIONNEMENT EN EAU POTABLE AUTONOME ET DURABLE DANS LA	060626C
Switzerland	SDC	14010	128	WATER RESOURCES POLICY/ADMIN. MGMT	2006000596
Switzerland	SDC	14030	28	BASIC DRINKING WATER SUPPLY AND BASIC SANITATION	1980000947
UNICEF		14010	128	WATER RESOURCES POLICY/ADMIN. MGMT	061684

Total aid for water in 2006 for   **Madagascar**   **USD thousand 80000**

And as a share of aid to all recipient countries   **1.23%**

Malawi

RECIPIENT / Donor	Agency	Sector	Amount USD thousand	Project description	CRS ID Number
Germany	BMZ	14010	15	TRANSNET - TRANSBOUNDARY MANAGEMENT OF NATURAL RESOURCES IN SADC REGION	2006003329
IDA		14010	11200	INFRASTRUCTURE SERVICES	060001b
Japan	MOFA	14030	3187	GROUNDWATER DEVELOPMENT IN LILONGWE WEST	050157
Norway	MFA	14020	45	PERSONNEL EXCHANGE	2006003729
Norway	MFA	14030	102	PERSONNEL EXCHANGE	2006002861
UNICEF		14010	218	WATER RESOURCES POLICY/ADMIN. MGMT	061727
United Kingdom	DFID	14010	121	WATER BASELINE	060734
United Kingdom	DFID	14010	1157	SAFE WATER SUPPLY & SANITATION PROVISION	060190

Total aid for water in 2006 for   **Malawi**   **USD thousand 16045**

And as a share of aid to all recipient countries   **0.25%**

Malaysia

RECIPIENT / Donor	Agency	Sector	Amount USD thousand	Project description	CRS ID Number
Japan	JICA	14020	77	WATER SUPPLY & SANIT. - LARGE SYST.	066066C
Japan	JICA	14040	85	RIVER DEVELOPMENT	066067C
UNICEF		14010	2	WATER RESOURCES POLICY/ADMIN. MGMT	061757

RECIPIENT Donor	Agency	Sector	Amount USD thousand	Project description	CRS ID Number

Total aid for water in 2006 for   Malaysia   **USD thousand 164**
And as a share of aid to all recipient countries   **0.00%**

Maldives

Canada	CIDA	14050	4364	TSUNAMI 2004 DEBRIS AND WASTE MANAGEMENT. PROG	060056
Japan	JICA	14020	57	WATER SUPPLY & SANIT. - LARGE SYST.	065761C
UNICEF		14010	24	WATER RESOURCES POLICY/ADMIN. MGMT	061766

Total aid for water in 2006 for   Maldives   **USD thousand 4445**
And as a share of aid to all recipient countries   **0.07%**

Mali

Belgium	DGCD	14030	529	GESTION INTEGREE DE L'EAU DELTA DU NIGER FONDS BELGE SURVIE	2006001311
Belgium	DGCD	14030	2193	APPUI À LA DÉCENTRALISATION DE LA GESTION DE L'HYDRAULIQUE	2004003683
Belgium	DGCD	14050	1706	ASSAINISSEMENT DE 2 VILLES MALIENNES	2002000358
Denmark	MFA	14010	1573	CAPACITY BUILDING AND SECTOR STRATEGIES	061202
Denmark	MFA	14020	7336	DECENTRALISED WATER AND SANITATION	061198
Denmark	MFA	14030	1077	REVIEWS, AUDIT AND PREPARATION OF NEXT PHASE	061382
France	AFD	14030	7657	ALIMENTATION EAU POTABLE SUD MALI	2006160100
Germany	GTZ	14010	3766	ADVISORY SERVICES FOR THE 'DIRECTION NATIONALE DE L'HYDRAULIQUE'	2006006035
Germany	KFW	14030	6418	BASIC DRINKING WATER AND BASIC SANITATION	2006000934
Germany	BMZ	14050	142	WASTE MANAGEMENT / DISPOSAL	2006005414
Italy	DGCS	14010	984	WATER RESOURCES POLICY/ADMIN. MGMT	060219
Italy	LA	14020	5	WATER SUPPLY & SANIT. - LARGE SYST.	061625
Italy	DGCS	14020	252	WATER SUPPLY & SANIT. - LARGE SYST.	060218
Japan	MOFA	14010	1	WATER RESOURCES POLICY/ADMIN. MGMT	063123C
Japan	JICA	14040	0	RIVER DEVELOPMENT	064107C
Luxembourg	MFA	14030	276	UNDP COMMUNITY WATER INITIATIVE,SÉNÉGAL,MALI,NIGER	061642
Luxembourg	MFA	14030	353	BASIC DRINKING WATER SUPPLY AND BASIC SANITATION	061108
Spain	MFA	14010	865	WATER IN ÁFRICA: IMPROVEMENT IN MALÍ BURKINA FASSO SENEGAL Y GUINEA	064611
Spain	MUNIC	14030	2	WATER COLLECTION	069784
Spain	MUNIC	14030	25	CONSTRUCTION OF A DWELL OF A BIG DIAMETER	069452
Spain	MFA	14030	162	WATER SUPPLY IN RURAL COMMUNITIES OF THE ECCLESIATIC UNIT OF SAN (MALI) - 2ª PHASE	064883
UNICEF		14010	121	WATER RESOURCES POLICY/ADMIN. MGMT	061788
UNICEF		14030	218	HYGIENE, SANITATION AND WATER SUPPLY PROGRAMMES	061789

Total aid for water in 2006 for   Mali   **USD thousand 35663**
And as a share of aid to all recipient countries   **0.55%**

Marshall Islands

Japan	JICA	14020	148	GRANT AID CO-OPERATION IN WATER SUPPLY	061574C

RECIPIENT Donor	Agency	Sector	Amount USD thousand	Project description	CRS ID Number
Japan	JICA	14050	23	WASTE MANAGEMENT/DISPOSAL	066402C

Total aid for water in 2006 for   **Marshall Islands**   **USD thousand 172**
And as a share of aid to all recipient countries   **0.00%**

Mauritania

RECIPIENT Donor	Agency	Sector	Amount USD thousand	Project description	CRS ID Number
France	AFD	14081	753	ASSISTANCE TECHNIQUE EN APPUI A LA DHA	2006155500
Ireland	DFA	14030	116	BASIC DRINKING WATER AND SANIT	060554
Japan	JICA	14040	11	RIVER DEVELOPMENT	064120C
Spain	MUNIC	14030	23	CONSTRUCTION OF A WATER WELL	069050
Spain	MUNIC	14030	4	TRAITEMENT DE L'EAU DES PUITS ET MÉTHODES DE CONSERVATION DANS LA LOCALITÉ DE KAÉDI'	068293
Spain	MUNIC	14030	12	POTABLE WATER SUPPLY FOR THE COMMUNITY OF NEBAGHIYA	066846
Spain	MUNIC	14030	4	PURCHASE OF PUMPS. MAURITANIA	066548
Spain	MFA	14030	157	POTABLE WATER SUPPLY MAGTA LAHJAR	064360
Spain	MFA	14030	301	IMPROVEMENT OF PUBLIC SOURCES OF NUAKCHOTT	064321
Spain	AG	14030	94	MANAGEMENT OF DESALINATION PLANTS OF THE BANC D´ARGUIN	061683
Spain	AG	14030	21	MOUSTAGHBEL. PHASE II. MAURITANIA	061680
Spain	AG	14050	8	REUSE OF DIFFERENT RECYCLING MATERIALS (ELECTRICAL APPLIANCE MATERIAL AND BATHROOM FITTINGS ETC) M	065948
UNICEF		14010	32	WATER RESOURCES POLICY/ADMIN. MGMT	061827

Total aid for water in 2006 for   **Mauritania**   **USD thousand 1536**
And as a share of aid to all recipient countries   **0.02%**

Mauritius

RECIPIENT Donor	Agency	Sector	Amount USD thousand	Project description	CRS ID Number
EC	CEC	14010	890	ACP EU WATER FACILITY : UNICEF - EAU ET ASSAIN EC PRIMAIRES, PROM HYG	2006100290
EC	CEC	14010	1792	ACP EU WATER FACILITY	2006100289
EC	CEC	14020	12552	WASTEWATER SECTOR POLICY SUPPORT PROGRAMME	2003100332
Japan	JICA	14020	22	WATER SUPPLY & SANIT. - LARGE SYST.	064133C

Total aid for water in 2006 for   **Mauritius**   **USD thousand 15256**
And as a share of aid to all recipient countries   **0.23%**

Mexico

RECIPIENT Donor	Agency	Sector	Amount USD thousand	Project description	CRS ID Number
Germany	BMZ	14010	25	REFORM OF THE WATER SECTOR IN THE MENA REGION	2006003349
Germany	BMZ	14010	56	INTEGRATED EXPERTS	2006002413
Germany	L G	14015	151	PROJECT STUDY FOR MONITORING OF REGIONAL WATER RESOURCES	2006012460
Germany	BMZ	14015	97	INTEGRATED EXPERTS	2006002421
Germany	BMZ	14020	37	INTEGRATED EXPERTS	2006002424
Germany	L G	14050	196	FEASABILITY STUDY	2006012240
IDA		14010	70	CDM TA FOR MEXICO	060029a
Japan	PRF	14010	15	WATER RESOURCES POLICY/ADMIN. MGMT	067271C
Japan	JICA	14020	89	WATER SUPPLY & SANIT. - LARGE SYST.	064760C

CRS AID ACTIVITIES IN SUPPORT OF WATER SUPPLY AND SANITATION – ISBN 978-92-64-05172-0 – © OECD/WWC 2008

RECIPIENT Donor	Agency	Sector	Amount USD thousand	Project description	CRS ID Number
Japan	JICA	14040	86	RIVER DEVELOPMENT	064761C
Japan	JICA	14050	90	WASTE MANAGEMENT/DISPOSAL	064762C
Spain	MUNIC	14030	20	WATER PROJECT	069817
Spain	MUNIC	14030	14	CONSTRUCTION OF THREE WATER CAPTATION WORKS	067503
Spain	MUNIC	14030	10	"RIGHT TO A PROPERTY AND WATER IN CHIAPAS"	067099
Spain	MUNIC	14030	102	CONSTRUCTION OF SYSTEMS TO GET PLUVIAL WATER IN THE COMMUNITIES OF XOCHITEPEC EL PLATINAR AND SANTA	066815
Spain	AG	14030	115	USE OF RAIN TO SATISFY BASIC NEEDS OF THE COMMUNITY OF GUADALUPE VISTA HERMOSA	062330
Spain	AG	14050	94	IMPLEMENTATION OF ECOLOGICAL BATHROOMS FOR HEATH AND ENVIRONMENTAL PROTECTION IN SAN LUIS POTOSÍ MÉ	062525
United States	Misc	14030	49014	OFFICE OF WATER (OW) / INFRASTRUCTURE ASSISTANCE: MEXICO BORDER	2006003019
United States	Misc	14050	4207	OFFICE OF INTERNATIONAL AFFAIRS (OIA): US MEXICO BORDER	2006003059

Total aid for water in 2006 for **Mexico** **USD thousand 54487**
And as a share of aid to all recipient countries **0.84%**

Micronesia, Fed. States

Japan	JICA	14050	58	WASTE MANAGEMENT/DISPOSAL	066417C

Total aid for water in 2006 for **Micronesia, Fed. States** **USD thousand 58**
And as a share of aid to all recipient countries **0.00%**

Middle East, regional

Canada	IDRC	14010	13	WATER RESOURCES POLICY/ADMIN. MGMT	060539b
Canada	IDRC	14010	176	SAFE USE OF WASTEWATER, EXCRETA & GREYWATER IN LOW-INCOME URBAN SETTINGS	060458b
Canada	IDRC	14010	246	WATER DEMAND INITIATIVE - WADIMENA	060162b
Germany	Fed Min	14010	5365	INTEGRATED WATER RESOURCE MANAGEMENT	2006011238
Germany	BMZ	14010	351	REFORM OF THE WATER SECTOR IN THE MENA REGION	2006003336
Germany	BMZ	14081	35	WATER EFFICIENT MANAGEMENT OF WASTEWATER, TREATMENT AND REUSE	2006003376
Sweden	Sida	14010	610	PREP. WATER MENA	2006003324
Sweden	Sida	14081	1587	244 TRANSBOUND WATER MENA 244 TRANSBOUNDARY MENA	2006008244

Total aid for water in 2006 for **Middle East, regional** **USD thousand 8384**
And as a share of aid to all recipient countries **0.13%**

Moldova

Austria	ADA	14020	816	WATER SUPPLY AND SANITATION FOR THE TOWN CANTEMIR	2006000227
Austria	ADA	14020	17	WATER SUPPLY & SANIT. - LARGE SYST.	20060000az
Greece	YPESDDA	14050	28	PROVISION OF A GARBAGE COLLECTION TRUCK AND FIFTY BINS	2006000328
Japan	JICA	14050	25	WASTE MANAGEMENT/DISPOSAL	063611C
Netherlands	MFA	14030	17	BASIC DRINKING WATER SUPPLY AND BASIC SANITATION	2003014589
Switzerland	SDC	14010	40	WATER RESOURCES POLICY/ADMIN. MGMT	2000002086
UNICEF		14010	13	WATER RESOURCES POLICY/ADMIN. MGMT	061876

RECIPIENT Donor	Agency	Sector	Amount USD thousand	Project description	CRS ID Number

Total aid for water in 2006 for   Moldova   **USD thousand 955**

And as a share of aid to all recipient countries   0.01%

Mongolia

Donor	Agency	Sector	Amount	Project description	CRS ID
Italy	LA	14020	100	WATER SUPPLY & SANIT. - LARGE SYST.	062188
Japan	JICA	14020	42	WATER SUPPLY & SANIT. - LARGE SYST.	066105C
Japan	JICA	14040	88	RIVER DEVELOPMENT	066106C
Japan	JICA	14040	134	RIVER BASIN MANAGEMENT MODEL CONSERVATION OF WETLAND AND ECO	061753C
Japan	JICA	14050	49	WASTE MANAGEMENT/DISPOSAL	066107C
Japan	JICA	14050	215	PROJECT FOR IMPROVEMENT OF SOLID WASTE MANAGEMENT IN ULAANBAATAR CITY	061022C
Korea	KOICA	14010	14	TRANINIG PROGRAM	2006861815
Korea	KOICA	14030	26	PROJECT FOR CONSTRUCTION OF NEW WATER WELLS IN MONGOLIA	2006862111
Netherlands	MFA	14010	301	WATER RESOURCES POLICY/ADMIN. MGMT	2006000792
Netherlands	MFA	14030	975	BASIC DRINKING WATER SUPPLY AND BASIC SANITATION	2006000522
Spain	ICO	14020	5937	WASTE WATER TREATMENT PLANT	060012

Total aid for water in 2006 for   Mongolia   **USD thousand 7880**

And as a share of aid to all recipient countries   0.12%

Montenegro

Donor	Agency	Sector	Amount	Project description	CRS ID
Austria	ADA	14020	1092	EASTE WATER DISPOSAL KOTOR	2006000240
Austria	ADA	14020	12	WATER SUPPLY & SANIT. - LARGE SYST.	20060000aq
Greece	YPEHO DE	14010	4	HIGHER EFFICIENCY OF WATER-RELATED DEVELOPMENT	2006000314
Luxembourg	MFA	14030	88	RECONSTRUCTION DU RÉSEAU HYDRAULIQUE À PETNJICA	060399
Luxembourg	MFA	14030	10	BASIC DRINKING WATER SUPPLY AND BASIC SANITATION	060021
United States	TDA	14020	476	WATER SUPPLY & SANIT. - LARGE SYST.	2006002967

Total aid for water in 2006 for   Montenegro   **USD thousand 1682**

And as a share of aid to all recipient countries   0.03%

Montserrat

Donor	Agency	Sector	Amount	Project description	CRS ID
United Kingdom	DFID	14020	2940	WATER SUPPLY & SANITATION - LARGE SYSTEMS	020299

Total aid for water in 2006 for   Montserrat   **USD thousand 2940**

And as a share of aid to all recipient countries   0.05%

Morocco

Donor	Agency	Sector	Amount	Project description	CRS ID
Belgium	DGCD	14020	189	WATER SUPPLY & SANIT. - LARGE SYST.	2004001135
Belgium	MPRF	14030	11	BASIC DRINKING WATER SUPPLY AND BASIC SANITATION	2006003647
Belgium	DGCD	14030	1074	ALIMENTATION EAU POTABLE AEP 5 PETITS CENTRES	2005007213
Belgium	DGCD	14030	2	BASIC DRINKING WATER SUPPLY AND BASIC SANITATION	2004001131

CRS AID ACTIVITIES IN SUPPORT OF WATER SUPPLY AND SANITATION – ISBN 978-92-64-05172-0 – © OECD/WWC 2008

RECIPIENT Donor	Agency	Sector	Amount USD thousand	Project description	CRS ID Number
Belgium	DGCD	14030	1408	TLAT LAKHSASS / ONEP ALIMENTATION EAU POTABLE	2002000344
Belgium	DGCD	14030	536	EAU POTABLE WILAYA AGADIR	2001001129
EC	EDF	14015	37655	ASSAINISSEMENT ET APPUI INSTITUTIONNEL	2006200445
EC	EDF	14020	12552	BONIFICATION D'INT‚R`TS : ASSAINISSEMENT DES VILLES MOYENNES	2006200434
France	MINEFI	14010	741	GESTION EFFLUENTS M'HAYA	2006003503
France	MINEFI	14015	591	RATIONNALISATION GESTION L'EAU	2006003517
France	AFD	14020	37655	PROGEA	2006109500
France	AFD	14020	37655	PROGRAMME DEPOLLUTION OUED SEBOU	2006102400
France	MINEFI	14020	261	MONTAGE D'UN PROJET MDP	2006003518
France	MINEFI	14040	486	DISPOSITIF D'ALERTES DES CRUES	2006003526
Germany	GTZ	14010	4393	WATER RESOURCES POLICY/ADMIN. MGMT	2006006076
Germany	BMZ	14010	462	REFORM OF THE WATER SECTOR IN THE MENA REGION	2006003341
Germany	KFW	14020	21966	WATER SUPPLY & SANIT. - LARGE SYST.	2006000998
Germany	KFW	14020	51462	PROGRAMME SECTORIEL D'AFP	2006000996
Germany	KFW	14030	2415	AEP RURALE II, TAROUDANT	2006000992
Germany	KFW	14030	16543	APPROVISIONNEMENT EN EAU DANS LA REGION DE LOUKKOS	2006000963
Germany	BMZ	14081	2	WATER EFFICIENT MANAGEMENT OF WASTEWATER, TREATMENT AND REUSE	2006003381
Greece	YPEHODE	14010	8	HIGHER EFFICIENCY OF WATER-RELATED DEVELOPMENT	2006000305
Italy	LA	14010	20	WATER RESOURCES POLICY/ADMIN. MGMT	061604
Italy	DGCS	14020	1139	WATER SUPPLY & SANIT. - LARGE SYST.	060056
Italy	LA	14050	105	WASTE MANAGEMENT/DISPOSAL	061908
Japan	JICA	14020	305	WATER SUPPLY & SANIT. - LARGE SYST.	063652C
Japan	JICA	14040	111	RIVER DEVELOPMENT	063653C
Japan	JICA	14040	99	REINFORCEMENT OF PUBLIC WORKS MATERIAL OF CENTRAL PARK FOR SUPPORT O	062498C
Japan	JICA	14040	337	STUDY ON WATER RESOURCE INTEGRATED MANAGEMENT PLAN IN HAOUZ PLAIN	060621C
Korea	KOICA	14010	14	TRANINIG PROGRAM	2006861188
Luxembourg	MFA	14030	1010	BASIC DRINKING WATER SUPPLY AND BASIC SANITATION	061114
Spain	MFA	14010	30	AT SUPPORT FOR THE ONEP	064249
Spain	ENV	14010	329	BEGINNING OF AN INTERNATIONAL CENTRE OF WATER MANAGEMENT	060485
Spain	MUNIC	14030	178	DRINKING WATER SUPPLY AND SANITATION FOR THE RURAL POPULATION OF BOUHAMED. CHEFCHAOUEN PROVINCE.	066222
Spain	MFA	14030	61	LONG-TERM CONSOLIDATION OF THE PILOT PROGRAM OF PHOTOVOLTAGIC PUMPS IN SOUTHERN MARRUECOS.	064366
Spain	AG	14030	60	IMPROVEMNET IN ENVIRONMENTAL AND SANITATION CONDITIONS IN THE COMMUNITY OF IHADDADEN	063218
Spain	AG	14030	114	WATER SUPPLY IN THE PROVINCE OF TIZANIT (MARRUECOS)	062752
Spain	AG	14030	113	FOUR DESALINATION PLANGTS WITH RENEWABLE ENERGIES IN MARRUECOS	061691
Spain	AG	14081	75	STRENGTHENING OF MUNICIPAL MANAGEMENT IN THE ENVIRONMENTAL AREA	063265
UNICEF		14010	1	WATER RESOURCES POLICY/ADMIN. MGMT	061916
United States	TDA	14050	322	WASTE MANAGEMENT/DISPOSAL	2006003062

Total aid for water in 2006 for   **Morocco**   **USD thousand 232492**

And as a share of aid to all recipient countries   **3.57%**

RECIPIENT / Donor	Agency	Sector	Amount USD thousand	Project description	CRS ID Number

Mozambique

Austria	ADA	14030	471	RURAL WATER SUPPLY AND SANITATION AT SOFALA	2006000062
Austria	ADA	14030	75	BASIC DRINKING WATER SUPPLY AND BASIC SANITATION	2003000092
Belgium	DGCD	14030	27	BASIC DRINKING WATER SUPPLY AND BASIC SANITATION	2004003286
EC	EIB	14020	38911	MAPUTO WATER SUPPLY	2006300046
Germany	BMZ	14010	45	TRANSNET - TRANSBOUNDARY MANAGEMENT OF NATURAL RESOURCES IN SADC REGION	2006003322
Ireland	DFA	14030	384	BASIC DRINKING WATER AND SANIT	060343
Italy	DGCS	14010	189	WATER RESOURCES POLICY/ADMIN. MGMT	060777
Japan	JICA	14020	4	WATER SUPPLY & SANIT. - LARGE SYST.	064144C
Japan	JICA	14040	78	RIVER DEVELOPMENT	064145C
Netherlands	MFA	14010	15627	WATER RESOURCES POLICY/ADMIN. MGMT	2003015511
Netherlands	MFA	14020	6406	WATER SUPPLY & SANIT. - LARGE SYST.	2003013876
Netherlands	MFA	14030	10459	BASIC DRINKING WATER SUPPLY AND BASIC SANITATION	2006000657
Portugal	ICP	14010	89	VISIT TO MOZAMBIQUE OF THE MINISTER OF ENVIRONMENT OF PORTUGAL	060873
Portugal	ICP	14010	201	INFRASTRUCTURE OF WATER SUPPLY, SANIT & WASTE SECTORS IN LUMBO	060814
Portugal	ICP	14010	56	PROTOCOL ON RIVER RESOURCES	060100
Portugal	ICP	14010	28	SUPPORT TO NAT DIRECTORATE OF WATER MOZAMBIQUE (AGREEMNT ON INT. RIVERS)	060092
Portugal	ICP	14030	76	WATER SUPPLY PROJECT - "BAIRRO DE MAZAQUENE" (MAXAQUENE QUARTER)	060452
Portugal	ICP	14040	3	RIVER DEVELOPMENT	060595
Spain	MUNIC	14030	36	RECONSTRUCTION OF A DWELL AND BOMB INSTALLATION. CONSTRUCTION OF 40 WC'S	069465
Spain	MUNIC	14030	24	WATER AND ELECTRICITY SUPPLY PROTECTION WALL AND SPORTS FIELD	067535
Spain	AG	14030	241	ENGINEERING WITHOUT BORDERS AGREEMENT	063398
Spain	AG	14040	354	CONTRIBUTION TO TH STRENGTHENING OF THE EXPLOITATION COUNTRYMEN MOVEMENT ALONG THE CHOKWE IRRIGATION	061013
Switzerland	SDC	14010	40	WATER RESOURCES POLICY/ADMIN. MGMT	2004004512
Switzerland	SDC	14010	399	CONTRIBUTION TO WATERAID ACTIVITIES	2004001544
UNICEF		14010	180	WATER RESOURCES POLICY/ADMIN. MGMT	061940
UNICEF		14030	217	SECTOR PLANS FOR HYGIENE, SANITATION AND WATER	061942
UNICEF		14030	1044	HYGIENE, SANITATION AND WATER SUPPLY PROGRAMMES	061941
United Kingdom	DFID	14020	90	WATER SUPPLY & SANITATION - LARGE SYSTEMS	990648
United States	MCC	14015	373	WATER RESOURCES PROTECTION	2006002925

Total aid for water in 2006 for Mozambique USD thousand 76125
And as a share of aid to all recipient countries 1.17%

Myanmar

Japan	JICA	14020	58	WATER SUPPLY & SANIT. - LARGE SYST.	065652C
Japan	JICA	14020	86	TC IN WATER SUPPLY	062951C
Japan	JICA	14020	207	THE PROJECT ON RURAL WATER SUPPLY TECHNOLOGY IN THE CENTRAL DRY ZONE	061062C
Japan	JICA	14040	41	RIVER DEVELOPMENT	065653C
Korea	KOICA	14010	5	TRANINIG PROGRAM	2006861592
Norway	MFA	14010	121	SMALLHOLDER IRRIGATION	2006004401
UNICEF		14010	35	WATER RESOURCES POLICY/ADMIN. MGMT	060438

CRS AID ACTIVITIES IN SUPPORT OF WATER SUPPLY AND SANITATION – ISBN 978-92-64-05172-0 – © OECD/WWC 2008

RECIPIENT Donor	Agency	Sector	Amount USD thousand	Project description	CRS ID Number
UNICEF		14030	466	HYGIENE, SANITATION AND WATER SUPPLY PROGRAMMES	060437
UNICEF		14030	317	COMMUNITY HYGIENE AND WATER SAFETY	060436

Total aid for water in 2006 for   **Myanmar**   **USD thousand 1334**
And as a share of aid to all recipient countries   **0.02%**

Namibia

RECIPIENT Donor	Agency	Sector	Amount USD thousand	Project description	CRS ID Number
Germany	BMZ	14010	15	TRANSNET - TRANSBOUNDARY MANAGEMENT OF NATURAL RESOURCES IN SADC REGION	2006003316
Germany	BMZ	14010	14	INTEGRATED EXPERTS	2006002415
Luxembourg	MFA	14020	2151	WATER SUPPLY & SANIT. - LARGE SYST.	061126
Luxembourg	MFA	14030	114	BASIC DRINKING WATER SUPPLY AND BASIC SANITATION	061128

Total aid for water in 2006 for   **Namibia**   **USD thousand 2294**
And as a share of aid to all recipient countries   **0.04%**

Nepal

RECIPIENT Donor	Agency	Sector	Amount USD thousand	Project description	CRS ID Number
Canada	CIDA	14030	71	RAMECHHAP RURAL HEALTH IMPROVEMENT PROJE	060187
Canada	CIDA	14030	13	KABHRE RURAL DEVELOPMENT	040681
Finland	MFA	14020	34	WATER AND SANITATION PROGRAMME III IN NEPAL	993097
Finland	MFA	14020	12179	RURAL WATER SUPPLY AND SANITATION PROJECT IN WESTERN NEPAL	060228
Finland	MFA	14020	5021	RURAL VILLAGE WATER RESOURCES MANAGEMENT PROJECT IN NEPAL	033095
Japan	JICA	14020	60	WATER SUPPLY & SANIT. - LARGE SYST.	065777C
Japan	JICA	14020	86	ADVISOR ON WATER SUPPLY MANAGEMENT	062933C
Japan	JICA	14020	375	GRANT AID CO-OPERATION IN WATER SUPPLY	060540C
Japan	JICA	14040	270	RIVER DEVELOPMENT	065778C
Japan	JICA	14050	33	WASTE MANAGEMENT/DISPOSAL	065779C
Japan	JICA	14050	193	THE STUDY ON THE SOLID WASTE MANAGEMENT FOR THE KATHMANDU VA	061164C
Japan	JICA	14050	510	TC IN URBAN SANITATION	060366C
Korea	KOICA	14010	109	TRANINIG PROGRAM	2006861637
UNICEF		14010	5	WATER RESOURCES POLICY/ADMIN. MGMT	061988
UNICEF		14030	136	HYGIENE, SANITATION AND WATER SUPPLY PROGRAMMES	061987
UNICEF		14030	238	COMMUNITY HYGIENE AND WATER SAFETY	061986
United Kingdom	DFID	14030	1187	WATER SUPPLY & SANITATION - LARGE SYSTEMS	990837
United Kingdom	DFID	14030	1656	RURAL WATER SUPPLY	950656

Total aid for water in 2006 for   **Nepal**   **USD thousand 22174**
And as a share of aid to all recipient countries   **0.34%**

Nicaragua

RECIPIENT Donor	Agency	Sector	Amount USD thousand	Project description	CRS ID Number
Austria	Reg	14030	4	BASIC DRINKING WATER SUPPLY AND BASIC SANITATION	2006008199
Austria	ADA	14030	60	BASIC DRINKING WATER SUPPLY AND BASIC SANITATION	20060022cu
Germany	BMZ	14020	17	LOCAL AGENDA 21 IN CENTRAL AMERICA	2006003368

RECIPIENT Donor	Agency	Sector	Amount USD thousand	Project description	CRS ID Number
Germany	BMZ	14030	14	SUPPLY WITH DRINKING WATER IN EL COYOLAR, PCHOTILLO, SAN DIEGO	2006004884
Germany	KFW	14030	15095	AGUA POTABLE Y SANEAMIENTO DE GRANADA	2006001177
Germany	L G	14050	46	ASSISTANCE FOR GARBAGE CLEARING	2006011426
Germany	BMZ	14050	70	INTEGRATED EXPERTS	2006002449
IDB Sp.Fund		14020	30000	POTABLE WATER AND SANITATION INVESTMENT PROGRAM	060091
Italy	DGCS	14020	3766	WATER SUPPLY & SANIT. - LARGE SYST.	060421
Italy	DGCS	14040	9	RIVER DEVELOPMENT	060489
Japan	JICA	14020	49	WATER SUPPLY & SANIT. - LARGE SYST.	064797C
Japan	JICA	14040	439	TC IN WATER RESOURCES DEVELOPMENT	060437C
Japan	JICA	14050	33	WASTE MANAGEMENT/DISPOSAL	064798C
Luxembourg	MFA	14030	1525	BASIC DRINKING WATER SUPPLY AND BASIC SANITATION	061134
Spain	AG	14010	9	INCIDENCE PROJECT ON THE RIGHT TO WATER AND SANITATION ACCESS NICARAGUA	061156
Spain	MUNIC	14030	21	VIABILITY STUDY FOR THE CREATION OF A DRINKING WATER NET IN VILLANUEVA	069955
Spain	MUNIC	14030	8	EXPANSION OF THE NETWORK AND WATER SUPPLY DUE TO GRAVITY IN LOS CHILES (NICARAGUA)	068994
Spain	MUNIC	14030	13	WATER SUPPLY AND SANITATION IN SEVEN COMMUNITIES IN THE DEPARTMENT OF MATAGALPA	068951
Spain	MUNIC	14030	23	IMPROVEMENT IN THE WATER LEVEL IN 28 COMMUNITIES IN THE MUNICPALITIES OF ESTELÍ PUEBLO NUEVO AND LI	068944
Spain	MUNIC	14030	63	REHABILITATION AND EXPANSION OF THE POTABLE WATER SYSTEM IN SAN JUAN OF THE COCO RIVER. DEPARTMENT O	068025
Spain	MUNIC	14030	6	WATER FOR YANKEE - 1.	068013
Spain	MUNIC	14030	5	PROMOTION OF SAFE WATER IN 5 RURAL COMMUNITIES	067650
Spain	MUNIC	14030	12	IMPROVEMENT OF THE AVAILABILITY AND USE OF WATER IN 10 RURAL COMMUNITIES IN CONDEGA MUNICIPALITY	067621
Spain	MUNIC	14030	23	SUSTAINABLE USE OF THE COCO -SOMOTO SUBBASIN	067607
Spain	MUNIC	14030	5	ENVIRONMENTAL IMPROVEMENT OF THE URBAN HELMET OF SAN MIGUELITO	067510
Spain	MUNIC	14030	26	PAVING AND SEWAGE PROJECT OF THE NEIGHBORHOOD SANDINO	067508
Spain	MUNIC	14030	17	VIABILITY STUDY TO CREATE A MUNICIPAL NETWORK OF POTABLE WATER SUPPLY	067000
Spain	MUNIC	14030	11	RESTORATION AND IMPROVEMENT OF A WELL . NICARAGUA	066557
Spain	MUNIC	14030	7	BUILDING OF A WELL AND A WATER TANK. NICARAGUA	066556
Spain	MUNIC	14030	8	COMMUNITY WOMEN DEFENDERS IN NICARAGUA	066553
Spain	MUNIC	14030	30	DRINKING FOR SEVERAL COMMUNITIES. NICARAGUA	066341
Spain	MFA	14030	126	PROJECT OF POTABLE WATER SUPPLY IN TWO COMMUNITIES OF THE MUNICIPALITY OF CASTILLO	065168
Spain	AG	14030	13	DEFINITIVE STUDIES OF POTABLE WATER SYSTEM IN MINA LIMÓN AND SANTA PANCHA	062893
Spain	AG	14030	25	LETRINIZATION AND SUPPORT FOR THE COMMUNAL ORGANIZATION IN THE COMMUNITY OF YAULE	062369
Spain	AG	14030	70	BASIC SERVICE PROVISION AND STRENGTHENING OF THE COMMUNAL ORGANIZATION TO IMPROVE HABITABILITY IN TH	062196
Spain	AG	14030	485	IMPROVEMENTS IN HEALTH CONDITIONS OF 226 FAMILIES IN 6 RURAL COMMUNITIES OF THE MUNICIPALITY OF TELP	061898
Spain	AG	14030	92	INSTALLATION OF DOMICILIARY HYGENIC SERVICES; A PROJECT TO IMPROVE BASIC SANTITATION CONDITIONS IN T	061160
Spain	Misc	14030	6	TECHNIOF ARSENIC ELIMINATION . NICARAGUA AREAS WITH PROBLEMS OF ANTURAL PRESENCE	0610269
Spain	MUNIC	14040	40	SUSTAINABLE USE OF THE COCO-SOMOTO SUBBASIN	068747
Spain	MUNIC	14040	15	MANAGEMENT OF THE COCO-SOMOTO BASIN	068347
Spain	MUNIC	14040	31	SUSTAINABLE USE OF THE COCO-SOMOTO SUBBASIN. NICARAGUA	066167
Spain	MUNIC	14050	38	SANITARY PADDING OF SOLID MUNICIPAL WASTE IN THE CITY OF LEÓN. NICARAGUA	067996

     CRS AID ACTIVITIES IN SUPPORT OF WATER SUPPLY AND SANITATION – ISBN 978-92-64-05172-0 – © OECD/WWC 2008

RECIPIENT Donor	Agency	Sector	Amount USD thousand	Project description	CRS ID Number
Spain	AG	14050	24	MANAGEMENT OF SOLID AND LIQUID WASTE IN THE BUS TERMINAL MARKET OF LEÓN NICARAGUA	061157
UNICEF		14010	3	WATER RESOURCES POLICY/ADMIN. MGMT	062019

Total aid for water in 2006 for   **Nicaragua**   **USD thousand 52378**
And as a share of aid to all recipient countries   **0.80%**

Niger

RECIPIENT Donor	Agency	Sector	Amount USD thousand	Project description	CRS ID Number
AfDF		14010	19123	PROJET DE VALORISATION DES EAUX	060008
Belgium	MPRF	14030	25	BASIC DRINKING WATER SUPPLY AND BASIC SANITATION	2006003655
Denmark	MFA	14010	2187	DANIDA LIASON OFFICE IN NIAMEY	061250
Denmark	MFA	14010	421	ADMINISTRATION, REVIEWS AND AUDITS	061248
Denmark	MFA	14010	505	DANIDA ADVISOR IN THE MINISTRY OF WATER AND ENVIRONMENT	061247
Denmark	MFA	14010	1703	SUPPORT TO CAPACITY STRENGTHENING AT NATIONAL LEVEL	061240
Denmark	MFA	14030	14703	WATER SUPPLY, HYGIENE AND SANITATION IN RURAL AND SEMI URBAN AREAS	061246
Denmark	MFA	14030	236	BASIC DRINKING WATER SUPPLY AND BASIC SANITATION	061245
EC	CEC	14010	1843	ACP EU WATER FACILITY : ETABLISSEMENT PROG INVEST & INVESTISSEMENTS	2006100302
EC	CEC	14010	1579	ACP EU WATER FACILITY : EAU & ASSAIN POUR DEV HUM DURABLE	2006100295
France	AFD	14030	13807	PROG. HYDRAULIQUE VILLAGEAOISE TAHOUA	2006171700
Germany	BMZ	14010	3	TRANSNET - TRANSBOUNDARY MANAGEMENT OF NATURAL RESOURCES IN SADC REGION	2006003308
IDA		14020	10000	WATER SECTOR PROJECT - ADDITIONAL FINANCING	060358
Japan	JICA	14020	70	WATER SUPPLY & SANIT. - LARGE SYST.	064169C
Luxembourg	MFA	14030	276	UNDP COMMUNITY WATER INITIATIVE,SÉNÉGAL,MALI,NIGER	061643
Luxembourg	MFA	14030	27	BASIC DRINKING WATER SUPPLY AND BASIC SANITATION	061143
Switzerland	SDC	14010	160	WATER RESOURCES POLICY/ADMIN. MGMT	2006000226
UNICEF		14010	147	WATER RESOURCES POLICY/ADMIN. MGMT	062045

Total aid for water in 2006 for   **Niger**   **USD thousand 66815**
And as a share of aid to all recipient countries   **1.03%**

Nigeria

RECIPIENT Donor	Agency	Sector	Amount USD thousand	Project description	CRS ID Number
Germany	BMZ	14010	1	TRANSNET - TRANSBOUNDARY MANAGEMENT OF NATURAL RESOURCES IN SADC REGION	2006003313
Germany	BMZ	14030	314	INTEGRATED COMMUNITY BASED AGRICULTURE AND WATER DEVELOPMENT PROGRAMME	2006008146
IDA		14010	50000	LAGOS METROPOLITAN DEVELOPMENT AND GOVERNANCE PROJECT	060010c
IDA		14020	74000	LAGOS METROPOLITAN DEVELOPMENT AND GOVERNANCE PROJECT	060010a
Ireland	DFA	14030	38	BASIC DRINKING WATER AND SANIT	061061
Ireland	DFA	14040	13	RIVER DEVELOPMENT	061193
Japan	JICA	14040	53	RIVER DEVELOPMENT	064190C
Japan	JICA	14040	170	THE PROJECT FOR RURAL WATER SUPPLY AND HEALTH IN YOBE STATE	061350C
Spain	MUNIC	14030	6	DRINKING WATER WELL PROSPECTION FOR HUMAN CONSUMPTION AND FOR THE AGRICULTURAL AND LIVESTOCK DEVELOP	068595
UNICEF		14010	9	WATER RESOURCES POLICY/ADMIN. MGMT	062098
UNICEF		14030	233	SECTOR PLANS FOR HYGIENE, SANITATION AND WATER	062099
UNICEF		14030	814	COMMUNITY HYGIENE AND WATER SAFETY	062097

RECIPIENT Donor	Agency	Sector	Amount USD thousand	Project description	CRS ID Number
United States	TDA	14020	251	WATER SUPPLY & SANIT. - LARGE SYST.	2006002972

Total aid for water in 2006 for   **Nigeria**   **USD thousand 125902**
And as a share of aid to all recipient countries   **1.93%**

North & Central America, regional

RECIPIENT Donor	Agency	Sector	Amount	Project description	CRS ID
Canada	IDRC	14010	20	WATER RESOURCES POLICY/ADMIN. MGMT	060999b
IDA		14010	420	CARIB-GEF-IMPLEMENTATION OF ADAPTATION MEASURES IN COASTAL ZONES	060145c
Spain	MFA	14030	119	SUPPORT FOR THE ENVIRONMENTAL PLAN ARAUCARIA XXI WITH CENTRALAMERICA AND DOMINICAN REPUBLIC	065041
UNICEF		14010	26	WATER RESOURCES POLICY/ADMIN. MGMT	060058

Total aid for water in 2006 for   **North & Central America, regional**   **USD thousand 586**
And as a share of aid to all recipient countries   **0.01%**

North of Sahara, regional

RECIPIENT Donor	Agency	Sector	Amount	Project description	CRS ID
Canada	IDRC	14010	220	GESTION LOCALE DE L'EAU À L'AIDE DE SYSTÈMES D'INFORMATION GÉOGRAPHIQUES	060988b
Canada	IDRC	14010	176	SAFE USE OF WASTEWATER, EXCRETA & GREYWATER IN LOW-INCOME URBAN SETTINGS	060457b
Canada	IDRC	14010	246	WATER DEMAND INITIATIVE - WADIMENA	060161b
Canada	IDRC	14050	13	WASTE MANAGEMENT/DISPOSAL	061229b
Germany	BMZ	14010	116	REFORM OF THE WATER SECTOR IN THE MENA REGION	2006003337
Spain	MFA	14010	9	MINOR CONTRACT ON ADVICING AND TECHNICAL ASSISTANCE TO LAUNCH THE PLAN OF ACTION AND PROMOTE ACCESS	065307
Spain	ENV	14010	752	TECHNICAL ASSISTANCE WATER ISSUES	060487
Spain	MUNIC	14030	26	WATER PUMPS	069399
Spain	MUNIC	14030	26	POTABLE WATER FOR WILAYA DE DAJLA	069057
Spain	MUNIC	14030	19	SUPPORT FOR THE POTABLE WATER SUPPLY PLAN FOR SCHOOL OF WAR VICTIMS (MUTILATED) IN SAHARAN REFUGEE C	068200
Spain	MUNIC	14030	20	ATTAINMENT AND POTABILIZATION OF HYDRIC RESOURCES IN THE LIBERATED ZONES OF WESTERN SAHARA AND SAHAR	068145
Spain	MUNIC	14030	170	SUPPORT FOR THE TRANSPORT INFRASTRUCTURE IN THE CAMPS WITH SAHARAN POPULATION; ACCESS TO POTABLE WAT	067400
Spain	MFA	14030	25	CONSTITUTION AND ORGANIZATION OF THE ALLIANCE FOR WATER	065441
Spain	AG	14030	58	AQUA-SAHARA PHASE II. ARGELIA	062758
Spain	AG	14030	50	"TRAINING AND EQUIPMENT OF A HYDROGEOLOGY UNIT (UH)	061302

Total aid for water in 2006 for   **North of Sahara, regional**   **USD thousand 1926**
And as a share of aid to all recipient countries   **0.03%**

Oceania, regional

RECIPIENT Donor	Agency	Sector	Amount	Project description	CRS ID
EC	CEC	14010	2445	ACP EU WATER FACILITY : PACIFIC	2006100330
EC	CEC	14010	6276	ACP - EU WATER FACILITY	2006100242
EC	CEC	14010	2887	ACP-EU WATER FACILITY	2006100236
Japan	JICA	14020	0	WATER SUPPLY & SANIT. - LARGE SYST.	066570C
New Zealand	NZAid	14010	481	"WATER T2: WATER DEMAND MANAGEMENT"	060637
New Zealand	NZAid	14010	459	"SOPAC WATER QUALITY MANAGEMENT"	060634

     CRS AID ACTIVITIES IN SUPPORT OF WATER SUPPLY AND SANITATION – ISBN 978-92-64-05172-0 – © OECD/WWC 2008

RECIPIENT Donor	Agency	Sector	Amount USD thousand	Project description	CRS ID Number
New Zealand	NZAid	14010	52	"WATER T2: ICU BULLETIN"	060269
New Zealand	NZAid	14015	68	"SOUTH PACIFIC RIVER CARE"	060330
New Zealand	NZAid	14030	132	"PIC DRINKING WATER QUALITY"	060451
New Zealand	NZAid	14050	2	"SPREP YEAR AGAINST WASTE"	060012

Total aid for water in 2006 for   Oceania, regional   **USD thousand 12803**
And as a share of aid to all recipient countries   **0.20%**

Oman

RECIPIENT Donor	Agency	Sector	Amount USD thousand	Project description	CRS ID Number
Japan	JICA	14020	51	WATER SUPPLY & SANIT. - LARGE SYST.	065383C
Japan	JICA	14040	6	RIVER DEVELOPMENT	065384C

Total aid for water in 2006 for   Oman   **USD thousand 57**
And as a share of aid to all recipient countries   **0.00%**

Pakistan

RECIPIENT Donor	Agency	Sector	Amount USD thousand	Project description	CRS ID Number
Canada	CIDA	14030	876	WATER AND SANITATION - PAKISTAN OXFAM WATSAN NWFP	060125
Germany	BMZ	14030	90	LOW COST SEWER SYSTEM FOR ORANGI PILOT PROJECT IN KARACHI	2006008147
Ireland	DFA	14030	402	BASIC DRINKING WATER AND SANIT	060272
Japan	JICA	14020	28	WATER SUPPLY & SANIT. - LARGE SYST.	065813C
Japan	JICA	14020	103	TC IN WATER SUPPLY	062379C
Japan	JICA	14040	39	RIVER DEVELOPMENT	065814C
Japan	JICA	14050	49	WASTE MANAGEMENT/DISPOSAL	065815C
Netherlands	MFA	14030	6778	BASIC DRINKING WATER SUPPLY AND BASIC SANITATION	2006000682
Netherlands	MFA	14030	14585	WATER: EARTHQUAKE	2006000317
Spain	MFA	14030	251	WATER AND SANITATION SUPPLY	064589
UNICEF		14010	80	WATER RESOURCES POLICY/ADMIN. MGMT	062147
UNICEF		14030	442	HYGIENE, SANITATION AND WATER SUPPLY PROGRAMMES	062146
UNICEF		14030	830	COMMUNITY HYGIENE AND WATER SAFETY	062145
United Kingdom	DFID	14050	37	SOUTH ASIAN CONFERENCE ON SANITATION	050854

Total aid for water in 2006 for   Pakistan   **USD thousand 24590**
And as a share of aid to all recipient countries   **0.38%**

Palau

RECIPIENT Donor	Agency	Sector	Amount USD thousand	Project description	CRS ID Number
Japan	JICA	14020	23	WATER SUPPLY & SANIT. - LARGE SYST.	066435C
Japan	JICA	14050	7	WASTE MANAGEMENT/DISPOSAL	066436C
Japan	JICA	14050	313	TC IN URBAN SANITATION	060982C
Japan	JICA	14050	765	IMPROVEMENT ON SOLID WASTE MANAGEMENT IN THE REPUBLIC OF PALAU	060208C

RECIPIENT Donor	Agency	Sector	Amount USD thousand	Project description	CRS ID Number

Total aid for water in 2006 for   **Palau**   **USD thousand 1108**

And as a share of aid to all recipient countries   **0.02%**

Palestinian Adm. Areas

Recipient/Donor	Agency	Sector	Amount	Project description	CRS ID
Austria	ADA	14010	74	WATER RESOURCES POLICY/ADMIN. MGMT	2006000055
Belgium	MPRW	14081	136	EDUC./TRNG:WATER SUPPLY & SANITATION	2002001402
Finland	MFA	14020	753	EMERGENCY WATER SUPPLY PROJECT IN SOUTHERN GAZA	050283
France	AFD	14020	15062	CONSTRUCTION STATION EPURATION NORD	2006151600
Germany	Fed Min	14010	609	INTEGRATED WATER RESOURCE MANAGEMENT	2006011240
Germany	BMZ	14010	139	REFORM OF THE WATER SECTOR IN THE MENA REGION	2006003354
Germany	KFW	14020	2510	WATER SUPPLY TULKAREM	2006001241
Germany	BMZ	14050	146	WASTE MANAGEMENT/DISPOSAL	2006005415
Germany	BMZ	14081	164	WATER EFFICIENT MANAGEMENT OF WASTEWATER, TREATMENT AND REUSE	2006003390
Greece	YPEHO DE	14010	8	HIGHER EFFICIENCY OF WATER-RELATED DEVELOPMENT	2006000309
Greece	YPEJ	14030	188	BASIC DRINKING WATER SUPPLY TO LAND LOCKED COMMUNITIES IN WEST BANK	2006000322
IDA		14010	3100	EMERGENCY SERVICES SUPPORT PROGRAM MULTI-DONOR TRUST FUND	060131e
IDA		14010	1350	GZ-INTEGRATED COMMUNITY DEVELOPMENT SUPPLEMENTAL	060125b
IDA		14010	4700	EMERGENCY MUNICIPAL SERVICES (REHAB. II)	060015a
IDA		14020	300	AVIAN AND HUMAN INFLUENZA PREVENTION AND CONTROL PROJECT	060083d
Japan	JICA	14050	186	GRANT AID CO-OPERATION IN URBAN SANITATION	061210C
Japan	MOFA	14050	5266	IMPROV. OF SOLID WASTE MANAGEMENT IN THE WEST BANK	060228
Norway	MFA	14010	39	CONSULTANT REPORT ON REGIONAL WATER ISSUES	2006003390
Norway	MFA	14010	1091	UN CONSOLIDATED APPEAL PROCESS UNDP	2006002652
Spain	MFA	14010	394	IMPROVE SOCIO-ECONOMIC CAPACITIES OF AGRICULTURES	064993
Spain	MUNIC	14030	38	CHANELLING MAIN LEAKS AND AWATER TANKS RECONSTRUCTION	066177
Spain	MFA	14030	628	IMPROVEMENT IN WATER SANITATION AND HYGIENE	064527
Spain	AG	14030	355	IMPROVEMENT IN SANITARY CONDITIONS & IN HABITABILITY OF DISABLED PEOPLE	060915
Sweden	Sida	14030	45	WATER/SANITATION ADV GAZA	2006002314
UNICEF		14030	637	HYGIENE, SANITATION AND WATER SUPPLY PROGRAMMES	063356
United Kingdom	DFID	14010	793	HYDROMETRIC MONITORING	060335
United States	AID	14010	1688	WATER RESOURCES POLICY/ADMIN. MGMT	2006002868
United States	AID	14010	4086	SUPPORT TO WEST BANK WATER PROJECTS	2006002860
United States	AID	14015	1172	GREATER ACCESS TO AND MORE EFFECTIVE USE OF SCARCE WATER RESOURCES	2006002924
United States	AID	14030	4440	SMALL WATER INFRASTRUCTURE PROGRAM	2006003023
United States	AID	14030	10000	SMALL WATER INFRASTRUCTURE PROGRAMME	2006003022

Total aid for water in 2006 for   **Palestinian Adm. Areas**   **USD thousand 60098**

And as a share of aid to all recipient countries   **0.92%**

Panama

Recipient/Donor	Agency	Sector	Amount	Project description	CRS ID
Germany	BMZ	14020	17	LOCAL AGENDA 21 IN CENTRAL AMERICA	2006003370

RECIPIENT Donor	Agency	Sector	Amount USD thousand	Project description	CRS ID Number
Japan	JICA	14020	40	WATER SUPPLY & SANIT. - LARGE SYST.	064832C
Japan	JICA	14040	94	RIVER DEVELOPMENT	064833C
Japan	JICA	14050	56	WASTE MANAGEMENT/DISPOSAL	064834C
Spain	MUNIC	14050	15	MANAGEMENT PROGRAMMES FOR SOLID WASTE IN KANKINTÚ	067757
UNICEF		14010	29	WATER RESOURCES POLICY/ADMIN. MGMT	063294

Total aid for water in 2006 for   **Panama**   **USD thousand 251**
And as a share of aid to all recipient countries   **0.00%**

Papua New Guinea

Japan	JICA	14020	118	WATER SUPPLY & SANIT. - LARGE SYST.	066459C
Japan	JICA	14050	18	WASTE MANAGEMENT/DISPOSAL	066460C
New Zealand	NZAid	14081	6	"LIVE & LEARN 2ND NATIONAL WATER EDUCATION CONFERENCE"	060040
UNICEF		14010	34	WATER RESOURCES POLICY/ADMIN. MGMT	062199

Total aid for water in 2006 for   **Papua New Guinea**   **USD thousand 176**
And as a share of aid to all recipient countries   **0.00%**

Paraguay

Japan	JICA	14020	89	WATER SUPPLY & SANIT. - LARGE SYST.	065130C
Japan	JICA	14050	105	WASTE MANAGEMENT/DISPOSAL	065131C
Spain	MUNIC	14030	8	TOILETS CONSTRUCTION	067638
Spain	MUNIC	14030	33	SUPPLY OF POTABLE WATER AND BASIC SANITATION FOR INDEGENOUS COMMUNITIES IN RURAL AND DRY AREAS OF PA	067435
Spain	MFA	14030	228	EXPANDING THE WATER PROVISION SYSTEM AND IMPROVING COVERAGE	064891
Spain	MUNIC	14050	7	PROCICLA. INPROVEMENT OF THE QUALITY OF LIFE OF WASTE RECYCLING	068840
UNICEF		14010	43	WATER RESOURCES POLICY/ADMIN. MGMT	062219

Total aid for water in 2006 for   **Paraguay**   **USD thousand 512**
And as a share of aid to all recipient countries   **0.01%**

Peru

Germany	BMZ	14010	9	STRENGHTENING THE MANAGEMENT CAPACITIES OF ECUADORIAN WATER UTILITIES	2006003360
Germany	BMZ	14010	101	STRENGHTENING THE MANAGEMENT CAPACITIES OF PERUVIAN WATER UTILITIES	2006003355
Germany	KFW	14020	15107	WATER SUPPLY & SANIT. - LARGE SYST.	2006001291
Germany	KFW	14020	7531	WATER SUPPLY & SANIT. - LARGE SYST.	2006001289
Germany	BMZ	14030	40	INTEGRATED EXPERTS	2006002437
Ireland	DFA	14030	15	BASIC DRINKING WATER AND SANIT	061169
Japan	JICA	14020	22	WATER SUPPLY & SANIT. - LARGE SYST.	065164C
Japan	JICA	14040	8	RIVER DEVELOPMENT	065165C
Korea	KOICA	14010	10	TRANINIG PROGRAM	2006861460
Korea	KOICA	14030	72	FEASIBILITY STUDY OF AMARILIS AREA WATER SUPPLY PROJECT IN PERU	2006862106
Spain	AG	14010	1956	COLLABORATION AGREEMENT BETWEEN THE GENERALITAT AND THE PETJADES ASSOCIATION	062638

RECIPIENT Donor	Agency	Sector	Amount USD thousand	Project description	CRS ID Number
Spain	MUNIC	14015	38	MANAGEMENT OF THE AGROSYSTEM IN THE MICROBASIN OF PAMPAHUASI	067072
Spain	AG	14015	43	USE OF THE HYDRIC RESOURCE IN THE COMMUNITIES OF CATCCAPAMPA AND QUERORA	062125
Spain	MUNIC	14030	24	BUILDING OF A DRINKING WATER SYSTEM AND LATRINES IN LA FLORIDA COMMUNITY	069834
Spain	MUNIC	14030	1	TECHNICAL STUDY AND DEV. OF INTEGRAL SYSTEMS OF POTABLE WATER SUPPLY	069109
Spain	MUNIC	14030	25	INFRASTRUCTURE CONSTRUCTION FOR A POTABLE WATER SYSTEM	068041
Spain	MUNIC	14030	25	HYDRIC SUPPLY WITH HEALTH PROMOTION	068036
Spain	MUNIC	14030	44	WATER SUPPLY FOR HUMAN CONSUMPTION FOR THE COMMUNITY OF QOTOWINCHO	067792
Spain	MUNIC	14030	41	EXPANSION AND IMPROVEMENT OF POTABLE WATER IN SANTA MARTE	067725
Spain	MUNIC	14030	119	POTABLE WATER AND SUSTAINABLE ENVIRONMENTAL SANITATION	067719
Spain	MUNIC	14030	13	INSTALLATION OF A DRINKING WATER SYSTEM IN JAYUA COMMUNITY	067645
Spain	MUNIC	14030	4	WATER SUPLY	067626
Spain	MUNIC	14030	10	RECONBSTRUCTION OF A WELL IN CHICLAYO PERU	067571
Spain	MUNIC	14030	3	INSTALATION OF SANITARY LETRINES FOR THE COMMUNITY OF CHIUT". CHOTA	067548
Spain	MUNIC	14030	11	POTABLE WATER SUPPLY PLAN POTABLE WATER PURIFICATION AND RECYCLING	067502
Spain	MUNIC	14030	22	IMPROVEMENT IN SYSTEMS OF WATER FOR HUMAN CONSUMPTION	067189
Spain	MUNIC	14030	8	CONSTRUCTION OF A POTABLE WATER SYSTEM AND LETRINES	066780
Spain	MUNIC	14030	16	POTABLE WATER SUPPLY FOR COMUNITIES IN THE MUNICIPALITY OF PILL COMARCA	066776
Spain	MUNIC	14030	41	IMPROVEMENT IN THE BASIC SANITATION OF THE COMMUNITY OF ARCAHUA	066775
Spain	MUNIC	14030	36	CONSTRUCTION OF SANITARY LETRINES FOR SIX ANDENIAN COMMUNITIES	066747
Spain	MUNIC	14030	55	WATER SUPPLY FOR HUMAN CONSUMPTION IN THE COMMUNITY OF QOTOWINCHO	066614
Spain	MUNIC	14030	23	SEWER SYSTEM FOR ARANJUEZ SETTLEMENT IN TALAVERA DE LA REYNA. PERU	066324
Spain	MUNIC	14030	496	INSTALLATION AND OPERATION OF WATERWORKS SYSTEM TO 24 HOURS SUPPLY (RURAL AREAS)	066224
Spain	MUNIC	14030	130	IMPROVEMENT OF SANITATION-ENVIRONMENTAL CONDITIONS IN JAQUI CARAVELÍ AREQUPA TOWNS. PERU	066212
Spain	MUNIC	14030	515	PROGRAM INTEGRATED MANAGEMENT OF WATER RESOURCES	066192
Spain	AG	14030	89	IMPROVEMENT OF BASIC SANITATION THROUGH A COMMUNAL SEWAGE SYSTEM	063872
Spain	AG	14030	24	IMPROVEMENT OF BASIC SERVICES IN TWO HUMAN ESTABLISHMENTS OF LIMA / PERU.	063721
Spain	AG	14030	202	POTABLE WATER AND SUSTAINABLE SANITATION FOR RURAL POPULATIONS	063468
Spain	AG	14030	228	IMPROVEMENT OF HEALTH AND BASIC SANITATION CONDITIONS OF THE DISTRICT OF CELENDIN PROVINCE OF CELEN	062562
Spain	AG	14030	190	POTABLE WATER CONDOMIAL SEWAGE SANITARY EDUCATION AND SERVICE MANAG.	062526
Spain	AG	14030	235	EXPANSION AND IMPROVEMENT OF THE POTABLE WATER SYSTEM	062510
Spain	AG	14030	122	CONSTRUCTION OF POTABLE WATER SYSTEMS IN CHOTA AND CUTERVO	062431
Spain	AG	14030	26	SUSTAINABLE SANITATION IN PERIURBAN NEIGHBORHOODS OF LIMA	062297
Spain	AG	14030	251	IMPROVING THE POTABLE WATER SYSTEM FOR THE CITY OF CHOTA	062266
Spain	AG	14030	97	INSTALATIONS OF POTABLE WATER SYSTEMS	062241
Spain	AG	14030	19	DONATION OF A POTABLE WATER SYSTEM	062130
Spain	AG	14030	350	IMPROVEMENTS IN SALUBRITY CONDITIOND IN HUMAN ESTABLISHMENTS	062003
Spain	AG	14030	215	CONSTRUCTION OF POTABLE WATER SYSTEMS IN 6 RURAL COMMUNITIES	061969
Spain	AG	14030	117	POTABLE WATER SYSTEMS INSTALATIONS IN ANDENIAN COMMUNITIES OF SANTA RITA TOCHEPAMPA LAS HOYADAS S	061944
Spain	AG	14030	753	DESALATION PROJECT IN PERU	061588
Spain	AG	14030	119	IMPROVEMENT INHEALTH & LIFE QUALITY IN FOUR SMALL VILAGES OF THE ANDES	061405
Spain	AG	14030	30	INSTALATION OF POTABLE WATER SYSTEMS FOR ANDENIAN COMMUNITIES	061247

CRS AID ACTIVITIES IN SUPPORT OF WATER SUPPLY AND SANITATION – ISBN 978-92-64-05172-0 – © OECD/WWC 2008

RECIPIENT Donor	Agency	Sector	Amount USD thousand	Project description	CRS ID Number
Spain	AG	14030	34	INSTALATION OF POTABLE WATER SYSTEMS FOR SOME RURAL COMMUNITIES	061128
Spain	AG	14030	54	CONSTRUCTION OF POTABLE WATER SYSTEM	061125
Spain	Misc	14030	24	TECHNICAL STUDY AND DEVELOPMENT OF A DRINKING WATER SUPPLY INTEGRAL SYSTEM	0610587
Spain	MUNIC	14040	35	STRENGTHENING OF THE MANAGEMENT OF HIERBA HUMA MICROBASIN PERÚ	068881
Spain	AG	14040	330	SUSTAINABLE INTEGRAL DEVELOPMENT OF SAN PABLO´S MICROSCIENCE	061012
Spain	AG	14040	376	HUATANAY RIVER´S RECOVERY V STAGE	061009
Spain	MUNIC	14050	79	INTEGRAL AND COMMUNITY URBAN SOLID WASTE MANAGEMENT . PERU	066431
Spain	AG	14050	82	PROMOTION OF CIVIC PARTICIPATION WITH EQUITY STARTING WITH TREATMENT	063843
Spain	Misc	14050	3	ASSESMENT OF THE SOLID WASTE IMPACT OF THE HUACARPAY VILLAGE (PERÚ)	0610393
Spain	Misc	14050	8	TSUSTAONABLE TECNOLOGIES FOR WASTE MANAGEMENT (II)	0610275
UNICEF		14010	3	WATER RESOURCES POLICY/ADMIN. MGMT	062240
United States	TDA	14020	5	WATER SUPPLY & SANIT. - LARGE SYST.	2006002986
United States	TDA	14050	5	WASTE MANAGEMENT/DISPOSAL	2006003067

Total aid for water in 2006 for　　　　　　**Peru**　　　　　　**USD thousand 30718**
And as a share of aid to all recipient countries　　　　　　**0.47%**

Philippines

RECIPIENT Donor	Agency	Sector	Amount USD thousand	Project description	CRS ID Number
Australia	AusAID	14010	18	WSS PERFORMANCE ENHANCEMENT PROJECT	1999001492
Australia	AusAID	14015	16	LAKE LAGUNA WATERSHED PROTECTION PROJECT	2006002160
Belgium	DGCD	14010	140	WATER RESOURCES POLICY/ADMIN. MGMT	2005000155
Finland	MFA	14040	4904	REHABILITATION OF WATERWAYS IN PHILIPPINES	060246
Germany	BMZ	14010	13	INTEGRATED EXPERTS	2006002416
Germany	BMZ	14020	143	INTEGRATED EXPERTS	2006002425
Germany	BMZ	14030	550	DEVELOPMENT OF AGRICULTURE AND WATER SUPPLY IN MINDANAO	2006004896
Germany	BMZ	14050	277	WASTE MANAGEMENT/DISPOSAL	2006005416
Germany	BMZ	14050	39	WASTE MANAGEMENT	2006003375
Germany	BMZ	14050	78	INTEGRATED EXPERTS	2006002450
Germany	BMZ	14081	81	INTEGRATED EXPERTS	2006002454
Italy	LA	14020	18	WATER SUPPLY & SANIT. - LARGE SYST.	061769
Japan	JICA	14020	140	WATER SUPPLY & SANIT. - LARGE SYST.	066149C
Japan	JICA	14020	600	SMALL WATER DISTRICTS IMPROVEMENT PROJECT	060405C
Japan	JICA	14020	556	TC IN WATER SUPPLY	060329C
Japan	PRF	14030	3	BASIC DRINKING WATER SUPPLY AND BASIC SANITATION	067403C
Japan	JICA	14040	709	RIVER DEVELOPMENT	066150C
Japan	JICA	14040	85	RIVER ADMINISTRATION IMPROVEMENT	062959C
Japan	JICA	14040	146	REHABILITATION OF FLOOD FORECASTING AND WARNING SYSTEM IN PA	061601C
Japan	JICA	14040	244	STUDY ON NATIONWIDE FLOOD RISK ASSESSMENT AND FLOOD MITIGATION PLAN	060879C
Japan	JICA	14040	247	TC IN RIVERS/SAND ARRESTATION	060859C
Japan	JICA	14040	268	TC IN WATER RESOURCES DEVELOPMENT	060793C
Japan	JICA	14050	1	WASTE MANAGEMENT/DISPOSAL	066151C
Japan	JICA	14050	347	THE STUDY ON RECYCLING INDUSTRY DEVELOPMENT IN THE PHILIPPINES	060597C

RECIPIENT Donor	Agency	Sector	Amount USD thousand	Project description	CRS ID Number
Korea	KOICA	14010	8	TRANINIG PROGRAM	2006861840
Norway	NORAD	14040	594	FLOOD CONTROL MASTER PLAN	2006001251
Spain	MUNIC	14030	6	PROJECT: DRINKING WATER CHANNELLING IN LANGGAL	068852
Sweden	Sida	14030	4408	WSP SANITATION FILIPPINES SANITATION FILIPPINES	2006003279
Sweden	Sida	14050	136	PHL. ADDITIONAL SURIGAO	2006002351
Switzerland	SDC	14015	14	WATER RESOURCES PROTECTION	1986000261
Switzerland	SDC	14030	11	BASIC DRINKING WATER SUPPLY AND BASIC SANITATION	1986000270
UNICEF		14010	5	WATER RESOURCES POLICY/ADMIN. MGMT	062265
United States	TDA	14020	5	WATER SUPPLY & SANIT. - LARGE SYST.	2006002985

Total aid for water in 2006 for   **Philippines**   **USD thousand 14810**
And as a share of aid to all recipient countries   **0.23%**

Rwanda

RECIPIENT Donor	Agency	Sector	Amount USD thousand	Project description	CRS ID Number
Austria	ADA	14030	6	BASIC DRINKING WATER SUPPLY AND BASIC SANITATION	2006000277
Belgium	DGCD	14020	46	WATER SUPPLY & SANIT. - LARGE SYST.	2005000668
Belgium	DGCD	14020	161	PROVISION D'EAU POTABLE ET DE L'ASSAINISSEMENT DANS LE DISTRICT DE NDIZA	2004000325
Belgium	DGCD	14030	32	BASIC DRINKING WATER SUPPLY AND BASIC SANITATION	2004000350
Belgium	MPRF	14081	32	EDUC./TRNG:WATER SUPPLY & SANITATION	2005009144
Germany	L G	14030	12	BASIC DRINKING WATER SUPPLY AND BASIC SANITATION	2006012061
Germany	KFW	14030	587	APPROVISIONNEMENT EN EAU POTABLE DANS 8 COMMUNES RURALES	2006001349
Japan	JICA	14020	46	WATER SUPPLY & SANIT. - LARGE SYST.	064244C
Japan	JICA	14020	283	GRANT AID CO-OPERATION IN WATER SUPPLY	060759C
Japan	MOFA	14030	4734	RURAL WATER SUPPLY	060111
Japan	JICA	14040	144	RIVER DEVELOPMENT	064245C
Luxembourg	MFA	14030	31	BASIC DRINKING WATER SUPPLY AND BASIC SANITATION	061593
Netherlands	MFA	14030	2471	BASIC DRINKING WATER SUPPLY AND BASIC SANITATION	2006000062
Norway	NORAD	14030	15	INTEGRATED WATER RESOURCES MANAGEMENT IN KIGALI NGALI	2006002241
Spain	AG	14030	187	PUBLIC WATER AND HEALTH IN THE DISTRICT OF KAMONYI	063485
UNICEF		14010	180	WATER RESOURCES POLICY/ADMIN. MGMT	062343

Total aid for water in 2006 for   **Rwanda**   **USD thousand 8966**
And as a share of aid to all recipient countries   **0.14%**

Samoa

RECIPIENT Donor	Agency	Sector	Amount USD thousand	Project description	CRS ID Number
Japan	JICA	14020	9	WATER SUPPLY & SANIT. - LARGE SYST.	066544C
Japan	JICA	14050	114	WASTE MANAGEMENT/DISPOSAL	066545C
Japan	JICA	14050	200	SOLID WASTE MANAGEMENT PROJECT FOR PACIFIC (SWAMPP)	062107C
New Zealand	NZAid	14050	10	"SAMOA DIGESTER"	060066

RECIPIENT Donor	Agency	Sector	Amount USD thousand	Project description	CRS ID Number

Total aid for water in 2006 for   Samoa   **USD thousand 332**
And as a share of aid to all recipient countries   **0.01%**

Sao Tome & Principe

Spain	MFA	14030	314	CO-OPERATION AGREEMENT TO PROMOTE THE POPULATION´S ACCESS TO BASIC SERVICES AND REDUCE INCIDENCE OF E	064695

Total aid for water in 2006 for   Sao Tome & Principe   **USD thousand 314**
And as a share of aid to all recipient countries   **0.00%**

Saudi Arabia

Japan	JICA	14020	46	WATER SUPPLY & SANIT. - LARGE SYST.	065400C
Japan	JICA	14040	36	RIVER DEVELOPMENT	065401C
Japan	JICA	14050	18	WASTE MANAGEMENT/DISPOSAL	065402C

Total aid for water in 2006 for   Saudi Arabia   **USD thousand 101**
And as a share of aid to all recipient countries   **0.00%**

Senegal

Belgium	DGCD	14030	5967	EAU POTABLE BASSIN ARACHIDIER (PARPEBA)	2002000510
Belgium	DGCD	14050	197	WASTE MANAGEMENT/DISPOSAL	2004001145
Canada	IDRC	14050	1058	DÉCHARGE DE MBEUBEUSS: AMÉLIORATION DES CONDITIONS DE VIE & DE L'ENVIR.	061171b
Germany	BMZ	14010	7	REFORM OF THE WATER SECTOR IN THE MENA REGION	2006003345
Germany	BMZ	14010	2	TRANSNET - TRANSBOUNDARY MANAGEMENT OF NATURAL RESOURCES IN SADC REGION	2006003310
Germany	BMZ	14030	314	BASIC DRINKING WATER SUPPLY AND AGRICULTURE SUPPORT	2006008148
IDA		14010	12513	SN-PARTICIPATORY LOCAL DEVELOPMENT PROGRAM	060055b
IDA		14010	16000	LOCAL AUTHORITIES DEVELOPMENT PROGRAM	060040d
Italy	LA	14020	25	WATER SUPPLY & SANIT. - LARGE SYST.	061551
Japan	JICA	14040	84	RIVER DEVELOPMENT	064271C
Japan	JICA	14040	890	TC IN WATER RESOURCES DEVELOPMENT	060166C
Luxembourg	MFA	14030	276	UNDP COMMUNITY WATER INITIATIVE,SÉNÉGAL,MALI,NIGER	061644
Luxembourg	MFA	14030	867	BASIC DRINKING WATER SUPPLY AND BASIC SANITATION	061155
Netherlands	MFA	14030	13493	ASSAINISSEMENT	2006000477
Spain	MFA	14010	865	WATER IN ÁFRICA: IMPROVEMENT IN MALÍ BURKINA FASSO SENEGAL Y GUINEA	064612
Spain	MUNIC	14030	38	SUPPORT FOR POTABLE WATER SUPPLY	067368
Spain	MFA	14030	218	IMRPOVING THE SANITARY AND ENVIRONMENTA SITUATION	064983
Spain	AG	14030	100	PILOT PROJECT WATER AND DEVELOPMENT IN SENEGAL	063328
Spain	AG	14030	75	IMPROVEMENT IN WATER ACCESS IN THE DEPARTMENT OF OUSSOUYE	063270
Spain	AG	14040	50	DINAMIZATION OF AN EXCHANGE CO-OPERATION	063189
UNICEF		14010	87	WATER RESOURCES POLICY/ADMIN. MGMT	062411
United States	TDA	14050	419	WASTE MANAGEMENT/DISPOSAL	2006003061

RECIPIENT Donor	Agency	Sector	Amount USD thousand	Project description	CRS ID Number
				Total aid for water in 2006 for Senegal	USD thousand 53547
				And as a share of aid to all recipient countries	0.82%

Serbia

RECIPIENT Donor	Agency	Sector	Amount USD thousand	Project description	CRS ID Number
Austria	ADA	14020	22	WATER SUPPLY & SANIT. - LARGE SYST.	2001009037
Germany	GTZ	14020	628	INFRASTRUCTURE PROGRAMME WATER SUPPLIES / SANITATION	2006006094
Germany	KFW	14020	12552	WATER SUPPLY & SANIT. - LARGE SYST.	2006001077
Germany	KFW	14020	3209	WATER SUPPLY & SANIT. - LARGE SYST.	2006001076
Germany	BMZ	14030	79	INTEGRATED EXPERTS	2006002432
Greece	YPEHO DE	14010	4	HIGHER EFFICIENCY OF WATER-RELATED DEVELOPMENT	2006000315
Japan	MOFA	14020	3900	SUPPLY NECESSARY EQUIPMENT FOR IMPLEMENTING WATER SUPPLY SYSTEM	060187
Japan	JICA	14040	48	RIVER DEVELOPMENT	063529C
Japan	JICA	14050	23	WASTE MANAGEMENT/DISPOSAL	063530C
Norway	MFA	14010	116	HYDROLOGICAL NETWORK DEVELOPMENT	2006004207
Norway	MFA	14010	75	SEMINAR	2006001009
Norway	MFA	14020	1247	FILTER AT THE MOJDEZ STATION	2006001175
Norway	MFA	14040	273	HYDROLOGY SMALL POWER PLANTS	2006002722
Norway	MFA	14040	122	HYDROLOGY: WATER BALANCE MAP	2006001058
Norway	MFA	14050	288	PUBLIC UTILITY SERVICES	2006003357
Sweden	Sida	14050	152	SOLID WASTE MANAGEMENT	2006003333
Switzerland	SDC	14030	519	INCORPORATION OF REG. WATER COMPANIES	2006000094
Switzerland	SDC	14030	519	RURAL WATER SANITATION AND SUPPLY SEK	2005000666
Switzerland	SDC	14030	6	BASIC DRINKING WATER SUPPLY AND BASIC SANITATION	2001002418
UNICEF		14010	4	WATER RESOURCES POLICY/ADMIN. MGMT	061564
				Total aid for water in 2006 for Serbia	USD thousand 23784
				And as a share of aid to all recipient countries	0.37%

Seychelles

RECIPIENT Donor	Agency	Sector	Amount USD thousand	Project description	CRS ID Number
EC	CEC	14050	3766	INTEGRATED SOLID WASTE MANAGEMENT PROGRAMME	2006100184
				Total aid for water in 2006 for Seychelles	USD thousand 3766
				And as a share of aid to all recipient countries	0.06%

Sierra Leone

RECIPIENT Donor	Agency	Sector	Amount USD thousand	Project description	CRS ID Number
EC	CEC	14010	821	ACP EU WATER FACILITY : SCALING UP WATER&SANIT CAPACIT IN SIERRA LEONE	2006100317
Ireland	DFA	14030	94	BASIC DRINKING WATER AND SANIT	060616
Japan	JICA	14020	28	WATER SUPPLY & SANIT. - LARGE SYST.	064315C
Japan	JICA	14040	54	RIVER DEVELOPMENT	064316C
Spain	MUNIC	14030	3	WELL CONSTRUCTION IN SIERRA LEONA	068975
Spain	MUNIC	14030	2	CONSTRUCTION OF A WELL IN THE TONKO LIMBA AND BRAMAIA AREA. CHIEFDOMS IN SIERRA LEONE	067618

CRS AID ACTIVITIES IN SUPPORT OF WATER SUPPLY AND SANITATION – ISBN 978-92-64-05172-0 – © OECD/WWC 2008

RECIPIENT Donor	Agency	Sector	Amount USD thousand	Project description	CRS ID Number
Spain	AG	14030	151	IMPROVEMENT IN POTABLE WATER ACCESS AND SANITATION	061400
UNICEF		14010	61	WATER RESOURCES POLICY/ADMIN. MGMT	062446
UNICEF		14030	155	HYGIENE, SANITATION AND WATER SUPPLY PROGRAMMES	062447
United Kingdom	DFID	14010	9200	WATER AND SANITATION IN FREETOWN	060904

Total aid for water in 2006 for **Sierra Leone** USD thousand **10568**
And as a share of aid to all recipient countries **0.16%**

Solomon Islands

Japan	JICA	14020	9	WATER SUPPLY & SANIT. - LARGE SYST.	066493C
Japan	JICA	14020	102	IMPROVEMENT AND REHAB. OF SOLOMON ISLANDS WATER AUTHORIT	062423C
Japan	JICA	14020	195	FOLLOW-UP CO-OPERATION OF WATER SUPPLY SYSTEM IN H	061140C
Japan	JICA	14020	509	TC IN WATER SUPPLY	060523C

Total aid for water in 2006 for **Solomon Islands** USD thousand **815**
And as a share of aid to all recipient countries **0.01%**

Somalia

Canada	CIDA	14030	1173	WATER AND SANITATION - OXFAM CAN TSUNAMI PROJECT/SOMALIA	060072
Ireland	DFA	14030	196	BASIC DRINKING WATER AND SANIT	060438
Luxembourg	MFA	14030	87	BASIC DRINKING WATER SUPPLY AND BASIC SANITATION	061044
Norway	MFA	14030	35	ASSESSMENT OF WATER PROJECT, MDUG REGION	2006002309
Norway	MFA	14030	731	WATER AND SANITATION	2006001363
UNICEF		14010	95	WATER RESOURCES POLICY/ADMIN. MGMT	062480
UNICEF		14030	154	HYGIENE, SANITATION AND WATER SUPPLY PROGRAMMES	062481

Total aid for water in 2006 for **Somalia** USD thousand **2470**
And as a share of aid to all recipient countries **0.04%**

South & Central Asia, regional

Finland	MFA	14010	502	ASSESMENT OF THE AMU-DARYA RIVER BASIN	060241
Sweden	Sida	14010	4747	WSP SANITATION MEKONG	2006003281
United States	TDA	14050	127	WASTE MANAGEMENT/DISPOSAL	2006003063

Total aid for water in 2006 for **South & Central Asia, regional** USD thousand **5376**
And as a share of aid to all recipient countries **0.08%**

South Africa

Belgium	MPRF	14030	555	WATER & SANITATION PROGRAM SEKHUKHUNE DISTRICT	2004007087
Canada	IDRC	14010	176	DEVELOPING COMMUNITY BASED GOVERNANCE OF WETLANDS IN CRAIGIEBURN VILLAGE	060982b
Denmark	MFA	14040	1346	REGIONAL WATER RESSOURCE STRATEGY FOR ZAMBEZI RIVER	061158
France	AFD	14020	50207	MISE A NIVEAU RESEAU EAU POTABLE SOWETO	2006115400

RECIPIENT Donor	Agency	Sector	Amount USD thousand	Project description	CRS ID Number
Germany	BMZ	14010	33	MANAGEMENT CAPACITIES FOR WATER AND SANITATION SERVICES IN LOCAL GOVERNMENT	2006003356
Germany	BMZ	14010	22	TRANSNET - TRANSBOUNDARY MANAGEMENT OF NATURAL RESOURCES IN SADC REGION	2006003333
Germany	Fed Min	14030	1245	DECENTRALIZED WATER SUPPLIES AND SEWAGE DISPOSAL	2006011294
Ireland	DFA	14010	511	WATER POLICY /ADMINISTRATION	060234
Ireland	DFA	14030	3964	BASIC DRINKING WATER AND SANIT	060026
Japan	JICA	14050	2	WASTE MANAGEMENT/DISPOSAL	063744C
Norway	MFA	14020	40	FREDSKORPSET PERSONNEL EXCHANGE	2006003413
Norway	MFA	14020	5	PERSONNEL EXCHANGE	2006001697
Norway	NORAD	14020	8	HEDMARK RENOVASJON & RESIRKULERING A.S	2006000853
Switzerland	SDC	14010	24	WATER RESOURCES POLICY/ADMIN. MGMT	1987000154

Total aid for water in 2006 for   **South Africa**   **USD thousand 58138**
And as a share of aid to all recipient countries   **0.89%**

South America, regional

Netherlands	MFA	14010	3130	DCO CONCERTATION	2006000184

Total aid for water in 2006 for   **South America, regional**   **USD thousand 3130**
And as a share of aid to all recipient countries   **0.05%**

South Asia, regional

Australia	AusAID	14015	15	WATER RESOURCES PROTECTION	2006002212

Total aid for water in 2006 for   **South Asia, regional**   **USD thousand 15**
And as a share of aid to all recipient countries   **0.00%**

South of Sahara, regional

Austria	ADA	14010	151	WATER RESOURCES POLICY/ADMIN. MGMT	2006000006
Belgium	DGCD	14030	123	RENFORCEMENT DES CAPACITÉS SUR LA GESTION DES EAUX DE PARTENAIRES	2004000316
Canada	IDRC	14010	220	GESTION LOCALE DE L'EAU À L'AIDE DE SYSTÈMES D'INFORMATION GÉOGRAPHIQUES	060989b
Denmark	MFA	14020	757	ACTION PLAN FOR CREPA 2006-2010	061238
EC	CEC	14010	1632	GESTION ADMINISTRTIVE DE LA FACILITE EAU ET FACILITE ENERGIE	2006100222
EC	CEC	14040	6276	PROGRAMME DE GESTION INTEGRE DES BASSINS TRANSFRONTALIERS	2006100159
EC	CEC	14040	6276	PROGRAMME DE GESTION INTEGREE DES BASSINS TRANSFRONTALIERS	2006100156
Germany	Fed Min	14010	16037	AVAILABILITY OF WATER, GLOBAL WATER CYCLE	2006011044
Germany	BMZ	14010	47	TRANSNET - TRANSBOUNDARY MANAGEMENT OF NATURAL RESOURCES IN SADC REGION	2006003304
Germany	L G	14030	46	PLANNING AND IMPLEMENTATION OF DECENTRALIZED WASTEWATER TREATMENT SYSTEM	2006011588
Germany	BMZ	14030	1497	IMPROVING COMMUNITY SANITATION, DECENTRALIZED WASTEWATER TREATMENT	2006004451
Germany	GTZ	14040	1883	RIVER DEVELOPMENT	2006006120
Germany	GTZ	14040	1883	CO-OPERATION AMONG RIVER BASIN ORGANISATIONS	2006006118
Italy	LA	14010	3	WATER RESOURCES POLICY/ADMIN. MGMT	061244
Netherlands	MFA	14030	59232	UNICEF WASH PROGRAMME	2006000685

CRS AID ACTIVITIES IN SUPPORT OF WATER SUPPLY AND SANITATION – ISBN 978-92-64-05172-0 – © OECD/WWC 2008

RECIPIENT	Agency	Sector	Amount USD thousand	Project description	CRS ID Number
Donor					
Netherlands	MFA	14030	32007	DMW AFDB TRUSTFUND RWSSI	2006000409
Norway	NORAD	14010	307	MARA RIVER BASIN MANAGEMENT	2006000602
Norway	MFA	14010	1325	WATER RESOURCES POLICY/ADMIN. MGMT	2006000293
Sweden	Sida	14010	7	STUDIES	2006003357
Sweden	Sida	14010	65	WORKSHOP MAY 2006	2006003346
Sweden	Sida	14010	1356	ZACPRO 6.2 CONT	2006003299
Sweden	Sida	14010	68	AUDITS, STUDIES, ETC.	2006003294
Sweden	Sida	14015	34	PROGR ASSESS (WDM III DESK APPRAISAL)	2006003466
Sweden	Sida	14030	10714	CREPA 2006-2010, CREPA CORE SUPPORT	2006003243
Switzerland	SDC	14030	38	BASIC DRINKING WATER SUPPLY AND BASIC SANITATION	1980000930
United States	TDA	14020	384	WATER SUPPLY & SANIT. - LARGE SYST.	2006002970
United States	AID	14040	61	RIVER DEVELOPMENT	2006003053
United States	AID	14040	2160	IMPROVED MANAGEMENT SELECTED RIVER BASINS	2006003051

Total aid for water in 2006 for   **South of Sahara, regional**   **USD thousand 144587**
And as a share of aid to all recipient countries   **2.22%**

Sri Lanka

RECIPIENT	Agency	Sector	Amount USD thousand	Project description	CRS ID Number
Australia	AusAID	14050	287	HIKKADUWA SEWAGE AND WASTE MANAGEMENT PR	2004002513
Austria	BMF	14020	365	EXPANSION OF WATER SUPPLY (INCREASE)	20040013g+
Canada	CIDA	14030	38	SANITATION FACILITIES FOR TSUNAMI	060639
Denmark	MFA	14020	9785	NUWARA ELIYA DISTRICT GROUP WATER SUPPLY PROJECT	061422
Denmark	MFA	14020	19865	TOWNS SOUTH OF KANDY WATER SUPPLY PROJECT	041043
Japan	MOFA	14010	0	WATER RESOURCES POLICY/ADMIN. MGMT	063338C
Japan	JICA	14020	255	WATER SUPPLY & SANIT. - LARGE SYST.	065688C
Japan	JICA	14020	119	TC IN WATER SUPPLY	062012C
Japan	JICA	14040	39	RIVER DEVELOPMENT	065689C
Japan	JICA	14050	55	WASTE MANAGEMENT/DISPOSAL	065690C
Korea	Kexim	14030	6440	GREATER GALLE WATER SUPPLY PROJECT PHASE II (SUPPLEMENTARY)	2006861148
Netherlands	MFA	14010	31	WATER RESOURCES POLICY/ADMIN. MGMT	2006001099
New Zealand	NZAid	14030	73	" RURAL COMMUNITY WATER SUPPLY AND SANITATION"	060344
New Zealand	NZAid	14030	32	"NUWARI ELIYA WATER SUPPLY"	060198
Norway	MFA	14010	28	FINAL REVIEW OF THIRD WATER PROJECT	2006000133
Sweden	Sida	14020	200	LKA, MONITORING WWT	2006002326
Sweden	Sida	14020	256	LKA WWT PROCUREM SUPPORT	2006002305
United States	TDA	14020	561	WATER SUPPLY & SANIT. - LARGE SYST.	2006002961

Total aid for water in 2006 for   **Sri Lanka**   **USD thousand 38431**
And as a share of aid to all recipient countries   **0.59%**

States Ex-Yugoslavia

RECIPIENT	Agency	Sector	Amount USD thousand	Project description	CRS ID Number
Italy	DGCS	14015	9	WATER RESOURCES PROTECTION	060874

RECIPIENT Donor	Agency	Sector	Amount USD thousand	Project description	CRS ID Number
Norway	MFA	14020	436	WATER SUPPLY & SANIT. - LARGE SYST.	2006002763

Total aid for water in 2006 for States Ex-Yugoslavia **USD thousand 444**
And as a share of aid to all recipient countries 0.01%

Sudan

RECIPIENT Donor	Agency	Sector	Amount USD thousand	Project description	CRS ID Number
EC	CEC	14010	2704	ACP EU WATER FACILITY	2006100299
Germany	BMZ	14010	1	TRANSNET - TRANSBOUNDARY MANAGEMENT OF NATURAL RESOURCES IN SADC REGION	2006003307
Greece	YPEJ	14030	60	BASIC DRINKING WATER SUPPLY AND BASIC NUTRITION TO THE DISPLACED PEOPLE	2006000324
Ireland	DFA	14030	386	BASIC DRINKING WATER AND SANIT	060345
Italy	DGCS	14020	632	WATER SUPPLY & SANIT. - LARGE SYST.	060303
Japan	MOFA	14030	1907	IMPROV. OF WATER-SUPPLY-RELATED FACILITIES	060260
Luxembourg	MFA	14081	60	EDUC./TRNG:WATER SUPPLY & SANITATION	061047
New Zealand	NZAid	14030	80	"OXFAM: RURAL WATER & LIVELIHOODS RESPONSE: SUDAN"	060357
Spain	MFA	14030	975	SUPPORT AND IMPROVEMENT IN SYSTEMS OF POTABLE WATER MANAGEMENT AND BASIC SANITATION AND HEALTHY HYG	064901
Spain	AG	14030	181	REPATRIATION AND REINTEGRATION OF REFUGEE ACCES TO WATER	062442
UNICEF		14010	113	WATER RESOURCES POLICY/ADMIN. MGMT	062526
UNICEF		14030	665	HYGIENE, SANITATION AND WATER SUPPLY PROGRAMMES	062527
United Kingdom	DFID	14010	2472	OXFAM PUBLIC HEALTH PROGRAMME IN DARFUR	060913
United Kingdom	DFID	14010	1268	OXFAM PUBLIC HEALTH PROGRAMME	060540
United Kingdom	DFID	14010	1582	SOLIDARITES EMERGENCY PROJECT	060537
United Kingdom	DFID	14010	2195	UNICEF-WATER SECTOR SUPPORT	060224
United States	AID	14030	739	BASIC DRINKING WATER SUPPLY AND BASIC SANITATION	2006003026

Total aid for water in 2006 for Sudan **USD thousand 16020**
And as a share of aid to all recipient countries 0.25%

Suriname

RECIPIENT Donor	Agency	Sector	Amount USD thousand	Project description	CRS ID Number
Netherlands	MFA	14020	84	WATER SUPPLY & SANIT. - LARGE SYST.	2003015101

Total aid for water in 2006 for Suriname **USD thousand 84**
And as a share of aid to all recipient countries 0.00%

Swaziland

RECIPIENT Donor	Agency	Sector	Amount USD thousand	Project description	CRS ID Number
Finland	MFA	14030	117	NGO SUPPORT / DRY SANITATION PROJECT IN MSUNDUZA TOWNSHIP	060166
Germany	BMZ	14010	15	TRANSNET - TRANSBOUNDARY MANAGEMENT OF NATURAL RESOURCES IN SADC REGION	2006003332
Italy	LA	14020	11	WATER SUPPLY & SANIT. - LARGE SYST.	062273
Japan	JICA	14020	6	WATER SUPPLY & SANIT. - LARGE SYST.	064365C
Japan	JICA	14020	107	TC IN WATER SUPPLY	062274C
Japan	JICA	14020	196	STRENGTHENING OF RURAL WATER SUPPLY SYSTEM IN RURAL AREAS	061131C
UNICEF		14010	26	WATER RESOURCES POLICY/ADMIN. MGMT	062581

CRS AID ACTIVITIES IN SUPPORT OF WATER SUPPLY AND SANITATION – ISBN 978-92-64-05172-0 – © OECD/WWC 2008

RECIPIENT Donor	Agency	Sector	Amount USD thousand	Project description	CRS ID Number
				Total aid for water in 2006 for Swaziland **USD thousand 477** **And as a share of aid to all recipient countries** 0.01%	

Syria

RECIPIENT Donor	Agency	Sector	Amount USD thousand	Project description	CRS ID Number
Germany	BMZ	14010	212	WATER RESOURCES POLICY AND ADMINISTRATIVE MANAGEMENT	2006005402
Germany	BMZ	14010	584	REFORM OF THE WATER SECTOR IN THE MENA REGION	2006003351
Germany	BMZ	14030	65	INTEGRATED EXPERTS	2006002438
Germany	BMZ	14081	966	EFFICIENT MANAGEMENT OF WATER SUPPLY AND WASTEWATER IN SYRIA	2006003391
Germany	BMZ	14081	5	WATER EFFICIENT MANAGEMENT OF WASTEWATER, TREATMENT AND REUSE	2006003386
Greece	YPEHO DE	14010	8	HIGHER EFFICIENCY OF WATER-RELATED DEVELOPMENT	2006000308
Greece	YPEJ	14030	126	PURIFICATION AND DISTRIBUTION OF SAFE DRINKING WATER & IMPROVEMENT	2006000321
Japan	JICA	14020	185	WATER SUPPLY & SANIT. - LARGE SYST.	065421C
Japan	JICA	14020	127	MANAGEMENT OF SEWER NETWORK	061872C
Japan	JICA	14020	520	THE STUDY ON SEWERAGE SYSTEM DEVELOPMENT	060353C
Japan	JICA	14040	437	RIVER DEVELOPMENT	065422C
Japan	JICA	14040	307	THE ESTABLISHMENT OF THE WATER RESOURCES INFORMATION CENTER	060696C
Japan	JICA	14040	520	TC IN WATER RESOURCES DEVELOPMENT	060354C
Japan	JICA	14050	96	WASTE MANAGEMENT/DISPOSAL	065423C
Japan	MOFA	14050	5009	IMPROV. OF EQUIPMENT FOR SOLID WASTE TREATMENT IN LOCAL CITIES	060127
UNICEF		14010	36	WATER RESOURCES POLICY/ADMIN. MGMT	062608
				Total aid for water in 2006 for Syria **USD thousand 9202** **And as a share of aid to all recipient countries** 0.14%	

Tajikistan

RECIPIENT Donor	Agency	Sector	Amount USD thousand	Project description	CRS ID Number
IDA		14010	9450	MUNICIPAL INFRASTRUCTURE DEVELOPMENT PROJECT	060022a
IDA		14020	5000	ADDITIONAL FINANCING FOR THE DUSHANBE WATER SUPPLY PROJECT	060332
IDA		14020	5550	MUNICIPAL INFRASTRUCTURE DEVELOPMENT PROJECT	060022b
Italy	DGCS	14020	12	WATER SUPPLY & SANIT. - LARGE SYST.	060858
Japan	JICA	14020	47	WATER SUPPLY & SANIT. - LARGE SYST.	065548C
Japan	JICA	14040	73	RIVER DEVELOPMENT	065549C
Japan	JICA	14040	125	TC IN RIVERS/SAND ARRESTATION	061898C
Japan	JICA	14040	1848	STUDY ON NATURAL DISASTER PREVENTION IN PYANJ RIVER IN REPUBLIC OF TAJIK	060072C
Spain	MUNIC	14030	22	ESTABLISHMENT OF WATER AND SANITATION SYSTEMS IN RURAL COMMUNITIES	068601
Spain	MUNIC	14030	25	INTEGRAL ASSISTANCE IN WATER AND SANITATION FOR RURAL COMMUNITIES	067762
Switzerland	SDC	14010	160	DISSEMINATION OF IWRM EXPERIENCES	2006000673
Switzerland	SDC	14010	160	REGIONAL RURAL WATER SUPPLY PROJECT CENTRAL ASIA	2006000607
UNICEF		14010	140	WATER RESOURCES POLICY/ADMIN. MGMT	062643

RECIPIENT Donor	Agency	Sector	Amount USD thousand	Project description	CRS ID Number

| | | | Total aid for water in 2006 for | Tajikistan | USD thousand 22611 |
| | | | And as a share of aid to all recipient countries | | 0.35% |

Tanzania

Donor	Agency	Sector	Amount	Project description	CRS ID Number
AfDF		14020	80906	RURAL WATER SUPPLY & SANITATION PROGRAM	060012
Austria	Reg	14030	46	BASIC DRINKING WATER SUPPLY AND BASIC SANITATION	2006008188
Austria	ADA	14030	28	BASIC DRINKING WATER SUPPLY AND BASIC SANITATION	20060022be
Belgium	DGCD	14020	51	WATER SUPPLY & SANIT. - LARGE SYST.	2006003692
EC	CEC	14010	2144	ACP EU WATER FACILITY	2006100310
Finland	MFA	14030	188	NGO SUPPORT / WELLS FOR THE MWASA DISTRICT IN TANZANIA	050187
Germany	BMZ	14010	98	TRANSNET - TRANSBOUNDARY MANAGEMENT OF NATURAL RESOURCES IN SADC REGION	2006003319
Germany	BMZ	14020	603	WATER SUPPLY AND SANITATION - LARGE SYSTEMS	2006005405
Germany	L G	14030	6	WATER SUPPLY AKERI VOCATIONAL CENTER	2006011442
Germany	L G	14030	13	WATER SUPPLY IN NYERE COMBRA YOUTH ECO-VILLAGE	2006011440
Germany	BMZ	14030	695	THE DIOCESAN DRINKING WATER PROGRAMME / MBULU	2006008151
Ireland	DFA	14030	116	BASIC DRINKING WATER AND SANIT	060634
Italy	LA	14020	6	WATER SUPPLY & SANIT. - LARGE SYST.	062064
Italy	LA	14030	10	BASIC DRINKING WATER SUPPLY AND BASIC SANITATION	062275
Italy	LA	14040	60	RIVER DEVELOPMENT	061677
Japan	MOFA	14010	0	WATER RESOURCES POLICY/ADMIN. MGMT	063143C
Japan	JICA	14020	198	WATER SUPPLY & SANIT. - LARGE SYST.	064382C
Japan	JICA	14020	188	RURAL WATER SUPPLY PROJECT IN COAST REGION AND DAR ES SALAAM PERI-URBAN	061193C
Japan	MOFA	14030	10567	ZANZIBAR URBAN WATER SUPPLY	060098
Japan	JICA	14040	79	RIVER DEVELOPMENT	064383C
Japan	JICA	14040	514	THE STUDY ON RURAL WATER SUPPLY IN MWANZA AND MARA REGIONS	060362C
Japan	JICA	14040	567	THE STUDY ON GROUNDWATER RESOURCES DEVELOPMENT AND MANAGEMEN	060321C
Japan	JICA	14040	2890	TC IN WATER RESOURCES DEVELOPMENT	060093C
Korea	KOICA	14010	14	TRANINIG PROGRAM	2006861328
Korea	KOICA	14030	20	DRINKING WATER DEVELOPMENT PROJECT IN EAST AFRICA	2006862186
Korea	KOICA	14030	1500	DEVELOPMENT OF GROUND WATER	2006862155
Spain	Misc	14010	74	STRENGTHENING FOR SUSTAINABLE MANAGEMENT BASIC SERVICES PLANNING AND SANITATION	0610343
Spain	MUNIC	14030	57	HYDROSANITARY PROGRAMME IN THE MANGOLA VALLEY (4ª PHASE)	068143
Spain	MUNIC	14030	54	INTEGRAL PROJECT OF WATER SYSTEMS FOR 8 HAMLETS IN THE DODOMA AND MOROGORO TANZANIA	066399
Spain	AG	14030	222	HRYDROHEALTH PROGRAM IN SAME DISTRICT PHASE II.	065905
Spain	AG	14030	20	HYDROHEALTH PROGRAM IN SAME DISTRICT. TANZANIA.	065854
Spain	MFA	14030	11	REPAIRMENT OF MANUAL PUMPS OF WATER SUPPLY OF WELLS IN TANZANIA	063966
Spain	AG	14030	251	HYDORSANITARY PROGRAM OF THE RURAL KIGOMADISTRICT (4ª PHASE)	061992
Spain	Misc	14030	4	TECHNICAL SUPPORT TO WATERSANITATION PROGRAMS IN TANZANIA	0610285
Switzerland	seco	14020	15353	PPP PROGRAM DODOMA / TABORAÉ TANZANIA	2006002754
UNICEF		14010	107	WATER RESOURCES POLICY/ADMIN. MGMT	062677
UNICEF		14030	148	COMMUNITY HYGIENE AND WATER SAFETY	062676

  CRS AID ACTIVITIES IN SUPPORT OF WATER SUPPLY AND SANITATION – ISBN 978-92-64-05172-0 – © OECD/WWC 2008

RECIPIENT Donor	Agency	Sector	Amount USD thousand	Project description	CRS ID Number

Total aid for water in 2006 for Tanzania **USD thousand 117806**
And as a share of aid to all recipient countries 1.81%

Thailand

RECIPIENT Donor	Agency	Sector	Amount USD thousand	Project description	CRS ID Number
Australia	AusAID	14015	3	THAILAND REPRESENTATION AT DELPHI DIALOG AND WORKSHOP	2006002727
Germany	L G	14050	2	WASTE MANAGEMENT/DISPOSAL	2006012403
Japan	JICA	14020	661	WATER SUPPLY & SANIT. - LARGE SYST.	066194C
Japan	JICA	14020	114	TC IN SEWERAGE	062119C
Japan	JICA	14040	44	RIVER DEVELOPMENT	066195C
Japan	Oth. MIN	14050	13	WASTE MANAGEMENT/DISPOSAL	067141C
Japan	JICA	14050	48	WASTE MANAGEMENT/DISPOSAL	066196C
Korea	KOICA	14010	4	TRANINIG PROGRAM	2006861868

Total aid for water in 2006 for Thailand **USD thousand 889**
And as a share of aid to all recipient countries 0.01%

Timor-Leste

RECIPIENT Donor	Agency	Sector	Amount USD thousand	Project description	CRS ID Number
Australia	AusAID	14010	64	RURAL WATER SUPPLY AND SANITATION	2006001443
Australia	AusAID	14020	1412	COMMUNITY WATER SUPPLY & SANITATION PROG	2006000278
Australia	AusAID	14030	14	COMMUNITY WATER SUPPLY & SANITATION PROG	2006002237
Ireland	DFA	14030	60	BASIC DRINKING WATER AND SANIT	060749
Japan	JICA	14050	11	WASTE MANAGEMENT/DISPOSAL	066239C
Portugal	ICP	14020	42	REHABILITATION OF THE ATAÚRO AQUEDUCT	060459
Spain	MFA	14030	355	SUSTAINABLE IMPROVEMENT IN TIMORENSES´HEALTH THROUGH POTABLE WATER ACCESS AND DEVELOPMENT OF LOCAL C	063979

Total aid for water in 2006 for Timor-Leste **USD thousand 1958**
And as a share of aid to all recipient countries 0.03%

Togo

RECIPIENT Donor	Agency	Sector	Amount USD thousand	Project description	CRS ID Number
Japan	JICA	14020	30	WATER SUPPLY & SANIT. - LARGE SYST.	064409C
Spain	MUNIC	14030	6	WATER FOR THE POPULATION OF NASSIEGOU	069035
Spain	MUNIC	14030	4	"DRILLING LABOUR AND CONSTRUCTION OF A WELL FOR THE POPULATION´S SUPPLY"	067097
Spain	AG	14030	377	IMPROVEMENT IN HEALTH AND SELF-MANAGEMENT CAPACITY OF THE POPULATION	061027
UNICEF		14010	42	WATER RESOURCES POLICY/ADMIN. MGMT	062751
UNICEF		14030	128	COMMUNITY HYGIENE AND WATER SAFETY	062750

Total aid for water in 2006 for Togo **USD thousand 588**
And as a share of aid to all recipient countries 0.01%

Tonga

RECIPIENT Donor	Agency	Sector	Amount USD thousand	Project description	CRS ID Number
Australia	AusAID	14030	29	AUST COMMUNITY ASSISTANCE SCHEME	2006001876

RECIPIENT Donor	Agency	Sector	Amount USD thousand	Project description	CRS ID Number
Australia	AusAID	14030	14	HA'APAI DEVELOPMENT FUND	1989000009
Australia	AusAID	14050	36	MONITORING	2006001748
Japan	JICA	14020	23	WATER SUPPLY & SANIT. - LARGE SYST.	066515C
New Zealand	NZAid	14030	286	"VILLAGE WATER SUPPLIES"	060550

Total aid for water in 2006 for   **Tonga**   **USD thousand 388**
And as a share of aid to all recipient countries   **0.01%**

Tunisia

RECIPIENT Donor	Agency	Sector	Amount USD thousand	Project description	CRS ID Number
Belgium	DGCD	14050	18884	PRET D'ETAT A ETAT - ASSAINISSEMENTET RÉHABILITATION DE LA BAIE DE SFAX	2006003003
France	MINEFI	14050	5422	ASSAINISSEMENT SITE TAPARURA VILLE DE SFAX	2006003004
Germany	BMZ	14010	133	REFORM OF THE WATER SECTOR IN THE MENA REGION	2006003343
Germany	BMZ	14010	2	TRANSNET - TRANSBOUNDARY MANAGEMENT OF NATURAL RESOURCES IN SADC REGION	2006003305
Germany	BMZ	14081	3	WATER EFFICIENT MANAGEMENT OF WASTEWATER, TREATMENT AND REUSE	2006003383
Greece	YPEHO DE	14010	8	HIGHER EFFICIENCY OF WATER-RELATED DEVELOPMENT	2006000307
Italy	DGCS	14020	4	WATER SUPPLY & SANIT. - LARGE SYST.	060147
Japan	JBIC	14020	46495	JENDOUA RURAL WATER SUPPLY PROJECT	063045
Japan	JICA	14040	470	STUDY ON INTEGRATED BASIN MANAGEMENT FOCUSED ON FLOOD CONTROL IN MEJERDA	060403C
Japan	JICA	14040	1645	TC IN WATER RESOURCES DEVELOPMENT	060085C
Japan	JICA	14050	81	WASTE MANAGEMENT/DISPOSAL	063679C
Japan	JICA	14050	93	PROMOTING OF PARTICIPATION OF JAPANESE CITIZEN IN URBAN SANITATION	062687C
Spain	MFA	14030	50	GUARANTEE POTABLE WATER SUPPLY IN THE POULATION THROUGH A DESALINATION WATER PLANT FROM WELLS NOURIS	064186
Switzerland	SDC	14050	56	WASTE MANAGEMENT/DISPOSAL	2003004862

Total aid for water in 2006 for   **Tunisia**   **USD thousand 73346**
And as a share of aid to all recipient countries   **1.13%**

Turkey

RECIPIENT Donor	Agency	Sector	Amount USD thousand	Project description	CRS ID Number
Finland	MFA	14050	24	WASTE WATER TREATMENT TRAINING	050309
Germany	BMZ	14020	24	INTEGRATED EXPERTS	2006002427
Germany	BMZ	14081	139	WATER EFFICIENT MANAGEMENT OF WASTEWATER, TREATMENT AND REUSE	2006003380
Greece	YPEHO DE	14010	8	HIGHER EFFICIENCY OF WATER-RELATED DEVELOPMENT	2006000306
Italy	LA	14020	13	WATER SUPPLY & SANIT. - LARGE SYST.	061295

Total aid for water in 2006 for   **Turkey**   **USD thousand 208**
And as a share of aid to all recipient countries   **0.00%**

Turkmenistan

RECIPIENT Donor	Agency	Sector	Amount USD thousand	Project description	CRS ID Number
UNICEF		14010	98	WATER RESOURCES POLICY/ADMIN. MGMT	062823

CRS AID ACTIVITIES IN SUPPORT OF WATER SUPPLY AND SANITATION – ISBN 978-92-64-05172-0 – © OECD/WWC 2008

RECIPIENT Donor	Agency	Sector	Amount USD thousand	Project description	CRS ID Number

Total aid for water in 2006 for   Turkmenistan   **USD thousand 98**

And as a share of aid to all recipient countries   0.00%

Uganda

RECIPIENT Donor	Agency	Sector	Amount	Project description	CRS ID
AfDF		14020	58841	RURAL WATER SUPPLY & SANITATION PROGRAM	060074
Austria	ADA	14010	180	WATER RESOURCES POLICY/ADMIN. MGMT	2006000166
Austria	Reg	14030	19	BASIC DRINKING WATER SUPPLY AND BASIC SANITATION	2006008195
Austria	ADA	14030	8786	SOUTH WESTERN TOWNS WATER AND SANITATION PROJECT	2006000015
Austria	ADA	14030	38	BASIC DRINKING WATER SUPPLY AND BASIC SANITATION	2003000030
Belgium	DGCD	14020	175	WATER SUPPLY & SANIT. - LARGE SYST.	2002000341
Belgium	MPRF	14030	20	BASIC DRINKING WATER SUPPLY AND BASIC SANITATION	2006003656
Belgium	DGCD	14030	176	GESTION DES EAUX DANS LE BASSIN DU LAC GEORGE	2005001208
Belgium	DGCD	14030	32	BASIC DRINKING WATER SUPPLY AND BASIC SANITATION	2004000351
EC	CEC	14010	4209	ACP EU WATER FACILITY : AFRIC MED&RES FOUND, WATER&SANIT PROG UGANDA	2006100309
EC	CEC	14010	471	ACP EU WATER FACILITY : MOBILIS LOC PRIV SECT FOR URB WATER&SANIT	2006100303
EC	CEC	14020	753	MID WESTERN TOWNS WATER SUPPLY	2006100200
Germany	BMZ	14020	507	WATER SUPPLY AND SANITATION - LARGE SYSTEMS	2006005406
Germany	BMZ	14020	77	INTEGRATED EXPERTS	2006002428
Germany	BMZ	14030	331	WATRE PROJECT KITAGWENDA	2006004899
Germany	BMZ	14030	28	IMPROVEMENT OF THE WATER SUPPLY IN MAKUKUULU	2006004894
Ireland	DFA	14030	139	BASIC DRINKING WATER AND SANIT	060680
Italy	LA	14010	263	IMPROVEMENT IN LIVING STANDARDS THROUGH FOOD SECURITY AND HYGIENE BY WAY OF TRAINING COURSES, SEED DISTRIBUTION AND SOLAR MECHANIZATION OF	061973
Italy	LA	14010	261	INTEGRATED WATER SUPPLY PROJECT IN THE DISTRICTS OF GULU, KITGUM AND PADERN	061959
Japan	JICA	14020	0	WATER SUPPLY & SANIT. - LARGE SYST.	064425C
Japan	JICA	14040	43	RIVER DEVELOPMENT	064426C
Norway	NORAD	14010	273	LAKE ALBERT EASTERN CATCHMENT MANAGEMENT INITIATIVE	2006002965
Norway	NORAD	14020	151	PLASTEC A.S	2006002303
Spain	MUNIC	14030	25	DEVELOPMENT OF THE ACCESS TOSAFE WATER AND BASIC SANITATION IN MARACHA COUNTY	069621
Spain	MUNIC	14030	8	IMPROVEMENT IN ACCESS TO SAFE WATER AND BASIC SANITATION IN THE COUNTY OF MARACHA	066778
Spain	MUNIC	14030	395	IMPROVEMENT OF DRINKING WATER ACCESABILITY AND ENVIRONMENTAL SANITATION BY THE CREATION OF INFRASTRU	066235
Spain	AG	14081	50	CHILDREN AND WATER EDUCATING WITH HEALTH	063191
Spain	AG	14081	194	IMPROVEMENT OF POTABLA WATER ACCESS AND SANITATION AND PROMOTION OF HYGIENIC HEALTHY HABITS IN 20 E	062128
UNICEF		14010	145	WATER RESOURCES POLICY/ADMIN. MGMT	062850
UNICEF		14030	742	HYGIENE, SANITATION AND WATER SUPPLY PROGRAMMES	062849
UNICEF		14030	357	COMMUNITY HYGIENE AND WATER SAFETY	062848

Total aid for water in 2006 for   Uganda   **USD thousand 77690**

And as a share of aid to all recipient countries   1.19%

Ukraine

RECIPIENT Donor	Agency	Sector	Amount USD thousand	Project description	CRS ID Number
Germany	BMZ	14050	38	INTEGRATED EXPERTS	2006002452
Germany	L G	14081	13	WORKSHOP 'INTEGRATED WATER AND RESOURCES MANAGEMENT'	2006012237
Sweden	Sida	14020	27	STUDY TOUR STHLM WATER	2006002542

Total aid for water in 2006 for   **Ukraine**   **USD thousand 77**
And as a share of aid to all recipient countries   **0.00%**

Uruguay

RECIPIENT Donor	Agency	Sector	Amount USD thousand	Project description	CRS ID Number
Canada	IDRC	14010	585	ECOPLATA: SUPPORT TO INTEGRATED COASTAL ZONE MANAGEMENT (URUGUAY)	060847b
IDA		14050	7000	URUGUAY - MONTEVIDEO LANDFILL GAS RECOVERY PROJECT	060309
Japan	JICA	14020	10	WATER SUPPLY & SANIT. - LARGE SYST.	065201C
Japan	JICA	14050	45	WASTE MANAGEMENT/DISPOSAL	065202C
Japan	JICA	14050	154	TC IN URBAN SANITATION	061508C
Japan	JICA	14050	310	THE STUDY ON CAPACITY DEVELOPEMENT FOR WATER QUALITY MANAGEM	060685C

Total aid for water in 2006 for   **Uruguay**   **USD thousand 8104**
And as a share of aid to all recipient countries   **0.12%**

Uzbekistan

RECIPIENT Donor	Agency	Sector	Amount USD thousand	Project description	CRS ID Number
Japan	JICA	14040	125	TC IN WATER RESOURCES DEVELOPMENT	061902C
Switzerland	SDC	14030	48	BASIC DRINKING WATER SUPPLY AND BASIC SANITATION	2003004856
Switzerland	SDC	14030	718	RURAL WATER SUPPLY AND SANITATION	2003001214
UNICEF		14010	2	WATER RESOURCES POLICY/ADMIN. MGMT	062916

Total aid for water in 2006 for   **Uzbekistan**   **USD thousand 893**
And as a share of aid to all recipient countries   **0.01%**

Vanuatu

RECIPIENT Donor	Agency	Sector	Amount USD thousand	Project description	CRS ID Number
Japan	JICA	14020	31	WATER SUPPLY & SANIT. - LARGE SYST.	066372C
Japan	JICA	14050	65	WASTE MANAGEMENT/DISPOSAL	066373C
Japan	JICA	14050	187	IMPROVEMENT OF BOUFFA LANDFILL	061199C
New Zealand	NZAid	14020	302	"RURAL WATER SUPPLY : VANUATU"	060563

Total aid for water in 2006 for   **Vanuatu**   **USD thousand 585**
And as a share of aid to all recipient countries   **0.01%**

Venezuela

RECIPIENT Donor	Agency	Sector	Amount USD thousand	Project description	CRS ID Number
Japan	JICA	14020	9	WATER SUPPLY & SANIT. - LARGE SYST.	065232C
Japan	JICA	14040	83	RIVER DEVELOPMENT	065233C
Spain	MUNIC	14030	3	WATER COOLER	068765
Spain	MFA	14030	126	PROJECT CONSTRUCTION OR IMPROVEMENT OF SMALL POTABLE WATER SUPPLY SYSTEMS IN RURAL AND INDEGENOUS C	065218
Spain	AG	14050	36	NATIONAL PROGRAM TO PROMOTE LOCAL INICIATIVES ON RECOVERY AND URBAN SOLID WASTE RECYCLING	061641

CRS AID ACTIVITIES IN SUPPORT OF WATER SUPPLY AND SANITATION – ISBN 978-92-64-05172-0 – © OECD/WWC 2008

RECIPIENT Donor	Agency	Sector	Amount USD thousand	Project description	CRS ID Number

Total aid for water in 2006 for   **Venezuela**   **USD thousand 256**
And as a share of aid to all recipient countries   **0.00%**

Viet Nam

RECIPIENT	Agency	Sector	Amount	Project description	CRS ID
AsDF		14010	52252	EMERGENCY REHABILITATION OF CALAMITY DAMAGE PROJECT	060055
AsDF		14010	78421	CENTRAL REGION WATER RESOURCES PROJECT	060027
Australia	AusAID	14020	36	DANANG URENCO SANITATION PROJECT	1996008103
Belgium	DGCD	14020	1883	EXTENSION ASSAINISSEMENT CANAUX DE HO CHI MINH VILLE, EXTENSION	2004001108
Belgium	DGCD	14030	651	BASIC DRINKING WATER SUPPLY AND BASIC SANITATION	2005000150
Denmark	MFA	14015	682	COMBATING WATER POLLUTION IN TRI AN	061261
Denmark	MFA	14020	1512	WATER SUPPLY, SANITATION, HYGIENE PROMOTION AND HEALTH.	001067
Denmark	MFA	14030	3163	WSPS II (WATER SUPPLY PROGRAMME) DANIDA ADVISERS	061136
Denmark	MFA	14030	9995	WSPS II (WATER SUPPLY PROGRAMME) UNALLOCATED FUNDS	061135
Denmark	MFA	14030	673	WSPS II RESEARCH	061134
Denmark	MFA	14030	1986	WSPS II (WATER SUPPLY PROGRAMME) MONITORING, REVIEWS, AUDITS	061133
Denmark	MFA	14030	3702	WSPS II (WATER SUPPLY PROGRAMME) TECHNICAL ASSISTANCE	061132
Denmark	MFA	14030	50311	WSPS II (WATER SUPPLY PROGRAMME) BUDGET SUPPORT	061131
Finland	MFA	14020	147	COMMUNAL WATER SYSTEM FEASIBILITY	050312
Finland	MFA	14050	93	WASTE WATER TREATMENT PROJECT IN VIETNAM	066009
Finland	MFA	14050	40	FEASIBILITY IN SEPTIC TREATMENT	050316
Germany	Fed Min	14010	140	INTEGRATED WATER AND RESOURCES MANAGEMENT	2006011288
Germany	GTZ	14020	5648	WASTEWATER MANAGEMENT IN PROVINCIAL URBAN CENTRES	2006006085
Germany	BMZ	14030	164	BASIC DRINKING WATER SUPPLY AND BASIC SANITATION	2006005412
Germany	BMZ	14030	19	INTEGRATED EXPERTS	2006002440
Germany	BMZ	14040	48	DEVELOPING POTENTIALS IN RURAL AREAS OF MEKONG RIPARIAN COUNTRIES	2006003372
Germany	Fed Min	14050	16	WASTE MANAGEMENT/DISPOSAL	2006011029
IDA		14010	8729	COASTAL CITIES ENVIRONMENTAL SANITATION PROJECT	060030d
IDA		14020	93525	COASTAL CITIES ENVIRONMENTAL SANITATION PROJECT	060030a
Italy	DGCS	14010	16	WATER RESOURCES POLICY/ADMIN. MGMT	060417
Italy	Art.	14020	6482	WATER SUPPLY AND DISTRIBUTION IN ME LINH	060016
Italy	Art.	14020	2840	WATER SUPPLY AND DISTRIBUTION IN QUANG NGAI CITY	060013
Italy	Art.	14020	4086	WATER SUPPLY AND DISTRIBUTION IN CA MAU CITY	060011
Japan	JICA	14020	91	WATER SUPPLY & SANIT. - LARGE SYST.	066262C
Japan	JBIC	14020	13376	2ND HOCHIMINH CITY WATER ENVIRONMENT IMPROVEMENT PROJECT	063036
Japan	JBIC	14020	26151	2ND HANOI DRAINAGE PROJECT FOR ENVIRONMENTAL IMPROVEMENT	063035
Japan	JICA	14020	366	GRANT AID CO-OPERATION IN WATER SUPPLY	060557C
Japan	MOFA	14020	301	GROUNDWATER DEVELOPMENT IN CENTRAL HIGHLAND PROVINCES	060189
Japan	JICA	14040	146	RIVER DEVELOPMENT	066263C
Japan	JICA	14050	29	WASTE MANAGEMENT/DISPOSAL	066264C
Korea	KOICA	14010	4	TRANINIG PROGRAM	2006861896
Korea	Kexim	14030	26000	EXPANSION OF THIEN TAN WATER PLANT PROJECT - PHASE II	2006861147

RECIPIENT Donor	Agency	Sector	Amount USD thousand	Project description	CRS ID Number
Korea	Misc	14040	632	DEVELOPMENT SUPPORT FOR HANOI'S HONG RIVER	2006010501
Luxembourg	MFA	14030	144	BASIC DRINKING WATER SUPPLY AND BASIC SANITATION	061171
Netherlands	MFA	14010	28	WATER RESOURCES POLICY/ADMIN. MGMT	2003007881
Netherlands	MFA	14015	8855	HAN_NDRM PROGRAMME	2006000488
Netherlands	MFA	14015	398	WATER RESOURCES PROTECTION	2003013703
Netherlands	MFA	14020	26045	WATER: SECTOR SUPPORT	2006001105
Netherlands	MFA	14081	10	EDUC./TRNG:WATER SUPPLY & SANITATION	2003015915
Norway	NORAD	14020	2	PUBLICATION TENDER DOCUMENTS HOI AN WATER SUPPLY	2006003144
Norway	NORAD	14020	1419	SONG CONG WATER SUPPLY PROJECT	2006002643
Norway	NORAD	14020	39	SONG CONG TENDERING PROCESS CONSULTANCY	2006000387
UNICEF		14010	157	WATER RESOURCES POLICY/ADMIN. MGMT	063060

Total aid for water in 2006 for   **Viet Nam**   **USD thousand 431452**
And as a share of aid to all recipient countries   **6.62%**

West Indies Unallocated

EC	CEC	14010	23215	ACP - EU WATER FACILITY	2006100235

Total aid for water in 2006 for   **West Indies Unallocated**   **USD thousand 23215**
And as a share of aid to all recipient countries   **0.36%**

Yemen

Germany	BMZ	14010	1089	REFORM OF THE WATER SECTOR IN THE MENA REGION	2006003353
Germany	BMZ	14010	86	INTEGRATED EXPERTS	2006002417
Germany	KFW	14020	40166	PROVINCIAL TOWNS PROGRAMME II	2006000644
Germany	KFW	14020	4393	WATER SUPPLY AND SANITATION IN SADAH	2006000638
Germany	KFW	14020	2510	SEWERAGE BAJIL AND BAIT AL-FAQIH	2006000635
Germany	KFW	14020	1004	SANITATION ZABID	2006000632
Germany	L G	14030	25	OVERHAULING AND REBUILDING OF CISTERNS FOR RAIN WATER, JANDALA DISTRICT	2006012185
Germany	BMZ	14030	268	BASIC DRINKING WATER SUPPLY AND BASIC SANITATION	2006005409
Germany	BMZ	14050	81	INTEGRATED EXPERTS	2006002453
Germany	BMZ	14081	3	WATER EFFICIENT MANAGEMENT OF WASTEWATER, TREATMENT AND REUSE	2006003388
Japan	JICA	14020	72	WATER SUPPLY & SANIT. - LARGE SYST.	065457C
Japan	JICA	14040	7	RIVER DEVELOPMENT	065458C
Japan	JICA	14040	318	TC IN WATER RESOURCES DEVELOPMENT	060659C
Japan	JICA	14040	571	STUDY FOR WATER RESOURCES MANAGEMENT AND RURAL WATER SUPPLY IMPROVEMENT	060317C
Netherlands	MFA	14020	1192	WATER SUPPLY & SANIT. - LARGE SYST.	2006000104
Netherlands	MFA	14030	17949	BASIC DRINKING WATER SUPPLY AND BASIC SANITATION	2006000720
UNICEF		14030	835	HYGIENE, SANITATION AND WATER SUPPLY PROGRAMMES	063103

CRS AID ACTIVITIES IN SUPPORT OF WATER SUPPLY AND SANITATION – ISBN 978-92-64-05172-0 – © OECD/WWC 2008

RECIPIENT Donor	Agency	Sector	Amount USD thousand	Project description	CRS ID Number

Total aid for water in 2006 for Yemen **USD thousand 70570**
And as a share of aid to all recipient countries 1.08%

Zambia

RECIPIENT Donor	Agency	Sector	Amount USD thousand	Project description	CRS ID Number
EC	CEC	14010	447	ACP EU WATER FACILITY:INTEGRATED COMMUNITY WATER&SANIT FOR URBAN POOR	2006100320
Germany	BMZ	14010	628	WATER RESOURCES POLICY/ADMIN. MGMT	2006006038
Germany	GTZ	14010	5120	WATER RESOURCES POLICY/ADMIN. MGMT	2006006018
Germany	BMZ	14010	17	TRANSNET - TRANSBOUNDARY MANAGEMENT OF NATURAL RESOURCES IN SADC REGION	2006003326
Germany	BMZ	14020	207	WATER SUPPLY AND SANITATION - LARGE SYSTEMS	2006005404
Germany	BMZ	14020	190	INTEGRATED EXPERTS	2006002429
Germany	BMZ	14030	260	BASIC DRINKING WATER SUPPLY AND BASIC SANITATION	2006005411
Germany	KFW	14030	1757	WATER SUPPLY AND SANITATION	2006001378
Germany	KFW	14030	12050	RURAL WATER SUPPLY	2006001377
IDA		14010	17250	WATER SECTOR PERFORMANCE IMPROVEMENT PROJECT	060009a
IDA		14020	4830	WATER SECTOR PERFORMANCE IMPROVEMENT PROJECT	060009b
Ireland	DFA	14010	4448	WATER POLICY /ADMINISTRATION	060027
Ireland	DFA	14030	4	BASIC DRINKING WATER AND SANIT	061380
Japan	JICA	14020	146	WATER SUPPLY & SANIT. - LARGE SYST.	064483C
Japan	JICA	14020	154	GRANT AID CO-OPERATION IN WATER SUPPLY	061513C
Japan	JICA	14020	244	THE PROJECT FOR GROUNDWATER DEVELOPMENT IN LUAPULA PROVINCE	060870C
Japan	JICA	14040	171	RIVER DEVELOPMENT	064484C
UNICEF		14030	783	HYGIENE, SANITATION AND WATER SUPPLY PROGRAMMES	063145

Total aid for water in 2006 for Zambia **USD thousand 48704**
And as a share of aid to all recipient countries 0.75%

Zimbabwe

RECIPIENT Donor	Agency	Sector	Amount USD thousand	Project description	CRS ID Number
EC	CEC	14010	1799	ACP EU WATER FACILITY	2006100324
Germany	BMZ	14010	15	TRANSNET - TRANSBOUNDARY MANAGEMENT OF NATURAL RESOURCES IN SADC REGION	2006003328
Italy	LA	14020	68	WATER SUPPLY & SANIT. - LARGE SYST.	061669
Japan	JICA	14020	65	WATER SUPPLY & SANIT. - LARGE SYST.	064214C
New Zealand	NZAid	14030	40	"COMMUNITY RAINWATER HARVESTING"	060231
Switzerland	SDC	14030	4	BASIC DRINKING WATER SUPPLY AND BASIC SANITATION	1980000949
UNICEF		14010	53	WATER RESOURCES POLICY/ADMIN. MGMT	063174

Total aid for water in 2006 for Zimbabwe **USD thousand 2044**
And as a share of aid to all recipient countries 0.03%

REFERENCE LISTS

LIST 1. Aid agencies

DAC MEMBERS

	AGENCIES	Acronym
AUSTRALIA	Australian Agency for International Development	AusAid
AUSTRIA	Austrian Development Agency	ADA
	Federal Ministry of Finance	BMF
	Provincial governments, local communities	Reg
BELGIUM	Official Federal Service of Finance	SPFF
	Directorate General for Co-operation Development	DGCD
	Flanders Official Regional Ministries	MPRF
	Walloon Official Regional Ministries	MPRW
CANADA	Canadian International Development Agency	CIDA
	International Development Research Centre	IDRC
DENMARK	Danish International Development Agency	DANIDA
	Ministry of Foreign Affairs	MFA
EUROPEAN COMMUNITIES	Commission of the European Communities	CEC
	European development Fund	EDF
	European Investment Bank	EIB
FINLAND	Ministry of Foreign Affairs	MFA
FRANCE	Ministry of Foreign Affairs	MAE
	French Development Agency	AFD
	Ministry of Economy, Finance and Industry	MINEFI
GERMANY	Bundesministerium für Wirtschaftliche Zusammenarbeit und Entwicklung	BMZ
	Deutsche Gesellschaft für Technische Zusammenarbeit	GTZ
	Kreditanstalt für Wiederaufbau	KFW
	Federal States & Local Governments	L.G.
	Federal Ministries	Fed. Min.
GREECE	Ministry of Foreign Affairs	YPEJ
	Miscellaneous	Alloi
	Ministry of the Environment, Land Planning & Public Works	YPEHODE
	Ministry of the Interior, Public Administration and Decentralisation	YPESDDA

CRS AID ACTIVITIES IN SUPPORT OF WATER SUPPLY AND SANITATION – ISBN 978-92-64-05172-0 – © OECD/WWC 2008

IRELAND	Department of Foreign Affairs	DFA
ITALY	Artigiancassa	Art.
	Direzione Generale per la Cooperazione allo Sviluppo	DGCS
	Local administration	LA
JAPAN	Ministry of Foreign Affairs	MOFA
	Other Ministries	Oth. MIN
	Prefectures	PRF
LUXEMBOURG	Ministry of Foreign Affairs	MFA
NETHERLANDS	Ministry of Foreign Affairs (DGIS)	MFA
NEW ZEALAND	New Zealand Agency for International Development	NZAid
NORWAY	Norwegian Agency for Development Co-operation	NORAD
	Ministry of Foreign Affairs	MFA
PORTUGAL	Portuguese Co-operation Institute	ICP
	Portuguese Government	GP
SPAIN	Autonomous Governments	AG
	Ministry of Foreign Affairs	MFA
	Ministry of Education, Culture and Sports	EDUC
	Municipalities	MUNIC
	Instituto de Crédito Oficial	ICO
	Ministry of Environment	ENV
SWEDEN	Swedish International Development Authority	Sida
SWITZERLAND	Swiss Agency for Development and Co-operation	SDC
	State Secretariat for Economic Affairs	Seco
	Swiss Agency for the Environment, Forests and Landscape	SAEFL
UNITED KINGDOM	Department for International Development	DFID
UNITED STATES	African Development Foundation	ADF
	Agency for International Development	AID
	Departement of Defense	DOD
	Department of Agriculture	AGR
	Trade and Development Agency	TDA
	Overseas Private Investments Corporation	OPIC
	State Department	STATE

MULTILATERAL ORGANISATIONS

Name	Acronym
United Nations Programmes and Funds	
United Nations Children's Fund	UNICEF
World Bank group	
International Development Association	IDA
Regional banks	
African Development Fund	AfDF
Asian Development Bank.Special Fund	AsDF

CRS AID ACTIVITIES IN SUPPORT OF WATER SUPPLY AND SANITATION – ISBN 978-92-64-05172-0 – © OECD/WWC 2008

LIST 2. DAC List of ODA Recipients

Effective from 2006 for reporting on flows in 2005, 2006 and 2007

Least Developed Countries	Other Low Income Countries (per capita GNI < $825 in 2004)	Lower Middle Income Countries and Territories (per capita GNI $826-$3 255 in 2004)	Upper Middle Income Countries and Territories (per capita GNI $3 256-$10 065 in 2004)
Afghanistan	Cameroon	Albania	* Anguilla
Angola	Congo, Rep.	Algeria	Antigua and Barbuda
Bangladesh	Côte d'Ivoire	Armenia	Argentina
Benin	Ghana	Azerbaijan	Barbados
Bhutan	India	Belarus	Belize
Burkina Faso	Kenya	Bolivia	Botswana
Burundi	Korea, Dem.Rep.	Bosnia and Herzegovina	Chile
Cambodia	Kyrgyz Rep.	Brazil	Cook Islands
Cape Verde	Moldova	China	Costa Rica
Central African Rep.	Mongolia	Colombia	Croatia
Chad	Nicaragua	Cuba	Dominica
Comoros	Nigeria	Dominican Republic	Gabon
Congo, Dem. Rep.	Pakistan	Ecuador	Grenada
Djibouti	Papua New Guinea	Egypt	Lebanon
Equatorial Guinea	Tajikistan	El Salvador	Libya
Eritrea	Uzbekistan	Fiji	Malaysia
Ethiopia	Viet Nam	Georgia	Mauritius
Gambia	Zimbabwe	Guatemala	* Mayotte
Guinea		Guyana	Mexico
Guinea-Bissau		Honduras	* Montserrat
Haiti		Indonesia	Nauru
Kiribati		Iran	Oman
Laos		Iraq	Palau
Lesotho		Jamaica	Panama
Liberia		Jordan	Saudi Arabia (1)
Madagascar		Kazakhstan	Seychelles
Malawi		Macedonia, Former Yugoslav	South Africa
Maldives		Republic of	* St. Helena
Mali		Marshall Islands	St. Kitts-Nevis
Mauritania		Micronesia,Fed. States	St. Lucia
Mozambique		Morocco	St. Vincent & Grenadines
Myanmar		Namibia	Trinidad & Tobago
Nepal		Niue	Turkey
Niger		Palestinian Adm. Areas	* Turks & Caicos Islands
Rwanda		Paraguay	Uruguay
Samoa		Peru	Venezuela
Sao Tome & Principe		Philippines	
Senegal		Serbia & Montenegro	
Sierra Leone		Sri Lanka	
Solomon Islands		Suriname	
Somalia		Swaziland	
Sudan		Syria	
Tanzania		Thailand	
Timor-Leste		* Tokelau	
Togo		Tonga	
Tuvalu		Tunisia	
Uganda		Turkmenistan	
Vanuatu		Ukraine	
Yemen		* Wallis & Futuna	
Zambia			

* Territory.

(1) Saudi Arabia passed the high income country threshold in 2004. In accordance with the DAC rules for revision of this List, it will graduate from the List in 2008 if it remains a high income country in 2005 and 2006.

As of November 2006, the **Heavily Indebted Poor Countries (HIPCs)** are : Benin, Bolivia, Burkina Faso, Burundi, Cameroon, Central African Republic, Chad, Comoros, Congo (Dem. Rep.), Congo (Rep.), Côte d'Ivoire, Eritrea, Ethiopia, Gambia, Ghana, Guinea, Guinea-Bissau, Guyana, Haiti, Honduras, Kyrgyz Republic, Liberia, Madagascar, Malawi, Mali, Mauritania, Mozambique, Nepal, Nicaragua, Niger, Rwanda, São Tomé and Príncipe, Senegal, Sierra Leone, Somalia, Sudan, Tanzania, Togo, Uganda and Zambia.

Acronyms

CRS	Creditor Reporting System
DAC	Development Assistance Committee
LDC	Least Development Country
LIC	Low Income Country
Macedonia (TFYR)	The Former Yugoslav Republic of Macedonia
MDG	Millennium Development Goal
ODA	Official Development Assistance
WSS	Water Supply and Sanitation

JOHN CLEARE'S
FIFTY BEST
HILL WALKS
OF BRITAIN

JOHN CLEARE'S
FIFTY BEST
HILL WALKS
OF BRITAIN

Webb & Bower

MICHAEL JOSEPH

For my ever-patient Joey
who always had the kettle on.

Half-title
Sunset over Llyn Fan Fawr, the lonely tarn that lies below the
steep east scarp of Bannau Brycheiniog. The Carmarthen
Fan is in the distance.

Frontispiece
Evening on Cherhill Down above the Vale of Calne, Wiltshire.

First published in Great Britain 1988 by
Webb & Bower (Publishers) Limited
9 Colleton Crescent, Exeter, Devon EX2 4BY
in association with Michael Joseph Limited
27 Wright's Lane, London W8 5TZ

Designed by Vic Giolitto

Production by Nick Facer/Rob Kendrew

Text and illustrations Copyright © 1988 John Cleare
Maps Copyright © 1988 Ordnance Survey

British Library Cataloguing in Publication Data
Cleare, John
John Cleare's fifty best hill walks of
Britain
    1, Mountaineering—Great Britain—
Guide-books   2. Great Britain—Description
and travel—1971– —Guide-books
I. Title
796.5'22   DA632

ISBN 0–86350–142–7

Typeset in Great Britain by Keyspools Limited, Golborne,
Lancashire

Colour reproduction by J Film Process, Bangkok, Thailand

Printed and bound in Spain by Graficromo, Cordoba

Contents

Introduction

'The true mountaineer is a wanderer . . .'

Albert Mummery 1893

Hill-walking – or fell-walking as it is often known in Britain – is the quintessential craft from which all other mountain sports are born. Without the basic ability to move safely over upland country there would be little rock-climbing, no fell-running or ski-touring and certainly no mountaineering. Climbing the snow and ice of a Himalayan giant is only an extreme extension of basic hill-walking techniques, an extreme to which relatively few aspire, while for thousands of folk hill-walking is a meaningful and enjoyable end in itself.

I was fortunate enough to become involved with mountains at a tender age and luckily the schoolmaster who led our climbing holidays – a real mountain man – ensured that though our days might start on steep rock, they would continue upwards to a summit and conclude with a circuit over the tops whatever the weather. Although rock was my early interest, I soon learnt to appreciate and then to cherish the walking. It was all part of the overall mountain game, and long days on the British hills have stood me in good stead among the mountains of six continents. For me too, hill-walking has become an end in itself.

Thus I make no apology that I, as an alpine and expeditionary mountaineer, should write a book on British hills, a book which includes such global trivia as the Wiltshire Downs between the same cover as Ben Nevis. Frank Smythe, the famous thirties mountaineer, considered the moody Surrey eminence of Holmbury Hill almost as much a mountain as many of his Himalayan summits. He knew that sheer height is no arbiter of what is a mountain.

In the British uplands the dividing line between hills and mountains may well depend on the weather or even the walker. Personally I enjoy my Wiltshire Downs even as I enjoy my Himalayan peaks, but I believe that reward is commensurate with effort, accomplishment with challenge. The walker who puts his all into an ascent of Bidean has achieved no less – in context – than the experienced climber on

Everest's summit. We play the same game, which is surely about delight rather than glory.

What then are the parameters of a *best* hill-walk? I have my own ideas, but the coverage of this book demanded routes to intrigue the experienced northern fell-walker with a week's holiday besides the aspiring chap from suburbia with a single day. I'm grateful to thirty friends – from Aberdeen to Kent – who gave me their opinions: a surprising concensus with some interesting gems. My initial list of one hundred walks included several *musts* if it was to be credible, the common denominator was ambience – the mood of the high places whatever the scale – while I ignored obviously 'coastal' walks. In eighteen months I walked, or repeated, them all. Practical considerations finally limited the present volume to just fifty routes, spread evenly over the British uplands.

Perhaps some words of explanation are pertinent. Each itinerary is a circuit from a reasonable parking place. The effort should go into the walking, not planning an expedition and it seemed pointless to recommend one-way routes requiring complex car shuttles or a non-walking partner. Every route in this book is a shorter or longer 'day's walk' save one – my own favourite in South Uist to which I was first introduced by the late great Tom Patey – and that actually requires a sleeping bag.

The main text, I hope, will enthuse, inform and inspire while the route itineraries are designed to be copied, as necessary, for use on the hill in conjunction with a compass and the relevant OS 1:50,000 or 1:25,000 map. For each route I have given only distances and height gained but never times. Time is subjective, folk walk at different speeds and I, for one, dawdle to take photographs, but all hill-walkers should be familiar with Naismith's Rule and should know how their own performance compares: they can thus estimate **their own** time. WW Naismith was a prominent Scottish mountaineer of the 1880s and 90s whose habit it was to walk, in a fairly direct line, from his home near Glasgow to the various Highland meets of the SMC, from this experience he concluded that his average speed was 3 mph with an extra half hour for every thousand feet he ascended, a formula that works surprisingly well. In modern metric terms read the Rule as: **5 kms/hour on the map plus 300 metres/$\frac{1}{2}$ hour on the ascent**. Allow a little longer for halts, bad weather, strong winds or really rugged ground.

Itinerary distances are from the map and are given to the nearest convenient half kilometre or quarter mile. For yards, of course, read metres. The meaning of the many abbreviations should be obvious: L for left, N for north – for instance – YDNP for Yorkshire Dales National Park, NCC for Nature Conservancy Council, RoW for Right of Way, and so on. The *true* bank of a stream is always that when facing downstream. Main routes on the maps are marked with a continuous line while a broken line indicates useful variations.

Summits, tops and significant eminences are named in capitals; an asterisk ★ against a Scottish height denotes a separate mountain in Munro's Tables. For those unfamiliar with the term, a Munro is a Scottish summit about 3,000 feet, first classified and listed by Sir Hugh Munro in 1891. The latest revision includes no less than 281 separate mountains plus a further 287 tops or significant lesser summits, the difference between the two being quite a complex matter! Today Munro bagging is almost a sport in itself and I would direct interested readers to both *The High Mountains* and *Munro's Tables* listed in the bibliography on page 206. *Bridge's Tables* is a similar summit catalogue for England and Wales listing also the County Tops.

Which leaves me only to thank, firstly those patient companions with whom I have shared these walks, who appear in my pictures and whose company I have greatly enjoyed: Caroline Aisher, Irvine Butterfield, John Chapman, Joss Cleare, Barry Cliff, Ronnie Faux, John and James Fowler, Carl Gilham, Emma van Gruisen, Ian Howell, Hywel and Caerwyn Lloyd, Joe Mitchell, Johnny Noble, Bill O'Connor, Alastair Stevenson and various folk from Glenmore Lodge. Secondly, for their understanding help with maps, Jim Page and his team at Ordnance Survey. And last but not least the colleagues whose advice, input and enthusiasm have kept the words flowing: Delian Bower, by publisher, Vic Giolitto, my designer, and my editor Alyson Gregory.

'Of Paradise can I not speak properly, for I have not been there,' wrote John Mandeville in 1360. But our British hills *are* accessible and the best of them are as beautiful as any in this world. I wish you as much delight walking among them as I have had.

Mid Craig Hill looks down on Loch Skeen, source of the Grey Mare's Tail.

The West Dartmoor Tors Devon

1″ Tourist Map, No. 1 'Dartmoor' (OS 1 : 50,000 Sheet 191)

Dartmoor is known for its ponies, its prison, the Hound of the Baskervilles and Widecombe Fair. Admittedly it is a moody place but as the largest and wildest tract of open country in southern Britain it allows a freedom to the walker unknown south of the Pennines. In 1951 the whole of Dartmoor, some 190 square miles of moorland, 128 square miles of farmland and 33 square miles of forest, was designated a National Park and the Dartmoor Commons Act of 1985 legalized free public access to almost all the wild country.

The moor itself is a great boss of granite from which time has stripped the overlying crust of softer Carboniferous rock. It has been eroded by rivers and frost – but was never glaciated. Today it forms two high, rolling plateaux mostly covered with a thick layer of peat, carpeted with heather, purple moor grass, bilberry and bent, studded with characteristic craggy tors and cut by fast-flowing streams in typically deep and narrow valleys.

Much of the larger northern plateau lies above 500 metres (1,500 feet) with the most dramatic heights rising in the north-west where two summits actually exceed 2,000 feet (620 metres) – High Willhays is the loftiest point in England south of Kinder Scout. Military use of Dartmoor dates back to 1870 and unfortunately the MOD still controls most of this area, although there is public access when training is not in progress (see the accompanying notes). Because of its height and the way the high ground drops abruptly to the rolling green chequer-board landscape of central Devon, this north-western sector of the moor is a favourite with many discerning walkers. This itinerary provides a good intro-

The River Tavy tumbles off the high moor through Tavy Cleave, an impressive defile originally carved by the torrential run-off of Ice Age thaws.

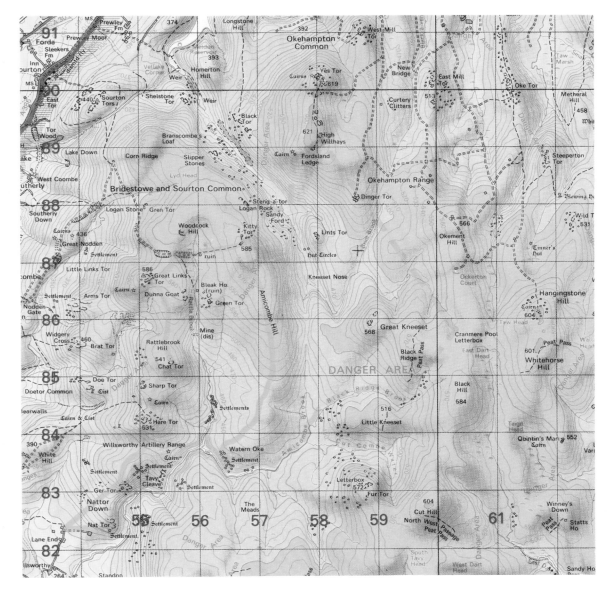

duction to the area while the longer variation visits its most interesting points.

Initially both routes ascend the picturesque defile of Tavy Cleave, where the rushing Tavy tumbles off the high moor in a series of rapids and small cascades between craggy granite outcrops. Such valleys – gorges almost – around the periphery of Dartmoor were originally carved by the torrential run-offs of Ice Age thaws and now contain rivers boisterous with an annual rainfall of around 260 centimetres (100 inches) on the high ground. Such rivers may not always be easy to cross. The path leads over green meads

and through scatters of lichen-covered boulders, so called 'clinter', prised from the jagged tors ranged along the skyline by the shattering action of Ice Age frost.

On a grassy terrace above the river lie the hut circles of Watern Oke. There are a couple of dozen of them – rings of rough stones mostly 3–5 metres across (10–15 feet) and rising about 1 metre from the ground. They date back some 3,500 years to a warmer and dryer climate than today and may mark the site of a summer pasture or sheiling. The huts would have been conical with a central post supporting a thatched

roof rising from the low granite walls. It is a thought-provoking place.

Northward a vague ridge of tussocky grass and stunted heather leads towards Kitty Tor. Alternatively, the river can be followed eastward and a careful route plotted across a boggy green desert to Cranmere Pool – reputed to be the middle of Dartmoor and its least accessible point. A century ago Cranmere Pool was a real pond but now it is merely a boggy hollow containing several puddles – and a Dartmoor letterbox. In 1854 James Perrot, a Chagford guide, started a custom whereby visitors to the

Pool would leave postcards in a jar which the next visitors would recover and post. Over the years the tradition grew and now a series of usually inconspicuous letterboxes are situated at remote points all over Dartmoor; they contain a visitor's book and a rubber stamp and enthusiasts set out to locate and visit them all. The modern letterbox is sited on the west 'shore'.

In good weather a bee-line can be made to High Willhays closeby the distinctive and sharper silhouette of Yes Tor; once the strange, pinnacled cairn is reached the most difficult navigation and worst going is over. Yes Tor, the

This is the most impressive of the several granite towers that comprise Great Links Tor.

Two Circuits on West Dartmoor

Length: short circuit 15.5 kms/9¾ miles
long circuit 24 kms/15 miles
Height gained: short circuit 430 m/1,410 ft
long circuit 722 m/2,370 ft
Difficulty: this is wild, rough and boggy moorland susceptible to quite fierce conditions. Competence with map and compass necessary even in clear weather. **Grid** bearings are given.
Warning: although usually accessible, this entire corner of moor is a military training area where live ammunition sometimes used. Red flags fly when training in progress – there is one near start. Check times with local post office, police, or phone Okehampton (0837) 2939

Start: Willsworthy Lane End, muddy parking area off farm track at 537823. (300 m/985 ft).

Short Circuit
(1) Track, then good path, leads E past barn to follow leat contouring into narrowing Tavy Cleave gorge. From weir rough path follows true R bank of river and gradually disappears. At major confluence cross L fork 100 m to N, continue 700 m along R fork to bend, then strike up steep bank to area of hut circles at 565834. (580 m/1,900 ft).
4 kms/2½ miles

(2) Follow high ground NW then N to flag post at summit of KITTY TORR (585 m/1,919 ft) and descend W to old workings before easy but boggy ascent WSW leads to summit and trig point GREAT LINKS TOR (586 m/1,923 ft).
6 kms/3¾ miles

(3) Descend S along high ground, crossing CHAT TOR (541 m/1,775 ft) to flag post among rocks HARE TOR (531 m/1,742 ft). Faint pony track leads SSW to flag post and rocks on GER TOR (448 m/1,470 ft). Descend SW, bridge crosses wide leat on direct line to car-park.
5.5 kms/3½ miles

Long Circuit
(1) As (1) above to river bank below hut circles, continue E along Amicombe Brook and Black Ridge Brook to final confluence at 593845. Strike NW via stream source over featureless ground to boggy hollow Cranmere Pool or follow compass bg 36° **grid** to reach Pool more surely. Letterbox on W bank. (560 m/1,837 ft).
9.5 kms/6 miles

(2) Cross undulating, featureless, often boggy moorland NW aiming at High Willhays visible just left of conspicuous Yes Tor. Or follow compass bg 326° **grid** to scattered rocks and tall narrow cairn on HIGH WILLHAYS summit. (621 m/2,039 ft).
4.5 kms/3 miles

(3) Descend SW into narrow steep-sided valley of West Okement river and ascend, steep at first, to summit flag pole KITTY TOR (585 m/1,919 ft). Continue as (2) above to rocky summit GREAT LINKS TOR (586 m/1,923 ft). Return to car-park as (3) above.
9.5 kms/6 miles

second 2,000-foot summit, is worth a short detour for its craggy top and wide views. The West Okement river must be crossed at Sandy Ford to reach Kitty Tor with its conspicuous flag post, while old tin workings are passed *en route* to Great Links Tor, the next summit. Great Links is actually several separate tors, each exhibiting interesting weathering patterns: another letter-box is located under the north-east corner. Onwards the wide ridge is easier ground though there are three more tors to ascend before the final rocks of Ger Tor above the mouth of Tavy Cleave are reached and the circuit completed.

Cheddar Gorge and the Mendip   Somerset

OS 1:50,000 Sheet 182

'. . . a stupendous chasm, quite through the body of the adjacent mountains. It appears as if the hill has been split asunder by some dreadful convulsion of nature.'

Thus was Cheddar Gorge described in 1784 by the New British Traveller, and indeed 200 years later it is still considered the most spectacular physical feature in southern Britain. This route makes a circuit of the gorge itself and continues over the Mendip Hills to the summit of Beacon Batch, their highest point.

An anticlinal outcroping of Mountain – or Carboniferous – limestone, the Mendip is a rather bleak and featureless plateau of drystone-walled pastures pitted with swallow-holes, scattered with farms and criss-crossed by narrow lanes and few roads. In several places the fold has eroded away to expose the more resistant old red sandstone beneath, forming several hills, such as

Blackdown, which rise above the general elevation of around 250 metres (800 feet). This is karst country where the water flows underground and where caves and pot-holes proliferate, attracting the 'spelunkers' who explore them. Along the plateau edge are several deep and narrow defiles originally cut by running water but left dry when the land rose and the water found a course underground: though less spectacular than the Cheddar Gorge, Burrington Coombe and Ebbor Gorge are such formations worth visiting.

Our route, shaped like a figure of eight, starts in the middle – the confluence of the several dry valleys which converge to form Cheddar Gorge

The route passes above these crags on the southern rim of the gorge known as 'the Pinnacles' which fall some 400 feet (over 120 metres) to the road at the Horseshoe Bend. Cheddar village and Brent Knoll on the Somerset Levels are seen in the distance.

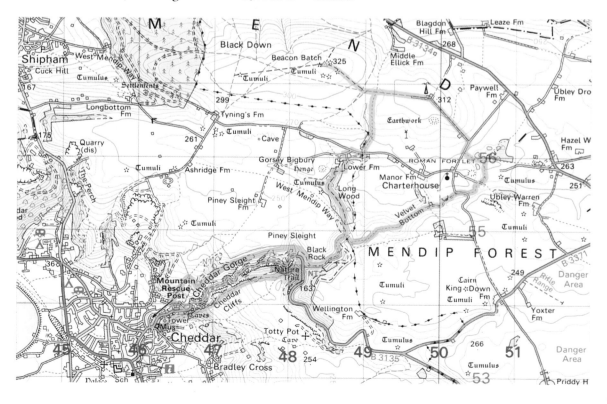

itself – allowing the walker to complete either circuit or both, depending on one's energy. The initial section climbs to open windswept downland on the southern side of the gorge where grey rocks protrude through the long grass and from where there are superb views southward over the broad Somerset Levels to Glastonbury Tor – the legendary Isle of Avalon – and the Quantock Hills beyond. The path runs close to the cliff edge, sometimes abrupt but often the highest of several tiers, the ledges bushy with ash and dogwood and the dark green of stunted yews. Several great gullies cut back into the cliff top and there are glimpses of the twisting road far beneath, but due to the strike of the limestone beds this is the steeper side of the Gorge, the northern side is less spectacular and little can be seen of the great cliffs below.

After descending into Cheddar village, in season a thriving tourist trap, the route ascends to National Trust territory on the northern side of the gorge from where the full majesty of the southern cliffs can be appreciated. The great wall of High Rock – a slightly overhanging 430 feet – is the tallest limestone cliff in Britain; Coronation Street, the now classic rock-climb pioneered by Chris Bonington, Tony Greenbank and the present author in 1965, runs straight up the middle. There are now many harder and more intimidating climbs hereabouts but most rise sheer above the road or various car-parks and climbing is banned between Easter and October. There are few easy routes however, the rock is loose and friable, and all Cheddar climbs are serious undertakings.

Apparently the ivy which today mantles much of the cliffs first arrived around the turn of the century when first the road was built. Where today cars and coaches grind round the tight bends and pedestrians take their lives in their hands, a narrow footpath once picked its way between the boulders beneath pristine pinnacles of gleaming limestone. A special feature of the limestone is its prolific and characteristic flora, best seen in early summer on this south-facing flank. This is the home not only of the rare Cheddar Pink but also of numerous adders – summer walkers should tread with care. The wild flowers are of course protected by law.

Returning to the start at Black Rock Gate, the route now follows the ancient drove road up the winding and attractive Black Rock valley. This is a Nature Reserve and interesting descriptive leaflets are available at a map board 200 metres up the track where a donation should be made. Access is permitted through the adjoining Long Wood Reserve where the entrances are passed to the important Longwood Swallet/August Hole cave system and the Rhino Rift pot-hole, and then a field path leads up onto the swelling heathland of Blackdown and through heather, gorse and, in summer, clumps of rippling pink willow-herb, to Beacon Batch. The wide panorama from the summit stretches from the Cotswolds to Exmoor and across the Bristol Channel to Flat Holm and Cardiff.

The return leg to Black Rock Gate follows the interesting Velvet Bottom dry valley, another Nature Reserve notable not only for its unusual ecology but also for the remains of the lead mines which were worked here from pre-Roman times until 1885 and are now scheduled as an Ancient Monument. This certainly is a fascinating and very varied walk.

This overgrown path on the slopes of Beacon Batch or Blackdown Hill – the highest point of the Mendip – leads over from Charterhouse to Burrington Coombe.

A Mendip Circuit

Length and height gained: Gorge circuit: 6 kms/3¾ miles, 300 m/1,000 ft
Blackdown circuit: +10.5 kms/6½ miles, +200 m/650 ft
Difficulty: easy walking but complete circuit involves considerable ascent
Warning: route often passes close to cliff tops where rock is loose and grass is slippery. On paths stones too can be very slippery when damp

Start: limited parking at Black Rock Gate layby 482545. (160 m/525 ft). Alternatively park in Gorge below High Rock or elsewhere.

(1) Good path waymarked 'Draycott W Mendip Way' ascends through woods from W side of road opposite layby. At junction above woods leave W Mendip Way taking RH fork to open grassland plateau (253 m/830 ft) whence path descends gradually close above cliff top to Jacob's Ladder tower. RoW now bears left through undergrowth to reach cottages, follow steep lane down into Cheddar village (15 m/50 ft).
2.25 kms/1½ miles

(2) Go 500 m up main road to top end of small lake and take RH of two lanes leading off left. After 200 m just before first cottage on RHS a marked RoW leads up to larger path ascending through wood, cross stile and take RH path to more open ground and wall near bush-lined cliff edge at 468543. Various paths now

lead eastward along gorge lip and converge to become narrow main trail signed 'Black Rock' which is followed along lip, via several stiles and a steep re-entrant, eventually to Black Rock Drove (lane) 200 m E of car-park.
3.75 kms/2¼ miles

(3) Follow drove road E through Black Rock Nature Reserve up dry valley past old quarry to stile and waymark 'W Mendip Way', 200 m onward stile by white gate is signed 'Longwood Nature Reserve', enter and continue through thick woods up narrow valley. Plank over stream just past August Hole cave entrance, then wooden steps lead up right to field edge and lane, tarmac drive leads R to road. Cross stile 100 m E opposite white bungalow onto RoW leading up little valley. At green lane on crest go W through gate onto open heathland of Blackdown. Ascend broad track to summit trig point of BEACON BATCH (325 m/1,067 ft).
4.25 kms/2½ miles

(4) Retrace route to green lane and continue past radio masts down metalled drive to road. 150 m S iron stile on LHS leads into Blackmoor-Charterhouse Nature Reserve. Cross 'dam' and follow track S to car-park before turning off left to reach road and crossing into Velvet Bottom Nature Reserve. Good track, then path, leads down dry valley which after 2 kms rejoins Black Rock Drove.
6.25 kms/4 miles

Wiltshire Downs – Cherhill and Oliver's Castle

OS 1:50,000 Sheet 173

'Wiltshire is a pleasant county, and of great variety. I have heard a wise man say that an ox left to himself would, of all England, choose to live in the north, a sheep in the south part hereoff, and a man in the middle betwixt both . . .'

Thomas Fuller,
History of the Worthies of England – 1675

Wiltshire has changed much in 300 years. Though still a pleasant and varied county, the oxen have gone and the great open chalk downs where once roamed myriad sheep have mostly surrendered to the plough – or the army. Fragments of these fragrant grasslands remain however, typically prolific with flowers in spring and summer, often where the scarp is too steep to cultivate. The questing walker can usually guarantee an interesting route by searching on maps for paths along the steepest scarps and among the most convoluted contours.

One will not have to look far for what the OS label as 'Antiquities', for Wiltshire has the greatest concentration in Britain. In prehistoric times these downs were the main inhabited region of our island and the scarp edges are scattered with the mounds and ditches of ancient hill forts, usually occupying – by no accident – the most spectacular sites. This particular route leads along the final westerly bastion of the chalklands of southern England, a scarp as convoluted as any and – though passing no famous monument or celebrated ruin – spanning some 5,000 years of Wiltshire history. It is perhaps epitomized by the delicate hang-gliders which on a summer's day soar out from the Iron

The great scarp of Oliver's Castle is seen to the south-east from the path that runs round below the hill from Heddington.

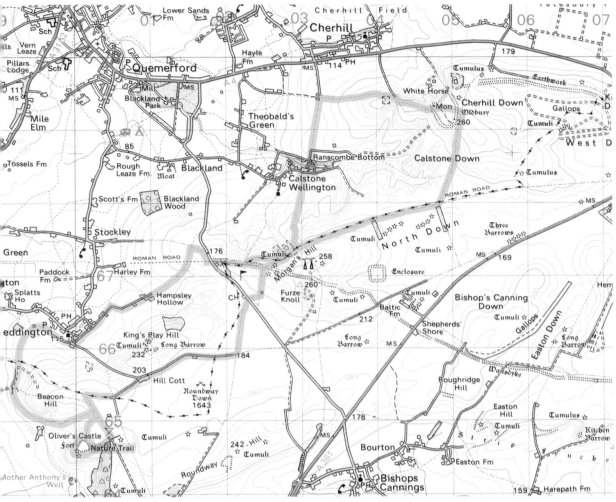

Age ramparts of Oliver's Castle, high over the 'Bloody Ditch' where the Parliamentary troopers perished, to float across the spreading cornfields of the Avon vale.

The Roman highway from Silchester to Bath and Wansdyke, the sixth century defensive ditch which stretches from Marlborough to the Bristol Channel, converge on a modern minor road on the slopes of Morgan's Hill where a convenient county council picnic site gives the start. The route can be broken if necessary into two circuits, one northward, one to the south. The objective of the former, Cherhill Down, is conspicuous right from the start for close to its summit rises the 37-metre (120-feet) obelisk of the Lansdown Column erected in 1845 and now in a somewhat dangerous state. The hilltop itself is crowned by Oldbury Castle, an impressive

Iron Age fort over a kilometre in circumference from where there are extensive views westward towards Calne and the Bristol conurbation: in mid-distance a distinctive beech clump caps the crest of the ascent ridge.

Cherhill is best known, however, for the White Horse which prances across its north-west-facing coombe and is best seen from the main A4 Bath road beneath. Alas, it is no ancient symbol for it was cut in 1780 on the instigation of a local Dr Alsop who directed his workmen by megaphone from over a mile away! The whole hilltop belongs to the National Trust. To the north-east the hill overlooks Yatesbury Field, until the early Sixties the site of a huge RAF training camp, no doubt remembered by many National Servicemen. Two kilometres of Roman Road lead back to Morgan's Hill.

Cherhill Down is best known for the white horse carved on its northern flank and which is familiar to motorists using the A4 road.

Below
Seen to the north-east from Morgan's Hill, Cherhill Down is crowned by the conspicuous obelisk of the Lansdown Column.

The circuit southward follows old and muddy rights of way under the hill, via the ancient village of Heddington with its strange dormer-windowed church, into the steep downland coombe between the swelling flank of Beacon Hill and the jutting prow of Oliver's Castle – a superb defensive site if ever there was one. Initially the return route lies along the scarp top before meeting the line of the Old Bath Road on Beacon Hill and following it eastwards between hedges of hawthorn, crab-apple and sloe onto Roundway Down. On old maps this shallow valley is named King's Play Down, possibly because of the bloody happenings here on 13th July 1643 when the King's army achieved its most decisive victory of the Civil War. Briefly, a small Royalist cavalry force, hoping to lift the siege of Devizes, tired after their hurried march from Oxford and outnumbered three to one, attacked a Parliamentary army of 2,000 horse and 2,800 foot who were drawn up across the valley. Avoiding the powerless infantry, they routed the cavalry and drove them from the field. Apparently the hapless Parliamentary horsemen fled south-west, only to find themselves plunging down very steep slopes into the coombe immediately south of Oliver's Castle where many perished. Meanwhile the Parliamentary infantry were rounded up by the Royalists aided by fresh troops who had sallied forth from Devizes.

Past Hill Cott the old road is metalled, a legacy of the Second World War when it became the artery of a large military camp of which little else now remains. A wide green lane leads to the North Wilts Golf Club beside an incongruous pig farm and so once more onto Morgan's Hill.

Cherhill Down and Oliver's Castle

Length: 21 kms/13 miles
Height gained: 300 m/1,000 ft
Difficulty: easy walking but paths may be very muddy or overgrown in places

Start: good parking at Smallgrain Plantation picnic site 020671. (190 m/623 ft).
Alternative start from Oliver's Castle picnic site 005648. (200 m/656 ft).

(1) Muddy lane at N side picnic site leads E to gate into Morgan's Hill Nature Reserve, take LH track 700 m to E corner of coppice where field path descends N to Calstone Wellington hamlet. At lane go 200 m R then fork L past farm to cross brook in Ranscombe Bottom. Narrow footpath leads uphill avoiding cottages to rejoin lane for 200 m before bridle-path strikes N across fields towards Cherhill. At path junction by barn 032694 footpath strikes E up hillside to reach lonely clump of beech trees. Several paths lead across National Trust property to conspicuous obelisk and Oldbury Castle hill fort (260 m/853 ft). From E ramparts path leads S down fields to join good track returning WSW to Morgan's Hill and road beyond car-park.
10 kms/6¼ miles

(2) Footpath immediately opposite leads between road and golf course 150 m NW to bridle-way signed 'Stockley', follow this 200 m to track junction and turn S down footpath descending to Hapsley Hollow stables – muddy. Metalled lane leads onwards under hill to Heddington village. At final muddy lane junction S of church go R through gate onto greenlane which becomes narrow track between high hedges behind village houses. At cottage 994663 high-hedged path contours S under hill to cross chalk lane leading steeply uphill. Path beyond completely overgrown but a few metres uphill a parallel tractor track continues its line, contouring below steep hillside into coombe between Beacon Hill and Oliver's Castle and climbs to hilltop. Turn R to Oliver's Castle and car-park at head of rough chalk lane. (200 m/656 ft).
6.5 kms/4 miles

(3) From Oliver's Castle, bridle-way leads NW to join chalk lane on Beacon Hill, follow it E to metalled lane bearing NE to Hill Cott. Continue to barn and 500 m beyond turn N onto wide green lane leading to golf clubhouse on road. Signpost on road indicates RoW N across golfcourse to Morgan's Hill, whence retrace route to Smallgrain Plantation car-park.
4.5 kms/2¾ miles

The Malvern Hills   Hereford and Worcester

OS 1:50,000 Sheet 150

'Mauborn hills or as some term
them the English Alps'

Celia Fiennes,
'Journeys' c 1690

They rise sheer from the flat Severn vale and to the unsuspecting visitor hurrying down the M5 at Worcester or breasting the Cotswold edge at Birdlip their long mountainous silhouette will seem improbable, a spectacular anomaly in this wide and gentle landscape. The Malvern Hills are certainly unusual, a narrow 12-kilometre (8-mile) crest of Precambrian volcanic rocks – mostly gneiss – found elsewhere only in the far north-west Highlands and as old as anything in Britain.

The Malvern ridge, steep-flanked and at its widest just over 1 kilometre ($\frac{3}{4}$ mile), holds fifteen named summits – ten of them rising above a thousand feet, and forms a natural boundary between the ancient shires of Worcester and Hereford – and Gloucester at its far southern point. Major gaps divide the crest into four distinct sections. Grassy and almost rugged, the higher more elegant northern tops are scattered with small rocky outcrops and in season with patches of rosebay willow-herb, gorse and broom, while the more intriguing southern hills rise above a woodland cloak of ash and oak and here especially exemplify the best of rural England. Certainly the poet William Langland thought so when he wrote the *Vision of Piers Plowman* towards the end of the fourteenth century: 'Ac on a may morwening/on Malverne hulles Me byfel for to slepe/for weyrynesse of wandryng.' and with the wide-spreading Severn plain below, the Cotswold edge lining the distance and hills and woods tumbling westward towards the dark shapes of the Black Mountains, he dreamt his famous dream of rustic England. Over the centuries many others have found inspiration here, most notably Sir Edward Elgar who was born nearby and was always closely associated with the hills. Malvern itself is a charming spa town with a strong cultural tradition.

The greater part of the hills and surrounding commons is under the jurisdiction of the Malvern Hills Conservators, a Body Corporate dedicated to protecting the land for the public benefit: they prevent encroachment, provide car-parks, check erosion and ensure free public access to some 3,000 acres. Obviously the traverse of the entire Malvern ridge is a classic walk but at first sight a one-way trip necessitating the use of two cars or barely existing public transport. The ridge crest itself involves considerable ascent but an almost level return route is suggested here which visits flanks of the hills usually unseen by the one-way walker while still being within the bounds of a fair day's hike.

Midsummer Hill, its bare summit crowned by a complicated Iron Age fort, rises above Holybush; careful navigation is required to find the best descent through steep woods directly to the Gullet, a deep defile where an imposing aban-

Herefordshire Beacon is crowned by the extensive ramparts of 'British Camp': in the distance beyond the shoulder of Broad Down stretches the Vale of Gloucester.

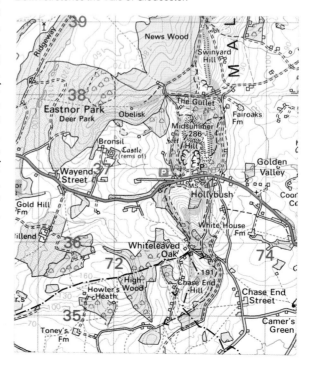

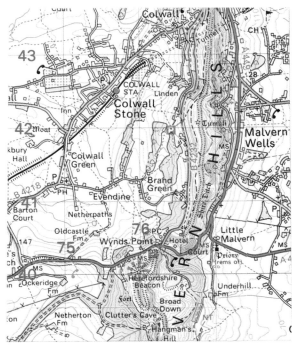

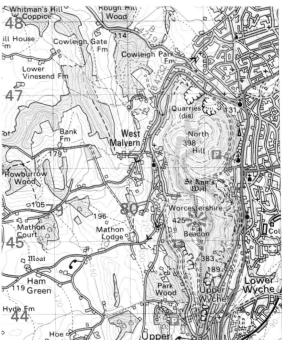

doned quarry holds a neat lake. A viewing ledge above its western lip displays the junction of the primeval Malvernian rocks with younger Silurian strata. From Swinyard Hill the route onwards becomes straightforward, following the prominent 'Shire Ditch' or 'Red Earl's Dyke' virtually the whole way. Constructed around 1290, the dyke was a boundary between the hunting preserves of the red-haired Earl of Gloucester and those of the Bishop of Hereford – it is said that the fence along its crest was so constructed that the Bishop's deer could leap eastward but the Earl's could not escape west! Herefordshire Beacon is the first 'thousand footer' and is crowned by British Camp, a very impressive thirty-two-acre hill fort built during the third century BC. 'One of the goodliest vistas in England', wrote the diarist John Evelyn of the summit panorama, and who would disagree?

North of the Wynds Point pass with its car-parks and café, the Pinnacle Hill massif is a particularly narrow switchback crest above almost precipitous eastern slopes. The Wyche pass beyond is the route of an ancient saltway from the Severn – only 7 kilometres ($4\frac{1}{2}$ miles) distant – into Wales: the actual cutting dates to 1840. The ascent to Worcestershire Beacon leads past the site of a long abandoned gold mine (at 769441), avoids with difficulty a metalled drive-way leading to an ugly café, and arrives at a sharp summit with a view that on a clear day extends over fifteen counties and must be one of the best in Britain. Now the crest curls round Green Valley to the final summit – North Hill – from where there are views down 800 feet (260 metres) into the town itself; with luck strains of a brass band will waft up on the wind.

The return to Hollybush is a long but gentle walk, often through woods and passing several points of interest including Clutter's Cave below Broad Down said to have been a hiding place of Owain Glyndwr. If energy permits, a short diversion to the conspicuous Eastnor obelisk is worth making. The circuit of the two southern-most hills, Ragged Stone and Chase End, is conveniently made from Hollybush. Certainly the most characterful of the Malverns, Ragged Stone Hill is surely the most shapely – though be careful! An old legend claims that they are cursed on whom its shadow falls!

The Malvern Circuit

Length and height gained:

Hollybush to North Hill	11.5 kms/$7\frac{1}{4}$ miles
	850 m/2,790 ft
North Hill to Hollybush	13 kms/8 miles
	175 m/575 ft
Ragged Stone Hill circuit	3.5 kms/$2\frac{1}{4}$ miles
	255 m/840 ft

Difficulty: easy walking on good paths

Start: parking beside track off N side A438 at Hollybush 759369. (160 m/525 ft)
'Pay + display' parking facilities at numerous points N.

(1) Ascend N up hillside to wood where well-defined path climbs diagonally L to gain crest and shelter MIDSUMMER HILL (286 m/938 ft). At rampart 200 m N take bridle-path descending back R. After 130 m turn L at second group of 3 ash trees onto narrow track through bracken following Shire Ditch then dropping steeply L into re-entrant descending to Gullet Quarry (150 m/490 ft). At E corner of lake ascend through fir copse onto brackeny hillside, climb steeply to rejoin Shire Ditch by fence on ridge. Follow crest over SWINYARD HILL (272 m/892 ft) and subsidiary summits of HANGMAN'S HILL and BROAD DOWN to British Camp earthworks and summit HEREFORDSHIRE BEACON (340 m/1,115 ft). Descend to road at Wynds Point pass. (236 m/774 ft).
4.5 kms/3 miles

(2) Go N 100 m up B4232 road, path leads up R behind hotel to ridge and Shire Ditch, follow crest N over BLACK HILL (308 m/1,010 ft), PINNACLE HILL (357 m/1,171 ft) and PERSEVERANCE HILL (325 m/1,066 ft) to road at Wyche Cutting (258 m/846 ft).
3.5 kms/$2\frac{1}{4}$ miles

(3) Take R fork ahead, Beacon Road, for 150 m then strike R to ridge crest and continue N over minor top SUMMER HILL to summit WORCESTERSHIRE BEACON (425 m/1,394 ft). Descend N over SUGARLOAF HILL (368 m/1,210 ft) and TABLE HILL to NORTH HILL (398 m/1305 ft).
3.5 kms/$2\frac{1}{4}$ miles
Short easy descents to Malvern (c 175 m/575 ft) if necessary.

(4) Return to Hollybush: Good paths contour E flank Worcestershire Beacon rejoining crest at Summer Hill. From road 50 m E of Wyche Cutting by cottage, path leads S contouring E flank Pinnacle Hill massif to road 300 m E of Wynds Point pass. From car-park path contours E flank British Camp to Broad Down, then contours W flank past Clutter's Cave and Hangman's Hill. At col beyond ('Pink Cottage') drop R to join wide track through News Wood to junction at 757380 and continuing S below W flank Midsummer Hill to Hollybush.
13 kms/8 miles

(5) Ragged Stone Hill and Chase End Hill: Track climbs S from A438 road just W of phone box. After 150 m path leads off R uphill behind cottage to summit RAGGED STONE HILL (254 m/833 ft). Descend narrow ridge S to track junction, continue SE over stile to lane. 70 m track leads onto CHASE END HILL, climb R to summit (191 m/627 ft). Path descends NW to lane and Whiteleaved Oak hamlet. Just N of Ragged Stone Cottage good contour track through gate on R returns to Hollybush.
3.5 kms/$2\frac{1}{4}$ miles

Below the slopes of Pinnacle Hill the chequer-board fields stretch towards the Severn.

Stiperstones – the Devil's Doing   Shropshire

OS 1 : 50,000 Sheets 126, 137

Once upon a time the Devil's apron strings broke and he was forced to drop the load of rocks he was carrying. Falling to earth they scattered along the crest of a high ridge of heathland that runs close to the Welsh border some ten miles from Shrewsbury. The rocks remain there to this day, a line of jagged quartzite outcrops standing along the spine of Stiperstones hill.

More akin to the wilder hills of North Wales than to the other more gentle hills of Shropshire, the Stiperstones is actually the second highest summit in the county and has a unique atmosphere engendered largely by its geology and resultant history. The hill is said to be the traditional meeting place of Shropshire witches; certainly the Devil's proximity will come as no

In this view from the Long Mynd the Stiperstones ridge is seen across the East Onny valley. Along the crest rise the outcrops of Cranberry Rocks (left), Manstone Rock and the Devil's Chair (right).

surprise to those who have traversed the Stiperstones on a misty winter's day! Geologically speaking however, the seven-mile-long Stiperstones ridge is the eroded crest of an anticline of Ordovician strata some 480 million years old, later folded by the 'Caledonian' earth movements responsible also for the Snowdonian mountains. A narrow band of light grey quartzite running almost vertically forms the hill crest and is flanked on the west by a steep band of standstones known as the Mytton and Tankerville Flags which contains veins of lead and barytes. From Roman times these minerals supported a mining industry which reached its zenith in the 1850s when no less than ten percent of Britain's lead originated in this small area, and which has so influenced the detail of the surrounding landscape.

Besides its quartzite spine, the other notable feature of the Stiperstones is the striking 'dingles', the narrow steep-sided valleys that cut deeply into its western flank and cradle long-abandoned mine shafts and adits. On the heathery western hillsides, linked by overgrown tracks and surrounded by crumbling walls and derelict hedges, stand relics of the smallholdings where the miners supplemented their livelihood. The main centres of mining activity were at The Bog, Pennerley and Snailbeach – where work ceased only in 1916: chimneys, the remains of engine houses, and tips of shiny calcite waste mark the old mine workings. Because of its geological and biological significance, the latter as an important acid moorland habitat, the Stiperstones has been protected as a National Nature Reserve since 1982 and the Nature Conservancy Council is intent on preserving its special wilderness quality.

The figure-of-eight circuit described here traverses the entire crest of the hill and visits its four contrasting flanks and thus gives much scope for individual route variations. The first section links the four main rock outcrops which were carved from the ridge by frost erosion during the last Ice Age when the Stiperstones stood as a nunatak above the surrounding glaciers: today they rise from blocky scree on a whaleback moor of heather and bilberry. Frost polygons and stripes – stone patterns characteristic of a permafrost landscape – are still discernable, particularly south-west of the second out-crop, Manstone Rock, which is the highest point. The remarkable panorama from its summit extends from Snowdon – some 93 kilometres (58 miles) distant – to the Brecon Beacons; from the Malverns to the Peak District. The famous Devil's Chair, a pinnacled knife-edge arete, is the third and steepest outcrop. Beyond the saddle north of Scattered Rock the character of the hill starts to change as the path, less frequented now and grassier, crosses the shoulder of the final rounded hilltop and gives an impressive glimpse down into the depths of Mytton Dingle. An ancient wind-blasted hedge appears as angles ease and the Shropshire Plain spreads out ahead, sweeping wide from Earl's Hill to the Wrekin. Green lanes, a copse, and sheep pastures lead to a pretty wooded valley with cottages and a chapel and the first hints of bygone mining activity, and past the derelict smelting mill itself to Snailbeach village. South of the village the road runs for a while beside the course of the abandoned narrow gauge Snailbeach Railway but after passing Crowsnest Dingle it can be avoided by a narrow right of way which contours the flanks of Oak Hill. Be prepared here for rough going between high hedges of bracken. The Stiperstones Inn may offer convenient refreshment before the ridge is regained by very steep slopes at the head of either of the two dingles, Mytton or Perkins Beach: both are interesting, the former containing traces of a wartime aircrash and the latter of an abandoned village. Once over the crest on the eastern flanks the character of the Stiperstones changes yet again as the final section of the route contours the edge of a gentler landscape of little tumbling hills that fall to the Onney Valley and rise beyond to the great bank of the Long Mynd.

If transport can be organized, an excellent traverse can be made along the Stiperstones crest to GR 383020 and then striking north-east to Habberley before climbing steep and imposing Earl's Hill with its Iron Age hill fort and descending to Pontsford: the route, of about 12 kilometres (7½ miles), follows rights of way except on Earl's Hill which is a nature reserve with permitted public access.

Manstone rock, the highest point on the Stiperstones crest, is seen here from the south.

A Traverse of the Stiperstones

Length of circuit: 13 kms/8 miles
Height gained: 415 m/1,360 ft
Difficulty: easy walking on good but sometimes rough and rocky tracks: one steep ascent
Note: this is a National Nature Reserve, please observe byelaws posted on car-park noticeboard which include strict control of dogs and prohibition of fires. Grouse shooting may take place between 12th August and 10th December
Warning: beware of old mine shafts

Start: Nature Reserve car-park at 369977. (420 m/1,380 ft).

(1) Take wide track to NW diagonally up hillside to first rocky tor, CRANBERRY ROCK, then follow ridge crest path past MANSTONE ROCK (trig pt 536 m/1,759 ft), DEVIL'S CHAIR and SCATTERED ROCK (pt 494 m) to track junction and cairn on broad saddle at 373999.
3 kms/1¾ miles

(2) Main track continues NE along ridge passing W of SHEPHERD'S ROCK (pt 502 m) gradually descending W flank of final rounded hill through several gates to track junction below power lines. Fork left to join metalled lane at gate by chapel at 380021 and descend into Snailbeach. Cut corner by lead smelter ruins, join road and continue SW to Stiperstones village.
6 kms/3¾ miles

(3) By Stiperstones Inn turn E into righthand of two lanes, leading first SE then S into deep valley of Perkin's Beach. Climb steeply from head of cwm to rejoin saddle at 373999.
1.5 kms/1 mile

(4) Descend path down E flank of ridge towards The Hollies until reaching jeep track 200 yards above farm, follow track S through forestry back to car-park.
2.5 kms/1½ miles

Variation
(2.A + 3.A) Follow (2) past Snailbeach and Crowsnest to 368014 opposite tower of old mine works. Ascend drive through gate to cottages, RoW passes in front of upper cottage and as a narrow path contours Oak Hill along intake fence into Mytton Dingle. Path ascends very steeply to ridge crest track below Shepherd's Rock. Continue S to saddle at 373999.
7.5 kms/4¾ miles

Pride of the Peak District – Kinder Scout

OS 1:50,000 Sheet 110

'A country beyond comparison uglier than any other I have seen in England, bleak, tedius, barren, & not mountainous enough to please one with its horrors.'

Thomas Gray: letter 1762

'Kinder Scout offers solitude and established stillness, older than the world'

E A Baker: Moors, Crags and Caves of the High Peak 1903

Time changes, and with it man's perception of the wilderness. And Kinder Scout, hardly a mountain, more a great mesa that holds the highest point in the Peak, is today a much loved wilderness. Together with Bleaklow, its slightly lower but even wilder neighbour, Kinder is the quintessence of the Dark Peak, that high and brooding gritstone moorland that sits astride the spine of England, the final link of the Pennine Chain. But Peak it is not: in Saxon times the region was known as Peaclond – literally 'Hill Country' – later travellers described the whole area as 'The Peak'; the Ordnance Survey in its first map of the region in 1864 misleadingly appended the name to the five-square-mile plateau surrounding its highest point while carefully lettering in the name 'Kinder Scout' along its western lip where the plunging stream of Kinder Downfall is the major feature. Cyn

Beyond the Kinder reservoir the moors sweep up to the western edge of the Kinder plateau: on the right is the dark gorge of Kinder Downfall and on the far left rises the shoulder of the so-called Upper Western Buttress.

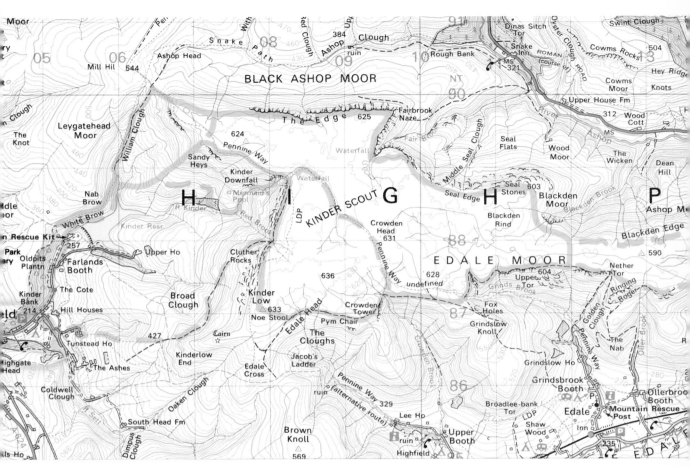

dwr Scwd in Anglo-Saxon means 'Hill of the Waterfall', so this at least seems an apt and proper title. Geologically, Kinder and Bleaklow are gritstone islands perched on the crest of the so-called Derbyshire dome, all that remains of the upper rock layer in the centre of this great anticline, the rest having been worn away to expose the layers beneath – notably the Limestone of the White Peak – but leaving gritstone verges to east and west. Practically, the Kinder Scout plateau is a desolation of shallow dunes of peat, sometimes heathery, often boggy, crisscrossed by 'groughs' – deep runnels – that make bee-line travel tedious and exhausting. Once clothed in broad-leaved forest after the permafrost melted, poorly drained, burned by man, grazed by animals, eroded and soured by atmospheric pollution, this is the landscape John Hillaby described in his *Journey through Britain* as '. . . land at the end of its tether . . .' likening the maze of chocolate-brown peat-hags to the

droppings of dinosaurs. Yet this monotonous plateau is fringed by steep edges, often rocky, sculpted and spectacular and if Kinder's heart is stern and uncompromising, its beauty lies round its perimeter.

Historically Kinder Scout played an important part in the development of both fell-walking and National Parks in Britain. Set midway between the great cities of Lancashire and Yorkshire and close to the industrial Midlands, the empty moors and challenging landscape had attracted hardy walkers as early as 1880 when the first rambling club was formed in Manchester, despite the fact that virtually the entire upland was jealously preserved grouse moor. An early test piece of the 'bog-trotters' – as they were soon known – was the now-classic Marsden to Edale walk across Black Hill, Bleaklow and Kinder in 1902. In the depression years of the Thirties rambling exploded as a natural escape from the frustrations of the cities and pressure for

Kinder Downfall and the Perimeter Circuit

Length: 22.5 kms/14 miles
Total ascent: 560 m/1,850 ft
Difficulty: a stout walk, rough often muddy paths along edges, short trackless sections across plateau rugged, strenuous. (1A) involves short stiff scramble
Note: this is serious country subject to fierce changeable weather. Typically featureless, navigational skills essential. Specific areas occasionally closed for shooting 12th August – 10th December, enquire PNP phone 062–981–4321

Start: PNP car-park at 048869, Bowden Bridge above Hayfield (215 m/705 ft).

(1) Ascend road 800 m, cross bridge continuing on path above river, crossing again to Pump House. Ascend L to contour path above reservoir, continuing E to N arm of lake whence descend to footbridge crossing outflow William Clough, ascend N of E to shoulder, hints of path, then up narrowing ridge – Upper Western Buttress – to bouldery promontory plateau lip. Follow well-defined path SE to head of cliffs Kinder Downfall (600 m/1,970 ft).
5 kms/3 miles

(2) Edge path continues S to OS pillar KINDER LOW (633 m/2,077 ft) whence strike 300 m SE, rough ground, to meet edge path near anvil-like Noe Stool rock, follow path E along southern edge past Pagoda Rocks, Woolpacks boulder field, above crag Crowden Tower to cross muddy swarth Pennine Way, swing N to craggy head Grindsbrook Clough (590 m/1,936 ft).
5.5 kms/3½ miles

(3) Following edge path to crags Upper Tor, continue 400 m E then strike N 400 m, rough ground, to join northern edge path above Blackden Brook. Follow path L, bearing first N to Seal Stones then W along Seal Edge, round craggy Fairbrook Naze and along Ashop Edge to conspicuous Boxing Glove stones (590 m/1,936 ft).
6 kms/3¾ miles

(4) Path continues along fading edge, descends to boggy saddle Ashop Head (500 m/1,640 ft), junction, whence descend path L down William Clough, rejoining ascent route above reservoir.
5.5 kms/3½ miles

Alternatives
(1.A) For experienced scramblers only: from William Clough bridge ascend to shoulder then contour rough hillsides due E 500 m, descending to W end small wood and Kinder River (direct route below wall crosses private non-access land). Ascend rugged valley beside river to scrambly boulders below waterfall/cliffs Kinder Downfall. Go 100 m L to corner whence stiff scramble up short vertical blocky wall into alcove leads to plateau above.

(2.A) Shorten route by using well-marked muddy Pennine Way linking Downfall to Grinsbrook Clough direct, or visa versa.
3 kms/2 miles

(4.A) If (1A) used in ascent, 700 m W of Boxing Glove stones strike S 300 m to join western edge path near summit Upper Western Buttress, descend (1) to reservoir.

access to the moors became intense for even by the mid-Thirties there was still not one public right of way on the upper fifteen square miles of Kinder Scout. Land that two centuries before had been criss-crossed by packhorse tracks and free for all was now barred by cordons of zealous gamekeepers, many of them not averse to strong-arm tactics, and walkers often developed commando methods to reach the tops. One unpleasant incident in 1932 was a Mass Trespass by some 600 ramblers up William Clough – a right of way above Hayfield since 1898 – and on towards the forbidden Kinder plateau: there was violence and five ramblers went to prison. Nevertheless, wheels were in motion and in 1939 Parliament examined the whole question of rights of way, finally passing an Access to Mountains bill which led eventually to the designation of Britain's first National Park – the Peak District – in 1951.

Today, with almost unrestricted access, there are many excellent walking routes on the Kinder massif but the finest is the lengthy perimeter circuit taking in the steep plateau edges where lies the best scenery. The approach suggested here allows competent scramblers to approach the Downfall, the most famous single feature on Kinder, up the wild gulch of the Kinder river from below where the 30-metre (100-foot) waterfall is seen at its most spectacular – especially in spate or when frozen. Non-scramblers can approach with caution but they cannot reach the plateau and must descend and detour via the Mermaid's Pool – where immortality is threatened for those who glimpse her on an Easter morning – to regain the recommended route up the pleasant ridge of the Upper Western Buttress. Going south the path approaches the marked 'summit' of Kinder Low, the unmarked highest place – point is hardly accurate – is some 800 metres distant towards the middle of the plateau and just three metres higher. Along the southern edges stands a sequence of grotesque boulders and outcrops such as the strange throne of Pym Chair, the tapering tor of the Pagoda, the scattered blocks of the weird Woolpacks – often known as 'Whipsnade' – and the imposing crags of Crowden and Grindsbrook Towers above the popular ascent routes from Edale.

Above Grindsbrook Clough energetic purists may consider extending the circuit a further 5½ kilometres (3½ miles) eastward to the spur-end of the plateau at Crookston Knoll before striking back westward to Blackden Edge. Here the northern edges are less frequented and more atmospheric and the path passes above the steep Chinese Wall on Seal Edge before striking north round the head of Fairbrook Naze and its sharp prow and west again over Black Ashop Edge – where stand the famous Boxing Glove Stones – to complete the circuit. A lasting impression of Kinder Scout in perfect weather is one of 'Big Sky Country' – as they say in Montana.

Kinder Downfall: this is a winter view of the famous falls which gave Kinder Scout its name.

White Peak and Dark Peak – A Hope Valley Horseshoe   Derbyshire Peak District

OS 1 : 50,000 Sheet 110

'The Peak of Derby being extraordinarily noted, I could not in my travels omit to visit it, especially upon the account of the dreadful cave called the Devil's Arse . . .'

William Lithgow,
Rare Adventures and Painful Peregrinations – 1614–32

Celebrated especially for its subterranean sights, the Hope Valley lies along the junction of the gentle limestone plateau of the White Peak and the frowning gritstone moorlands of the Dark Peak. Peak Cavern – as William Lithgow's cave is now more genteelly known – is just one of a

number of impressive caves that open into the valley close to the village of Castleton. The surrounding surface landscape is as dramatic and interesting as any in the Peak and the walk suggested here samples something of the contrasting countryside on both sides of the valley.

Castleton, a charming village and something of a tourist venue in season, is both a convenient place to start and a refreshing place to finish the walk. Once a medieval model town, it is dominated by the imposing ruins of Peveril Castle perched on a great crag rearing between the twin clefts of Peak Cavern and Cave Dale, a stronghold built by William Peveril, bastard son of the Conquerer, when Steward of the surrounding royal hunting preserve of Peak Forest.

After crossing farmland and fields the ascent of the graceful cone of Lose Hill is straightforward and the excellent summit panorama stretches from the gritstone edges beyond the Derwent to the moody folds of the Kinder Scout massif at the head of the Edale sanctuary below. Across the portal of Edale and crowned with a rocky 'Pike' stands shapely Win Hill which can be added to the walk for the penalty of a few more miles. Apparently a bloody Dark Ages battle took place hereabouts – possibly on the Roman Road that runs along the western flanks of Win Hill – between Edwin of Northumbria and Cuicholm of Wessex. On the eve of battle the former king camped on Win Hill and won the day, the latter king on Lose Hill and lost it!

Of a mountainous style almost unique in the Peak District, the crest westward from Lose Hill is comparatively narrow, runs over two intermediate tops and is generally known as the Great Ridge. From Back Tor, the first top, the path picks its way down the edge of shattered cliffs, a manifestation of the shales – neither gritstone nor limestone – of which the ridge is composed, and which appear again on the culminating point of the ridge, Mam Tor. Its east face scooped out to form a huge striated cliff, Mam Tor – the so-called 'Shivering Mountain' – is a dramatic landmark. The triangular cliff, falling around 100 metres (300 feet) from near the summit but not particularly steep, is the result of incessant landslips which have exposed layer upon layer of crumbling shales and siltstones. Solidified under snow and ice the face provides an esoteric winter climb though first ascended as long ago as 1898 in summer by the redoubtable J W Puttrell! The hilltop is circled by the ramparts and ditches of a large and important Bronze Age hill fort which obviously predates the cliff over which a sizable section has fallen away: as recently as 1977 landslides permanently cut the main A625 road below the face.

The ruins of eleventh-century Pevril Castle guard the mouth of the dry limestone valley of Cavedale.

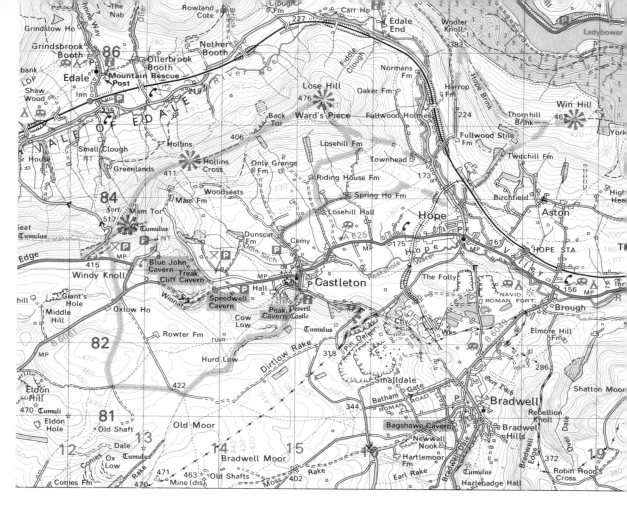

A Hope Valley Horseshoe

Length: 13 kms/8 miles
Total ascent: 435 m/1,425 ft
Difficulty: mostly good paths but a trifle rough in places
Note: try not to exacerbate footpath erosion on Great Ridge Section (2)

Start: most aesthetic start is Castleton car-park 149830. (290 m/950 ft) but other convenient car-parks are marked on map below Mam Tor and Hope, fees usually payable.

(1) From village centre take Hollowford Rd N 500 m to lane junction, follow farm track R, then field path to lane junction at Losehill Hall PNP Centre. Continue 100 m E where concessionary field path over stile leads N to pass R of Riding House Farm whence RoW ascends E across hillside almost to Lose Hill Farm, go up L to signpost from where path climbs to grassy spur and summit topograph LOSE HILL (476 m/1,562 ft).
3.5 kms/2 miles

(2) Follow well-worn path SW along ridge crest crossing BACK TOR (435 m/1,427 ft) with steep rocky descent and BARKER BANK (426 ft/1,398 ft) to track junction at Hollins Cross saddle (390 m/1,280 ft) then long gradual climb to summit trig pt MAM TOR (517 m/1,696 ft). Descend path and steps SW to lane at Mam Nick then footpath S to road.
4 kms/2½ miles

(3) RoW opposite leads 500 m over Windy Knoll to another road, go R to track junction whence path strikes S across pasture to wall continuing over low hill to track junction 126814. Go E beside wall to join stony lane at gate, continue E 200 m to gate by sheep pens where path strikes L across fields 400 m to fork in shallow depression, go R, through iron gates, deepening depression becomes narrow Cave Dale leading directly to Castleton.
5.5 kms/3½ miles

Variation to include Win Hill Pike

(1.A) From Losehill Hall continue track E to Hope Village, cross Edale Rd and river at Killhill Br, lane ascends under rly, go R then NE up track to Twitchill Farm whence RoW climbs to crest and trig pt, rocky outcrop WHIN HILL PIKE (462 m/1,516 ft). Descend ridge 2 kms W then NW to track junction almost at edge woods, descend L track across steep hillside S to lane crossing rly, then Edale Rd crossing river. After 50 m turn R up Townhead lane for 300 m, fork L, track then path climbs to grassy spur and Losehill Pike summit.
Extra 8 kms/5 miles, plus 325 m/1,0670 ft

The imposing east face cliffs of Mam Tor fall towards the head of the Hope Valley, the remains of the main road, finally obliterated in 1977, are seen below. In the distance, eastward, the Great Ridge stretches towards Lose Hill.

Mam Tor and a large surrounding area belong to the National Trust and the popular summit is usually reached from the lane on the narrow col of Mam Nick. A steep, rough and badly eroded (and currently closed) but infinitely more scenic route descends south-west down the cliff edge to the abandoned road which can be followed round to the entrance to Winnats or 'Wind Gates' Pass, now the only road, steep, narrow and twisting, down to Castleton: it is sobering to learn that in 1938 there were serious government proposals to blast a modern highway through this delightful limestone gorge! However, our route now gently ascends onto the limestone plateau, and dark gritstone dykes give way to walls of white rocks. Near its lowest point the

path passes over Windy Knoll where excavations in Bone Cave below the old quarry have revealed bones of bison, wolf and bear. Further on the route crosses pastures riddled with the shafts and workings of lead mines abandoned some 200 years ago. Eventually the path drops into the interesting dry valley of Cave Dale, a narrow, twisting and rocky defile probably carved by glacial melt water after the Ice Age and the scene of mass rallies in the 1920s of ramblers protesting at the then forbidden access to 'Old Kinder Scout and the Moors roundabout' as the well known song puts it. Peveril Castle rises sheer above the craggy lower end of the Dale and the centre of Castleton is only a couple of hundred metres further on.

Along the Derwent Crest Peak District

OS 1: 50,000 Sheet 110

'... where Derwent rolls his dusky floods
Through vaulted mountains, and a night of
woods.'

Erasmus Darwin: *The Botanic Garden* 1789

The Peak District, remarkable for its dearth of natural lakes, is scattered liberally with over fifty made by man of which the largest and least incongruous are the three that impound the headwaters of the River Derwent — the Howden, Derwent and Ladybower reservoirs. Rising on the desolate Bleaklow plateau with its annual rainfall of some 150 centimetres (60 inches) the Derwent had little chance of escaping the East Midlands water engineers: the first two dams were completed in 1916 and the Ladybower, ten years in construction, in 1945. Derwent derives from the Celtic *derwen* meaning oak, and Derwent Dale must once have been a peaceful and attractive wooded valley, the surrounding moors were always sheep country and Ladybower alone drowned no less than ten farms besides the pretty villages of Ashopton and Derwent Chapel. Tragic though it was, the flooding of the dale and the subsequent reafforestation of its slopes — albeit with pine and larch — has created a comely lakeland landscape in harmony with the surrounding hills and high moors, an extremely popular 'lung' for the nearby Sheffield conurbation and a walking venue unique in the Pennines.

For serious walkers the area holds several excellent itineraries of which the most interesting is probably the circuit of the tops and edges — noted for their extraordinary rocky outcrops — that rise above the eastern shores. At Fairholme the National Park facilities provide a good starting point close to the awesome 35 metres (115 feet) castellated wall of the Derwent Dam. During the war the hills hereabouts echoed to the roar of 617 Squadron's Lancasters as they swept down the narrow valley to perfect their dangerous 'dambusting' techniques, but the lakeside is now tranquil and a good foresty track leads along it towards the Howden Dam that

blocks the northern end of the valley. But before the dam is reached the route turns eastward to follow an ancient packhorse track up the narrow clough of the Abbey Brook: with its lively stream, slopes scattered with oaks and leading deep into the heart of the moors this is one of the most charming small glens in the Peak District. Walkers striding happily up this grassy track, secure in the midst of National Trust territory, should know that it was not always so peaceful. Although well used from times immemorial, by an Enclosure Act of 1816 this track was closed from the Derwent up to the watershed while the descent eastward to the Don valley remained a right of way, a ridiculous situation which rankled the militant hikers of the Twenties and Thirties when the surrounding moors were closely patrolled by gamekeepers: many a fracas occurred. In 1932 a mass trespass down the valley was confronted by a waiting band of keepers and police and a battle and several arrests ensued.

Lost Lad is the first summit and all routes up to it are rough and more or less boggy, it commands a wide view and is crowned with a useful topograph but it tells a sad tale. One springtime many years ago a shepherd noticed the words 'lost lad' scratched on a boulder here and found nearby the body of a young boy lost in bad weather the previous autumn. On the skyline rises the spectacular silhouette of a ruined castle which on closer aquaintance proves to be Back Tor — the highest point of the walk. 'The Spirit of the Moors has his home on Back Tor ...' wrote John Derry in his classic book *Across the Derbyshire Moors* in 1926, and surely this spiky gritstone palisade standing atop the wide and gently swelling moorland and guarded by grouse, curlew and mountain hares, is an imposing — and particularly in mist — an eerie place. In clear conditions however it is one of the best viewpoints in the Peak, not only towards the

At the southern end of the Derwent Edge, Whinstone Lee Tor looks down on the long ribbon of the Ladybower Reservoir: in the distance rears the Kinder Scout Plateau.

looming plateaux of Bleaklow and Kinder, but into Edale and to distant Axe Edge, while eastwards can be seen the towers and chimneys of Sheffield, Barnsley and the cities beyond the Peak.

The terrain ahead is almost flat but the path is often boggy and badly eroded. Near the Cakes of Bread a rocky scarp materializes along the eastern lip of the moor and a diversion here provides better going with glimpses of the lakes and the green vale below. To locate anything resembling a dove among the boulders littering the craggy edge of Dovestones Tor requires a powerful imagination but the weird Salt Cellar beyond is a conspicuous landmark. Now a great henge rears on the moorland horizon, the Wheel Stones, a clutch of bizarre towers which could well have been designed by Michelin men though from other angles the alternative name of Coach and Horses seems apposite: the scramble to the summit is easier than it looks. The Hurkling Stones, said to resemble crouching warriors, are the final strange outcrops before the path passes above the cliffs of Whinstone Lee Tor and starts descending towards Grindle Clough and moody Ladybower. There, as the lake-shore lane passes little Mill Brook Bay, it seems appropriate to give a poignant salute to Derwent village beneath the dark waters.

Right
Standing high on the Derwent Edge, the celebrated gritstone pillar of the Salt Cellar looks eastwards over the Ladybower Reservoir towards the distant plateau of Kinder Scout.

Like the ruins of a great castle, rocky battlements of Back Tor r from the moorland skyline: view from the north near Lost Lad.

A Circuit of the Derwent Crest

Length: 16 kms/10 miles
Total ascent: 380 m/1,250 ft
Difficulty: mostly easy, though sometimes boggy, paths but one section rough, damp and trackless
Note: almost entire route to Salt Cellar on National Trust land, many variations, short cuts, possible

Start: Fairholme PNP car-park/picnic site 173893 between Ladybower and Derwent reservoirs.

(1) Minor road leads E below Derwent Dam whence path ascends L to E end Dam, joining wide track leading N up shore Derwent Reservoir. At bend 200 m before bridge first major tributary, gated track leads R above plantation wall into Abbey Brook valley, grassy track continues ascending gradually across S slopes deep glen. Cross second stream to follow grassy path more steeply over shoulder, leave path at highest point striking R up steep grassy nose, continue over rough tussocky plateau ascending slightly R to join poor path leading S along wide ridge crest. Join bigger boggy path, ascend to junction, small cairn, go L, ascend to skyline cairn, topograph, LOST LAD (518 m/1,699 ft), boggy path continues SE to skyline rocks, OS cairn BACK TOR (538 m/1,765 ft).
6.5 kms/4 miles

(2) Follow wide path S, cross signposted junction, then near Cakes of Bread bear R along plateau edge, passing over rocky Dovestone edge to obvious Salt Cellar pillar. Path continues past conspicuous towers Wheelstones, across shallow col, signpost, to follow ridge curving R to rocky edge Whinstone Lee Tor (c 410 m/1,345 ft).
4.5 kms/2¾ miles

(3) Continue 100 m to gully on RHS shoulder, path cuts steeply back R under hill to join boggy contour track along wall, continue N to signpost 'Derwent/Moscar', continue 50 m then strike L on good path, pass plantation edge, ford stream to Grindle Clough barns. Go L through yard, gate, descend steep pasture to join lakeside lane, continuing N past Derwent cottages, etc, to Ladybower Reservoir head and Fairholme.
5.0 kms/3¼ miles

Variation

(1.A) Continue Abbey Brook track round shoulder to ascend valley third stream, Sheepfold Clough, towards Lost Lad. Path fades into bogs, reappears higher. This variation 1.2 kms longer, wetter, more gentle.

Stanage and the Hathersage Heights

Peak District

OS 1 : 50,000 Sheets 110, 119

Probably the best known and certainly the most extensive of the gritstone edges that fringe the Peak District is Stanage, a stark battlement of coarse brown rock that lines the escarpment of the Hallam Moors high above the village of Hathersage in the verdant Derwent valley. Eastward from the bouldery edge-top a flat peaty plane of heather, crowberry and moorgrass wide to the wind ripples towards the Sheffield suburbs while westward, beyond steep bracken and green depths, the swelling uplands of the White Peak smile towards the brooding plateaux of the Dark Peak that line the horizon. Stanage Edge is an atmospheric place and its crest

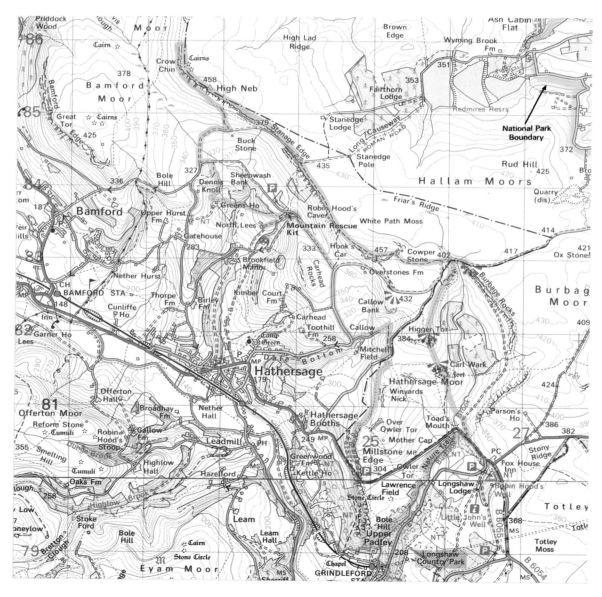

Long-abandoned millstones litter the slopes beneath the gritstone crags of Stanage High Neb.

a popular walk while its 4 kilometres (2½ miles) of almost continuous cliffs between 12 and 25 metres (40–80 feet) high make it the busiest rock-climbing venue in the Peak with well over 500 guide-booked climbs dating back to 1890 and of every kind and standard.

But there is more on the heights above Hathersage than merely Stanage Edge and this route links interesting local features into an infinitely variable circuit. Completing the major ascents early on, the walk starts at the cwm-like head of the Burbage Brook, and descends over the prow of Higger Tor to the strange kopje of Carl Wark rising in the middle of the heathery expanse of Hathersage Moor. Higger Tor is variously claimed to mean 'Hill of God' or 'Higher Tor' and is celebrated among climbers for its imposing Leaning Block. Carl Wark however is an ancient hill fort where steep rocky edges on three sides and a stone-revetted rampart still 3 metres high (10 feet) enclose a small plateau: it has been described as 'a natural fortress improved by art' and is probably of Dark Ages origin. Leaving the wild moorland the route follows the pretty alder-lined Burbage Brook down a shallow green valley, part of the National Trust's 1,000-acre Longshaw Park estate, formerly home of the Duke of Rutland. Suddenly woods close in as the stream pours over a series of foaming cataracts and into a steep and narrow defile. This is Padley Gorge, deep and rocky but never precipitous, hung with stunted oaks, floored with mossy boulders and the habitat of uncommon birds, it is a beautiful place and a Nature Reserve.

Below the gorge the route climbs to the long-abandoned Lawrencefield Quarry hidden in a jungle of whispering birches, its slabby faces rising from a silent pool while rows of completed millstones nearby await transport that will never come. The famous Surprise Pass on the A625, noted for its sudden view across the Derwent valley towards Kinder, lies just above the cliff top and beyond rear the dark walls of Millstone Edge, its conspicuous quarried escarpment dominating the Hathersage dale. Here the most interesting of several tracks follows the base of the formidable cliffs, in places 40 metres

(140 feet) high, which were an important training ground for aid climbers in the '50s – though the old piton cracks and the blank walls between are today climbed free. A good path, initially steep and rough, leads through the cliffs, round the lip of Hathersage Moor and down into the hollow below Callow Bank which is a favourite launch site for radio-controlled gliders.

The steep Dale Bottom road climbs to Hook's Car where it joins the line of the old turnpike which runs below Stanage Edge and once linked the Woodlands Valley to Sheffield, continuing to a weathered plantation where the narrow road forks. Round the corner a right of way leads down through the trees towards North Lees Hall and although our route bears off almost immediately, the Hall is worth a short diversion: a battlemented tower-house originally built around 1410, it is the original of Thornfield House in Charlotte Brontë's novel *Jane Eyre*. At Dennis Knoll above lonely Greens House the ancient packhorse track over Stanage Edge meets the old turnpike while a short way north stands High Neb, almost the end of the Edge and the loftiest point in the area. Flaunting a series of impressive buttresses and the distinctive overhanging beak (or 'neb') of 'Qietus', for long one of the boldest climbing leads on gritstone, High Neb was, like all these moors, closely keepered and theoretically inaccessible until comparatively recently, and is still subject to access agreements between the National Park and the landowners.

Over a hundred millstones lie scattered beneath the edge hereabout, until the advent of synthetic carborundum killed the industry they would have been carved on site, fitted to wooden axles and rolled, two by two, to waiting transport. Now an easy path stretches along the edge-top, past the packhorse track, across the line of the 'Long Causeway' Roman road, above the sandy ledge of Robin Hood's cave – a favoured bivouac of early climbing pioneers – and past the tops of famous rock climbs busy every weekend. Then as the cliffs fade away the path rises easily to the final rocky tor above the Cowper Stone and the finish is in sight.

Stanage Edge and the Padley Gorge

Length: 20.5 kms/12½ miles
Total ascent: 570 m/1,870 ft
Difficulty: easy walking on good though sometimes stony paths
Note: route described may be changed at will with wide choice of obvious variations

Start: route described from car-park at 263829, Burbage Rocks, on Ringinglow – Hathersage road (410 m/1,345 ft). Other convenient parking places include: Hook's Car (244830) or PNP car-park Hollin Bank, toilets closeby, (237837).

(1) From Burbage Rocks follow road 300 m W, strike S on moorland path above edge to rocky nose Higger Tor, path descends over Carl Wark hill fort to Burbage Brook near A625 bridge (Toad's Mouth). Beyond road follow good path R bank Brook S to edge woods/Nature Reserve continuing on rocky main path down through forested Padley Gorge to grassy clearing. Strike up R past water-works building 50 m to gate whence narrow path climbs into woods, zigzags up through overgrown spoil tips to level birch woods below old Lawrencefield Quarry, good track continues to A625 road at 248801 (280 m/920 ft).
6 kms/3¾ miles

(2) Cross road, go R to take higher of two stony tracks leading N below cliffs Millstone Edge to good path contouring moor beyond to RoW sign on minor road 253817. Go R 100 m to RoW sign, path drops L across slope to valley bottom track, go L to join minor road, ascend R to Hook's Car junction, take road L below Stanage Edge, descending gradually to fork at small wood (305 m/1,000 ft).
5 kms/3 miles

(3) Follow R fork 100 m to track leading L through pines, grassy path soon breaks R across bouldery pastures below oak wood to stream, stepping stones, mill-pond. Path beyond continues, gates, stiles, to Greens House whence track ascends R to rejoin road at Dennis Knoll plantation. From corner 200 m L take good moorland track R until obvious path leads L below craggy Edge to High Neb. Continue N 500 m (or sooner) to ascend easily to cliff-top, whence good path leads R to OS pillar HIGH NEB (458 m/1,502 ft).
5 kms/3 miles

(4) Follow path SE above cliffs Stanage Edge to OS pillar near Cowper Stone (457 m/1,500 ft). Path continues E across moor to road at Burbage Rocks.
4.5 kms/2¾ miles

Stanage Edge, seen here above the Plantation, stretches away northward towards the distant prow of High Neb.

The Roaches Crest Staffordshire Peak District

OS 1: 50,000 Sheets 118, 119

'Here are also vast Rocks which surprise with Admiration, called the Henclouds and Leek Roches. They are of so great a Height and afford such stupendous Prospects that one could hardly believe they were anywhere to be found but in Picture'

Compleat History of Staffordshire: 1730

Stand on the gritstone battlements of the Roaches and look down upon the wide spreading Cheshire Plain fading away towards the distant soft Midland shires. Turn behind and gaze at rolling moors and serried hills – the sterner stonier upland landscape of the North. There can be few places in Britain where the frontier between lowland and highland is so manifest as this long gritstone ridge that dominates the Staffordshire border country north of Leek.

Ideally the traverse of the Roaches ridge should start from its far north-western end where Danebridge nestles in a smiling landscape of rounded hills, but it is only a small village and one should park with discretion in the narrow lane by the bridge. Field paths lead past a four-square farmhouse to the strange beak of Hanging Stone jutting from the grassy hillside. A plaque set into the rock commemorates *Burke, a noble mastiff . . . a gun and a ramble his heart's desire with the friend of his life the Swythamley Squire. 1874.* Here the corner can be cut by a narrow path – not actually a right of way – past a small tarn directly to the ridge crest. The first landmark beyond the bare whaleback ridge of Back Forest is the pass at Roach End with a remote keeper's cottage closeby. Now the moorland is more rugged with boulders scattered over the carpet of heather and bilberry, and rocks sometimes eroded into weird shapes such as the conspicuous Bearstone. Although the Roaches summit hardly justifies its name – obviously derived from the Norman-French *roche* – rock becomes very much in evidence thereafter as the path approaches the scarp-like lip of the almost unbroken line of crags that edge the ridge for

two kilometres: an ancient larch plantation below the cliffs and the sharp shape of Hen Cloud now visible ahead add a real mountain ambience to the wide view. Doxey Pool is passed, a peculiar grass-fringed tarn on the ridge crest and mentioned in Domesday Book, and then the path drops into the straggly trees below the rearing crags. And what crags! This is one of the most celebrated climbing areas in the Peak and the gritstone walls and buttresses are laced with dozens of routes up to 30 metres (100 feet) high. Most famous is surely Sloth, a fierce and uncompromising line out over a huge overhanging neb, a truly spectacular lead pioneered by the great Don Whillans in 1952. 'Was it 'ard?' Asked an onlooker who watched the first ascent. 'Not if yer use yer loaf,' replied Whillans dryly.

At the edge of the wood stands Rockhall, a bizarre cottage built into the cliff, its rooms hewn from the living rock. Not surprisingly it has been the abode of a succession of eccentric characters, from highly aggressive gamekeepers to the recent Doug – self-proclaimed 'Lord of the Roches' – but none stranger than Bowyer of the Rocks, a notorious moss trooper and local terror, and his daughter Bess who sheltered thieves, smugglers and deserters here. Her own lovely daughter had a beautiful voice and would sing among the rocks in a unknown tongue: but she was abducted and old Bess was left to die of a broken heart. If you are alone and unlucky you may see the spectre of a singing woman who haunts the Roaches. If you are lucky you may see a wallaby. Several escaped from a local menagerie years ago – together they say with a few musk oxen – and after travelling as far as Macclesfield (!) the wallabies settled in hereabouts: many unsuspecting visitors have been astonished to see these outlandish marsupials hopping around among the boulders.

From near Doxey Pool on a misty day, the Roaches escarpment takes on a strangely alpine character as it stretches away towards the distant prow of Hen Cloud.

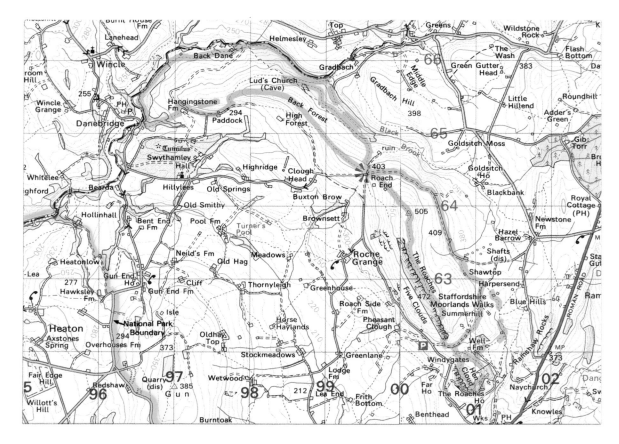

The Roaches Crest

Length: 18.5 kms/11½ miles
Total ascent: 580 m/1,900 ft
Difficulty: easy walk on good paths
Note: most aesthetic start to route is Danebridge, but parking problems may dictate alternative starts

Start: limited roadside parking at Danebridge 965652 (190 m/623 ft), better but still limited at Roach End 996645 (407 m/1,335 ft) or PNP car-park below Roaches at 004621 (340 m/1,115 ft).

(1) From road 100 m S of bridge waymarked path leads up L past cottages, continues NE through wood, up fields, through farmyard and L of Hangingstone Farmhouse to join contouring track below prominent 'Hanging Stone'. Go E past Paddock house and gate to track junction, turn N up to moorland, at ridge crest waymarked concessionary path strikes E along broad Back Forest Ridge, gradually bending SE to join rough track leading eventually to lane crossing ridge at Roach End.
4 kms/2½ miles

(2) Path continues along ridge crest, climbing past eroded Bearstone Rocks to summit trig point THE ROACHES (505 m/1,657 ft). Gradual descent onwards leads along edge of craggy scarp to Doxey Pool. 350 m beyond Pool by fence at shallow saddle easy gully drops R through gap in crags to path descending through woods, down steps, to Rockhall cottage and moorland track leading L to wide col (350 m/1,150 ft). Now concessionary path through stile leads across field to well-defined ridge, ascend to rocky summit HEN CLOUD (410 m/1,345 ft).
4 kms/2½ miles

(3) Path drops steeply SE to plantation, then R to wide track, go 50 m R to narrow trail branching up R to contour steep slopes below W face cliffs and rejoin ascent path. At col follow track NE for 1 km to gate, now strike NW to stile, pass Shawtop ruins to lane at Shaw House, follow lane L climbing gradually round hillside to Roach End.
4.5 kms/2¾ miles

(4) Descend N down track 100 m, cross stile L, follow path descending to edge of woods where lesser path branches L contouring through woods. Lud's Church lies on L some 250 m after slight ascent/descent. Path continues 200 m to junction where fork cuts back R descending to river and footbridge below Gradbach. Now take path L through woods along S bank river, then through fields above river and finally woods again to Danebridge.
6 kms/3¾ miles

The entire 1,000-acre Roaches Estate was acquired by the Peak National Park in 1979, so perhaps the wallabies are safe now.

Hen Cloud is one of the more imposing 'peaks' in the Peak District, a real mini-mountain standing aloof beyond the end of the Roaches ridge. The odd name comes from the Celtic *clud* meaning a rock. Our route now traverses Hen Cloud before turning back below the north-eastern slopes of the Roaches towards Danebridge. Do not miss Lud's Church as you descend into the secluded Dane valley on the return journey. Here, hidden in the depths of Back Forest, is a deep dark chasm in the gritstone, the result of an ancient landslip. A mossy and mysterious place, it has been identified as the Green Chapel of the classic medieval poem *Sir Gawain and the Green Knight* (the atmosphere is certainly appropriate) and also with the Lollards, precursors of the Reformation, who held illicit religious services here in the fourteenth century, indeed it is probably named from one of their number, Walter de Ludank. It was later a popular 'sight' for energetic Victorian pedestrians.

The final stretch leads through pretty woods and meadows along the banks of the tumbling Dane, here the border between Staffordshire and Cheshire.

The mysterious cleft of Lud's Church high above the upper reaches of the River Dane, was a meeting place of the fourteenth-century Lollards and has associations, it is said, with Sir Gawain and the Green Knight.

Ingleborough and the Karst Country

Yorkshire Dales

OS 1:50,000 Sheet 98

Ingleborough has been described as the most interesting mountain in England and though a comparison with, say, Scafell would be invidious, the mountain does have a unique appeal and is justly famed. Its massif dominates the countryside of west Craven and although the distinctive ramp-shaped summit is often invisible from the surrounding dales, Ingleborough projects its singular personality over a large area of the Pennines. It is the third highest summit in Yorkshire – though at one time held to be the highest in England. While unspectacular in mountain terms, the Ingleborough massif is scenically remarkable, a geological treasure trove besides having considerable botanic and archaeological significance.

Much of the mountain's special character results from the unique limestone landscape of its lower slopes, a major reason for the designation of the Yorkshire Dales National Park in 1954. Basically the massif is a huge moorland plinth of Great Scar or mountain limestone some 250 metres (800 feet) thick, on top of which repeated layers of shales, sandstones and limestones – the Yoredale Beds – form the final stepped 300-metre (1,000 foot) summit structure capped with a thin layer of millstone grit. Hillside streams flow downwards only to disappear underground on reaching the limestone. Gaping Gill is the most celebrated of more than a hundred such sinks or potholes which surround the mountain, the water often reappearing at resurgences along the base of the limestone above a lower impervious strata. Twenty thousand years ago deep glacier ice covered Ingleborough and many of its features are due to glacial action, not only did the ice grind out the scars and the vast areas of limestone pavement so characteristic of the Ingleborough moors but the powerful melt-water torrents carved valleys and gorges in the limestone when the ground was still frozen and the cave systems blocked, valleys long since abandoned by the water. It is textbook karst country and the finest cave region in Britain.

Our route leaves the pretty village of Clapham by the private Clapdale valley, its beautiful lake and delightful woods created in the early 1800s by members of the Farrer family of Ingleborough Hall at the top of the village. Various exotic shrubs and plants in the wood were planted by Reginald Farrer, the famous plant collector who died in 1920. Clapdale narrows beyond the woods and the path, now a right of way, passes the craggy portal of Ingleborough Cave, a fine show cave, and the closeby Beck Head resurgence where the Clapham Beck emerges from the inner recesses of the Ingleborough/Gaping Gill system beyond. Thereafter Clapdale becomes a classic dry valley, the path passing the Foxholes neolithic rock shelter on the left before turning ninety degrees into Trow Gill.

Probably cut by a retreating melt-water waterfall, Trow Gill is a small but imposing limestone gorge, its walls over 25 metres (80 feet) high and its head narrowing to a mere slot through which the path climbs over a steep boulder slope. Beyond the slot the Ice Age torrent once roared through a narrow green valley which twists through the limestone onto the plateau near the rocky hollow of Bar Pot where a black hole beneath a gnarled ash tree is the caver's regular entrance to the Gaping Gill system. Gaping Gill itself lies a little further on, where the Fell Beck, chattering down over the moor, suddenly plunges into a dark fern-hung abyss. It is the head of Britain's highest unbroken waterfall, albeit in darkness, and all of 104 metres (340 feet) high. It is given to few to visit the awesome base of the falls in a huge cathedral-like chamber over 150 metres (500 feet) long and 30 metres wide where the water – glinting with hints of daylight – pours through a shaft in the vaulted roof. Reaching this sanctum demands several hours strenuous caving or a descent

On the south-eastern slopes of Ingleborough the Fell Beck plunges into the vertical shaft of Gaping Gill – though underground, the highest unbroken waterfall in Britain.

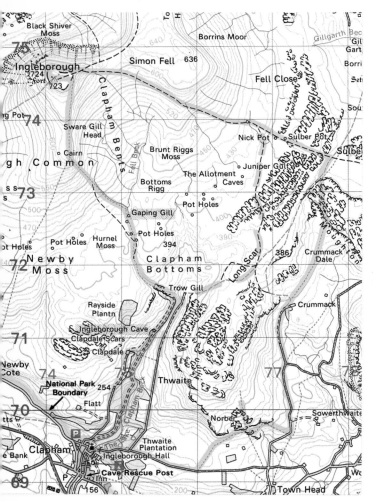

A Circuit over Ingleborough

Length: 17.5 kms/11 miles
Total ascent: 670 m/2,200 ft
Difficulty: easy walking on paths, often boggy, sometimes faint. Mountain is high and exposed, navigation awkward, especially stage (3), compass advisable and 1:25,000 map, showing walls, useful
Note: RoW to Clapdale Farm avoids Clapdale Wood toll. Ingleborough Cave open daily summer, weekends winter. YDNP visitor centre in Clapham car-park

Start: large YDNP car-park at 745692 near Clapham village centre, charge payable. Toilets, etc (160 m/525 ft).

(1) Go N to church, L across bridge, R to corner then L whence sign indicates route to Ingleborough Cave leading R through wood-yard past cottage, 10p payable. Follow good path N along lake shore, through woods into narrow valley. Pass Ingleborough Cave, continue through gate, sharp bend, into Trow Gill. Path emerges into shallow valley leading to stile and rocky depression Bar Pot. Muddy path bears N over flat moorland 400 m to fenced pothole Gaping Gill. (405 m/1,330 ft).
4.5 kms/2¾ miles

(2) Ascend steepening path over boggy moorland then gritstone scree NW to flat shoulder of Little Ingleborough. Muddy path continues N along plateau to second flat shoulder then ascends across rocky slopes on RHS to summit plateau, go 300 m W to

down the main shaft by the bosun's chair set up during summer bank holidays by local cave clubs. The intrepid French speleologist, Edouard Martel, was the first man to enter Gaping Gill, lowered down this main shaft in 1895, while in 1983, after more than a decade of concerted attempts, cave divers finally entered Gaping Gill to emerge in Ingleborough Cave.

The route up Ingleborough is now straightforward, crossing the shoulder of Little Ingleborough before climbing to the extensive summit plateau graced with its topograph and shelter-wall built to commemorate the '53 Coronation. Traces of ramparts encircle the plateau and nineteen hut circles have been located here, for this is Britain's highest Iron Age hill fort. It seems to have been built by the local Brigantes tribe in reaction to the Roman occu-

pation rather than to enjoy the wide view which in good conditions stretches from Pendle Hill and the Isle of Man to Scafell.

Initially the descent follows the badly eroded route of the famous 'Three Peaks Race' across the flanks of Simon Fell to an area of weird limestone landscape known as Sulber Scars. Here a desert of white pavement fretted by clints and grikes – blocks and fissures – stretches in every direction, though close examination will reveal a wealth of plant life flourishing in the protective grikes. Several routes now lead back to Clapham but navigation hereabouts can be confusing: the suggested route descends via mysteriously named Thieves Moss and Beggar's Stile, over two strange craggy amphitheatres and past a deserted Iron Age settlement into the head of charming little Crummack Dale. Down the dale lie the famous Norber Erratics, scores of slaty boulders plucked by the ice from an outcrop up the valley and deposited here on a limestone bench. The final stretch, Thwaite Lane, is a continuation of celebrated Mastiles Lane above Kilnsey, a section of that medieval monastic highway that crossed the southern Dales to link the lands of Fountains Abbey.

topograph, shelter wall, OS cairn INGLEBOROUGH (724 m/2,374 ft).
2.5 kms/1½ miles

(3) From NE corner plateau, path descends R of ridge crest then across S flank Simon Fell, eroded, boggy, to ruined shooting hut 767740. (425 m/1,395 ft). Level path continues over stile to signpost whence follow single pointer 'Footpath' W through expanse limestone pavement to lonely signpost at 778735 (Sulber) whence go R signed 'Clapham'. At stile 400 m drop L through wicket gate to Thieves Moss below rocky amphitheatre, follow faint path through rocky pavement, cairns, to Beggar's Stile notch in Moughton Scars escarpment. Cross stile, descend to grassy terrace traversing W side Crummack Dale to stile above farm. (300 m/985 ft).

(4) Join track W of Crummack farm, continue c 1.5 kms S to tarmac, at bend strike R through stile to Nappa Scars craglet, stile and signpost beyond at 766697 amid Norber Erratics, best just up slope above. Follow sign 'Clapham' W below crags (Robin Proctor's Scar), then SW to stile, lane leads W (Thwaite Lane) eventually through tunnels to Clapham church.
5 kms/3 miles

Alternative
(3.A + 4.A) At first signpost past shooting hut faint path strikes S, over area limestone pavement, to join ancient drove road at Long Scar. Continue SW over hillside to gate and walled lane (Long Lane) leading S to Thwaite Lane and Clapham.
Saves 2 kms/1¼ miles

The 'grikes' – or fissures – between the limestone 'clints' are often sanctuaries for small plants. Our route crosses this area of characteristic limestone pavement at Sulber below the slopes of Simon Fell, Ingleborough's eastern satellite.

The Wharfedale Heights – Great Whernside   Yorkshire Dales

OS 1 : 50,000 Sheet 98

> 'Our maps are candid charts of desolation
> And wear the Pennine weather on their sleeve'
>
> Ivor Brown: 'The Moorland Map' from *Landmarks*: 1943

Great Whernside is the monarch of the Eastern Dales. Not to be confused with Whernside, one of the famous 'Three Peaks' of Craven, Great Whernside rears its massive whale-back above upper Wharfedale, the fifth highest summit in Yorkshire, the most easterly two-thousand-footer in Britain north of the Peak. A typically unspectacular Yorkshire mountain, its gritstone summit sits on layers of shales, sandstones and limestones – the Yoredale beds – which rise from a plinth of Great Scar limestone. Accordingly,

from high desolate moorland of peaty grass and dark boulders, its terrain changes to smiling hillsides slashed with white limestone scars and seamed with caves, while myriad shafts riddle the middle slopes, dotted with the abandoned workings of old lead mines to lend the mountain a hint of times-past, of melancholy even. Nevertheless its summit is windblown, aloof and atmospheric, far higher it seems than a mere 2,310 feet. Great Whernside is a worthy mountain.

One of the gems of Wharfedale, the picturesque small village of Kettlewell is mentioned in Domesday Book, it has a long history of marketing, mining and farming and still boasts no less than three inns, a reflection of one-time importance as a stage on the old coaching route

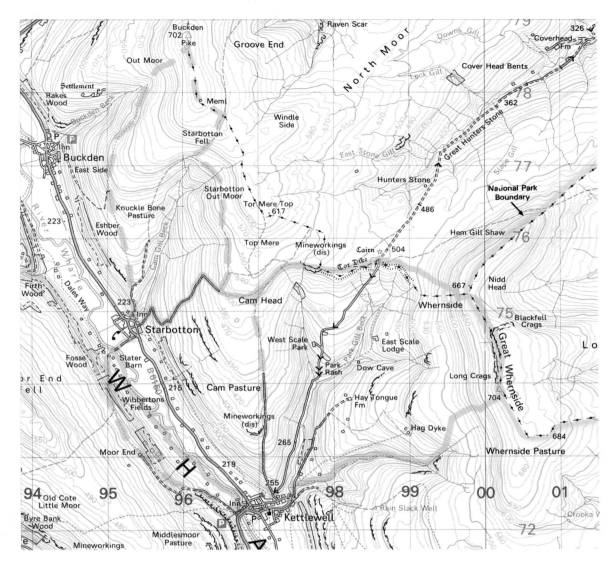

The route from Kettlewell onto Great Whernside follows Dowber Gill, carved deeply into the limestone on the western flanks of the hill.

from Keighley to Richmond. Our walk starts through the village and enters the deep rugged Dowber Gill, V-shaped and not quite a gorge but set with limestone outcrops and initially scattered with gnarled hawthorns and ash trees. The tumbling beck pours over impressive little cascades and waterslides but in summer its waters often disappear and resurge in its rocky bed. Then the limestone ends, the rocks darken and the valley forks. High up on the right stand the ruins of the old Providence lead mine, once one of the largest hereabouts, while in the often-dry stream bed below a metal manhole covers Providence Pot. Here starts one of the classic Pennine caving expeditions, following the water through the extraordinary Dowbergill Passage for over 1,500 metres – nearly a mile – to its resurgence in the depths of Dow Cave in Caseker Gill, the next valley northwards on the flanks of Great Whernside.

Climbing out of Dowber Gill the route crosses tussocky moorland and a terrace littered with aircraft wreckage to a low edge of strange gritstone boulders, crinkly surfaced and almost rectangular, crowned by the huge cairn of Great Whernside. Almost ringed by a horizon of rolling Pennine tops, the best views are eastwards towards the distant Cleveland Hills. Continuing almost level over the bouldery outcrop of Blackfell Top and alternating between boggy patches and a shallow rocky edge, the crest drops towards Black Dike End and a view down to the twin Nidderdale reservoirs – a lake view unusual in the Yorkshire Dales. Then the path plunges down the massive rounded shoulder of Great Whernside to meet the lonely Coverdale road, once the Richmond stage coach route and unmetalled until 1953,. The rumpled flanks of Park Gill and Cam Gill frame a green glimpse of Wharfedale far below.

From the road a boggy route strikes across featureless moorland towards Tor Mere Top and Buckden Pike, but our 'edge' route – and a more interesting approach to the Pike – follows an old green lane along the line of the enigmatic Tor Dike. Nearly a mile of still-discernible Iron Age ramparts, the Dike follows the lip of the moorland plateau encompassing the natural limestone scars: it appears to defend the pass over to Coverdale and may have been built around 70 AD by the local Brigantes tribe, possibly against the Roman invader. The lane is the ancient trackway from Wharfedale, Arncliffe and Settle to Nidderdale; the Cam Head fork was once a busy junction and from it a direct descent can be made down the medieval 'Top Mere' track to Kettlewell. Meanwhile the walled lane, wide enough for a sheep or cattle drove, drops steeply to the dale floor at Starbotton, a now quiet hamlet but once an important crossroads which, all but destroyed by the catastrophic flash flood of June 1686, later became a flourishing mining village until the industry declined in the 1840s: the name means 'Stony Valley'. A modern footbridge takes the ancient route over the Wharfe, a noted trout stream, and the gentle riverside path now followed by the modern 'Dales Way' leads back southwards to Kettlewell while along the western flank of the valley a slightly longer alternative path shows the glacial trough of the dale to good advantage – although it existed before the ice – while one of Britain's finest lynchet or 'Celtic field' systems should be visible across the dale if the light is low.

On the wide saddle between Great Whernside and Tor Mere Top the route follows the line of the ancient Tor Dike. The Cam Gill valley beyond leads down towards Kettlewell.

A Great Whernside Circuit

Length: 15.5 kms/9½ miles
Total ascent: 550 m/1,800 ft
Difficulty: easy walking mostly on good paths/tracks, Great Whernside summit high and exposed, compass useful in poor conditions
Note: route lies within YDNP car-park, nearest Visitor Centre at Grassington

Start: large YDNP car-park at 968723 near Kettlewell bridge, charge payable (210 m/690 ft).

(1) From car-park go L, first R, pass toilets, church, continuing beside stream to rough track, Dowber Gill Bridge. Turn R into camp-site field, keep R, stile, on beckside path into narrow Dowber Gill following good path N bank of beck. At Providence Pot cave entrance beck forks, climb steeply N out of L fork, hints

of path, to open hillside above Hag Dike Scout centre, join well-defined path ascending easily NE to crest, OS and summit cairns GREAT WHERNSIDE (704 m/2,310 ft).

5 kms/3 miles

(2) Strike N 600 m along shallow ridge crest, poor path, boggy and rocky, past shelter ring to bouldery outcrop Blackfell Top. Path descends gradually NW to wall, stile, dropping steeply beyond to boggy plateau and unfenced road at hill lip. Across road green lane contours W round hill lip to Cam Head track junction, signpost, continue R round hillside, stiles, lane becomes walled, wide, stony, zigzagging steeply down to Starbotton village.

7 kms/4½ miles

(3) At S end village turn R off road through gate, lane leads to River Wharfe footbridge. Cross, go L joining Dales Way along W bank to Kettlewell.

3.5 kms/2¼ miles

Continuation to Buckden Pike

(2.A) At gate in walled lane 600 m beyond Cam Head junction strike up R, follow track 300 m, gate, then contour hillside above wall, past old mine workings, to junction above steep valley head, cairn. RH track climbs to boggy plateau meeting hillcrest path by wall. Follow path N onto wide ridge, past memorial cross, to OS cairn BUCKDEN PIKE (702 m/2,032 ft) extra ascent 230 m/760 ft).

8.5 kms/5¼ miles from Great Whernside

(3.A) Descend SW from cairn into hollow above Buckden Beck, traces of path, aiming at abandoned mine workings whence footpath contours S round hillside, gradually descending to join drove road above Starbotton. Continue to Kettlewell as for (3).

8 kms/5 miles

The Cleveland Crest   North York Moors

OS 1:50,000 Sheets 93, 100

Cleveland, they say, was originally 'cliff-land' – a region of steep and craggy scarps looking down on the flatlands of the twisting Tees and the green vale of its tributary the little River Leven. But the modern county of Cleveland comprises the industrial cities of Teesside and little more, leaving the closeby Cleveland Hills, as the north-western extremity of the North York Moors, to remain as they should be – part of North Yorkshire.

The North York Moors contain England's largest expanse of heather-covered upland – some 500 square kilometres (196 square miles) – one reason for their designation as a National Park in 1952, and although some areas are conserved as grouse shoots there is a tradition of

de facto access to the open moors and several long distance moorland walks have become renowned in recent years. Longest is the Cleveland Way which follows the western and northern scarps from Helmsley before taking the precipitous east coast southward to Filey – a distance of 150 kilometres (93 miles). Most classic is the Lyke Wake Walk from Osmotherly to the sea at Ravenscar, a test piece of 64 kilometres (40 miles) to be completed in under twenty-four hours. Originally a gruelling expedition, more than thirty years of use have worn a muddy trail

The sandstone crags of the Wainstones rise beside the Cleveland Way on the edge of Hasty Bank and overlook the Vale of Cleveland: across Garfitts Gap rise the slopes of Cold Moor.

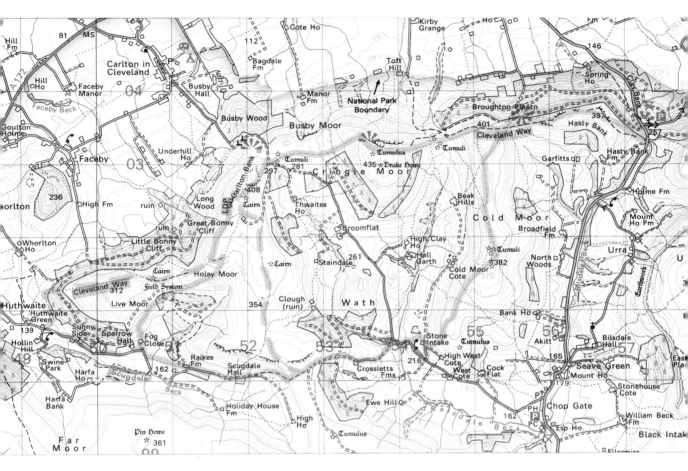

across expanses of once trackless moorland and erosion has now become a frequent problem.

Coincident with one of the best sections of both walks, the circuit suggested here savours all three intrinsic aspects of the moors: the imposing edge, the heather plateaux and the charming dales. Hasty Bank, the first summit, is easily reached and its long crest, supported by the dark crags of Raven's Scar, provides a magnificent view out over the Vale of Cleveland. Eastward the forest-hung scarp extends from Urra Moor to the strange pinnacle of Rosebury Topping above Captain Cook's boyhood home, the village of Great Ayton, with a distant glimpse of the ocean beyond. A thousand feet below the patchwork of fields sweeps northward to a horizon of towers, spires and chimneys, of steam, clamour and smoke – the Teesside conurbation.

Dropping off Hasty Bank the route passes the Wainstones, an inviting area of shattered sand-

stone boulders and grotesque towers which catches the sun and is the most important climbing crag in the area: the name is said to derive from a Danish chieftain who was slain here. Cold Moor rises beyond Garfitts Gap and its wide heathery crest is followed south to Point 382 m where the right of way leading directly down into Raisdale becomes lost in knee-deep heather – not overly strenuous if taken in **descent**. Some walkers may prefer to continue the traverse of the northern edges from Garfitts Gap in one continuous section, although that necessitates a strenuous **ascent** of these slopes on the return. Other indirect paths do exist on this grouse moor.

Raisdale is a smiling and pretty valley with moors swelling high on either side but its narrow lane is a through route and the dale is not as secluded as Scugdale which is reached by a long but gentle ascent over Barker's Ridge. A deserted cottage stands in a blind coombe below craggy bracken-cloaked hillsides at the head of Scugdale and rough pastures dotted with stunted oaks and hawthorns fall to a burbling beck. A line of grey spoil heaps scattered across the hillside are among the few signs of the jet mining industry that once flourished in this charming and venerable-seeming dale.

The secret oak-forested hollow above Fog Close is known as Snotterdale. The long climb above it on to heathery Holey Moor is never steep and the Cleveland Way is soon joined. One can continue down Scugdale to join the Way — indicated by acorn symbol waymarks — at Huthwaite, but the price is an extra two kilometres. The upper section of Holey Moor where it narrows towards the summit of Carlton Bank is the home of the Newcastle and Teesside Gliding Club and while the soaring gliders are graceful creatures of the wind, the discordant paraphernalia of hangers and huts, winches and wires, adds nothing to the ambience of this fine hill.

Cringle Moor, the next top, is the second highest point on the North York Moors but the path avoids the actual summit and follows the scarp edge. According to the memorial topograph agreeably sited at the head of the northern spur the superb vista extends as far as Cross Fell and Ingleborough. Signs of long-abandoned alum and jet workings are common among the hollows as the edge continues over the northern end of Cold Moor and drops once more to Garfitts Gap from where a forestry track — once an old jet miner's trail — leads back to Clay Bank avoiding Hasty Bank. If it is evening however, connoisseurs will return along the high edge to savour the Teesside lights glinting beyond the wide vale.

The Cleveland Crest

Length: 19.5 kms/12 miles
Total ascent: 940 m/3,080 ft
Difficulty: mostly good paths, RoW through trackless heather on Cold Moor
Note: high moors subject to changeable weather. Route sometimes badly eroded on N scarps, follow NP erosion control signs. Grouse shooting may take place Cold Moor 12th Aug to 10th Dec, enquire locally

Start: good FC car-park Clay Bank 572035. (240 m/787 ft).

(1) Cross road to woods, steps lead to wide forestry track, ascend L 200 m to corner, cross stile on to open moor, join well-used path climbing steeply W to flat top and edge HASTY BANK

From Barker's Ridge our route drops into the secluded head of Scugdale. In the distance the nose of Lime Kiln Bank rises above Swainby village.

(397 m/1,302 ft). Continue along edge, descend past Wainstones crags to wide saddle Garfitts Gap (300 m/984 ft) now poor path ascends steeply SW across hillside to wide sandy track on crest Cold Moor, continue 1.3 kms S to tumuli (382 m/1,253 ft) then strike SW down wide shoulder, RoW overgrown, to gate, wall, Stone Intake Cottage. Descend lane W to Raisdale bridge 540005 (185 m/607 ft).

(2) 100 m W track forks L past Raisdale Mill up to Crossletts Farm and Barker's Ridge saddle beyond (330 m/1,083 ft) where path forks R descending across hillside by power lines to narrow lane at Scugdale Hall. Descend lane to Raikes Farm, RoW cuts corner, take drive to Fog Close farm where sunken trackway above intake wall leads up around hillside, through wood to stile at N corner, now join sandy track ascending NE across Holey Moor, after 800 m strike L to join Cleveland Way along W lip of moor, continue N – beware gliders, winch cables – to OS trig pt and finger stone CARLTON BANK (408 m/1,338 ft).
7 kms/4½ miles

(3) Descend NE round old quarry edge then down through old alum workings to road 523030 (290 m/951 ft), go R 30 m to Cleveland Way path and climb to stone seat, topograph on spur CRINGLE MOOR (c 410 m/1,345 ft), continue round N edge, descend steeply to Raisdale saddle (310 m/1,017 ft) and ascend over COLD MOOR (401 m/1,316 ft) to meet outward route Garfitts Gap. Cross stile N to wide forestry track contouring round E below Hasty Bank scarp to start.
6 kms/3½ miles

Variation avoiding Scugdale
(2.A) From Barker's Ridge saddle (527001) path strikes NW between craglets 800 m to track junction by small Brian's Pond. Continue N to gliding access track and summit Carlton Bank.
Saves 2.5 kms/1½ miles and 165 m/550 ft ascent

High Cup Nick and the Northern Pennines

OS 1:50,000 Sheet 91

Nowhere is that backbone of England, the Pennine Chain, more impressive that at its northern end where for nearly 50 kilometres (30 miles) between the Stainmore and Tyne Gaps a great west-facing escarpment falls steeply from the barren moors into the green and fertile Vale of Eden. Rising to almost 3,000 feet (900 metres) on Cross Fell, the Pennine's loftiest summit, this mountain wall echoes – though more forcefully – the jumbled hills of the Lake District of similar height but so very different in style, that face it across the Vale. Indeed, a clear view of this formidable rampart from one of the higher Lakeland fells, especially in winter, is possibly the most remarkable upland vista in England. This northern section of the Pennines – the 'Alston Block' – is formed by a thick platform of mountain limestone, capped by gritstone and uptilted westward to where the scarp edge lies along a great fault and in the aeons since the slippage occurred the Vale beyond has been covered with a layer of new red sandstone.

Craggy outcrops of igneous Whin Sill dolerite intruded into the rocks of the scarp form High Cup Nick, the unique feature of our route.

Dufton is one of several picturesque villages that lie beneath the scarp, owing much of their prettiness to the colourful sandstone so characteristic of the older buildings. The Pennine Way drops to the village on its journey from upper Teesdale to Cross Fell and our walk initially follows its route from Dufton back up to the Pennine crest. Set on gently rising pastures, Bow Hall farm is the last habitation and offers teas and accommodation to weary hikers before the walled green lane beyond climbs high onto the hillside and views open up across the smiling Vale. Wild Boar Fell and the Howgills appear on the horizon, the lane becomes a path and the dark, shapely pyramid of Murton Pike looms

This weird pinnacle stands among the dolerite cliffs that line the slopes of High Cup Nick close below the Pennine Way path.

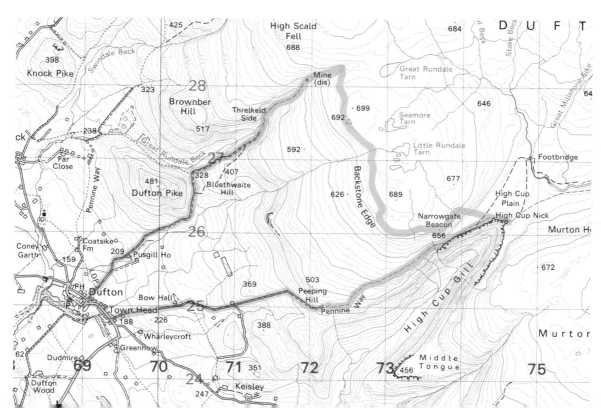

across the glinting river twisting below in the deep mouth of High Cup Gill: the dull crump of artillery is often heard from behind the Pike. As the Gill narrows lines of cliffs appear around its lip and the path crosses a couple of cascading becks – awkward when frozen – to follow a gentle terrace of greensward round the crag-hung hollow to the wide saddle of the Pennine watershed ahead. Precipitous grass slopes fall beneath the crags into the blind and secluded head of this curious valley where there is only the odd sheep to hint that the scale is less than it looks. One should traverse the cliff top below the path to admire the shattered columnar structure of the dolerite and several spectacular pinnacles. This is High Cup Nick, cutting right back into the spine of the great escarpment, a place to pause before climbing northwards to the conspicuous rocky cairn of Narrowgate Beacon that has already overlooked so much of the route.

Beyond the Beacon a shallow and intermittent gritstone edge lines the lip of the empty windswept moors of what is loosely known as Dufton Fell and the stern and almost featureless foreground presents a remarkable contrast to the green chequer-board of Edenside far below. A trickling beck issues from a miniature canyon in the peat, gushing over the edge on its journey to

the Solway and flowing from lonely Little Rundale Tarn nearby where just 200 metres distant another shallow pool empties to the North Sea. Ahead the white radar dome on Great Dun Fell points the way to the Backstone Edge OS pillar while the cairn of the true summit, invisible from afar, stands a short way into the moor. Hidden until the last moment, the deep re-entrant of Great Rundale has recently been scarred with a rocky jeep track; a ramshackle bothy lies at its head beside Great Rundale Tarn but our route follows the track down, past the long abandoned workings where lead and barytes were won in the mid nineteenth century, to the ugly devastation of the modern mine. Rugged and crag-ringed, Great Rundale is not open like High Cup Gill but strangely enclosed, its mouth choked by the rearing cone of Dufton Pike, a delectable valley until recently when, at its most imposing level, its bottom was ripped out in the quest for barytes, a mineral used in paint and by the oil industry. Ironically, at the time of writing, the bottom has dropped out of the barytes market and the valley is again silent. The track leads easily down through the attractive lower valley and round Dufton Pike back to the village.

High Cup Nick and Backstone Edge

Length: 15.5 kms/9¾ miles
Total ascent: 550 m/1,800 ft
Difficulty: easy walking on good paths or tracks, fairly rough but flat on high exposed moorland where compass essential
Note: empty moorland section (2) is regularly walked but is not RoW. Live firing at military training area centred Hilton 5 kms S. This route and Pennine Way do not pass through Danger Area

Start: no formal car-park Dufton village (690250), limited parking possible with discretion by village green, at gravel dump 688253 or in Bow Hall Lane 695249. Do not inconvenience residents.

(1) From village green follow Appleby road SW 400 m to hollow, Bow Hall lane strikes L waymarked 'High Cup Nick'. Ascend metalled lane past Bow Hall farm, green lane beyond, gated, to open fell. Beyond last gate good path continues, climbing easily to horizontal terrace above steep flanks High Cup Gill leading eventually to wide saddle above Gill head. High Cup Nick (580 m/1,900 ft).
6 kms/3¾ miles

(2) Return 300 m to signpost inscribed 'footbridge', strike W, no path, to ascend easy grass slopes R of screes joining narrow path on plateau lip above. Follow path L to prominent cairn NARROWGATE BEACON (656 m/2,152 ft) continuing round edge W then N, poor path, to cairn/shelter walls. Path fades, edge soon disappears, strike R 200 m, boggy, to higher less pronounced edge followed by faint path to Little Rundale beck, and on to OS cairn 692 m, summit cairn BACKSTONE EDGE 200 m NE (699 m/2,292 ft). Continue N, descend boggy trackless slopes to join jeep track in deep re-entrant 725283.
4 kms/2½ miles

(3) Descend track W, through mine workings, down S side Gt Rundale valley, L round Dufton Pike, crossing fords, to reach centre Dufton village.
5.5 kms/3½ miles

A little brook trickles over the steep lip at the very head of High Cup Nick, a peculiar valley that cuts deeply into the great western scarp of the Pennines.

The Roman Wall

Northumberland

OS 1:50,000 Sheet 86

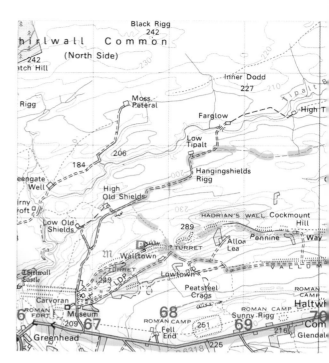

'Old men who have followed the Eagles since boyhood say nothing in the Empire is more wonderful than first sight of the Wall!'

Puck of Pook's Hill: Rudyard Kipling

This walk along the Wall will set you thinking. Thinking of the Roman legionnaires who built it over 1,800 years ago; of the Imperial auxiliaries who guarded it for two and a half centuries and the Pictish barbarians who finally overran it as the Dark Ages began. This was the stormy North-West Frontier of the Roman Empire, it marked the edge of civilization. What do we share with those soldiers of Rome as we pace their Wall? There are still larks in the summer sky and curlew on the hill, the autumn bracken was as golden then as now and the winter wind as sharp as it sweeps in from Scotland with snow in its teeth. Their Wall may be shrunk and broken and its garrison long gone but it is still the most impressive boundary on the map of Britain.

The Wall – Hadrian's Wall – spans England at its narrowest point and the best section of its seventy-three-miles length is where it crosses the backbone of England as the Pennines blur towards the Cheviots. Not only is this the most scenic countryside, but up here the Wall and its supporting roads, ditches and forts have been best preserved from centuries of stone-stealing vandals and the march of progress and lie within the Northumberland National Park. Walls do not naturally lend themselves to meaningful circuits but the one suggested here covers much of the best of the Wall itself besides sampling a little of the surrounding countryside it so dominates. The route described gives a reasonable day's hike but is easily broken into shorter outings or extended to provide a 20-mile (32-kilometre) marathon, as desired.

Canny military engineers, when they built the Wall between AD 120 and 130, took advantage of a remarkable geological feature, a rippling ridge of igneous rock – a quartz-dolerite called Whin Sill – intruding into the surrounding

The most photogenic section of Hadrian's Wall is this stretch over Cuddy's Crags looking eastward towards Housesteads Crags. The Wall continues over the distant prow of Sewingshields Crags.

sedimentary strata and characteristically forming a craggy scarp frowning steeply to the north. The Wall follows the scarp crest to form a superb defence line. The Steel Rigg to Housesteads section is typical, running above a sequence of steep cliffs of which one – Highshield Crags rising over 35 metres (80 feet)) vertically above the dark waters of Crag Lough (pronounced 'loff') – is the most famous rock-climbing venue in Northumberland.

At the time of writing National Trust archaeologists are working on the Steel Rigg section and it is fascinating to watch them at work, both excavating and renovating, carefully replacing each stone in its original position with new mortar. The Wall hereabouts was originally some 5 metres (15 feet) high of which the top third – the breast-work – has fallen while the lower third is buried in the detritus of centuries. The turf has been cut back in places on the northern side to expose as many as eight previously hidden courses of stonework and here the Wall rears an imposing ten feet and more, though throughout most of our route it rises only half that height and has sometimes even been entirely dismantled and replaced at some juncture by a dry stone shepherd's wall. Three Milecastles – 39, 38 and 37 – stand between Steel Rigg and Housesteads, there was one every Roman mile (1,481 metres/1,620 yards) with two 'turret' guard posts between them. Just west of impressive Milecastle 37 is Cuddy's Crags – so named from St Cuthbert whose body was supposed to have been laid here *en route* from Lindisfarne to Durham – the most popular and photographed section of the entire Wall, leading shortly to the major fort of Vercovicium at Housesteads where the 1,000-man garrison has been replaced by a car-park, a visitor centre and a small museum under the care of English Heritage. The site is well worth a diversion and if one descends from the Wall to enter it the usual fee should be paid. However, the Knag Burn Gateway closeby, a unique fourth-century customs post virtually in the centre of England, is on the Wall itself and can be visited for free.

The return route westward along old tracks and footpaths running below the northern scarps is not without interest for this is the viewpoint of the Wall that would have discouraged would-be attackers, though rough pasture and grassy fields have largely replaced the moorland of heather, bracken, birch and alder scrub that lay below the Wall in times past. Look to the walls of field and farmhouse for army-issue quarried stones, obviously filched long ago from the Wall itself. At reclaimed Cawfields Quarry our route rejoins the Wall and returns eastward over fairly rugged country to Steel Rigg. Beyond the deep Shield Gap the Wall, unfortunately not here at its structural best, climbs towards its highest point on Winshields summit. From here the view is extensive, the dark conifers of Wark Forest roll northward towards the Cheviot outliers and the Scottish Liddesdale Hills, southward the Pennines rise towards Cross Fell while the giant cooling towers of Annan power station stand westward above the Solway flats: no, it is no longer quite as the Romans saw it!

It is worth diverting a few yards from the Wall at Housesteads to explore the ruins of the Roman fort of Vercovicium: this is the remains of the headquarters or administration block.

A Circuit of Hadrian's Wall

Length of circuit: 21 kms/13 miles
Total ascent: 570 m/1,870 ft
Difficulty: easy going on the Wall, some muddy fieldpaths elsewhere
Note: background knowledge is essential to appreciate fully any walk along Hadrian's Wall. Several guides and handbooks are available, OS special 2 inch: mile map 'Hadrian's Wall' is useful. Especially interesting are nearby musems at Vindolanda Fort (771664) with full-size replica of Wall section, and Museum of the Roman Army (667658) near Greenhead. County Council plan Theme Park at reclaimed Walltown Quarry (668659)

Start: good car-park at Steel Rigg 751676 (280 m/920 ft) – toilets.

(1) At SE corner car-park gate leads onto Wall, follow it up and down E above Peel Crags, over Steel Rigg, past Milecastle 39, over Highshield Crags (330 m/1,085 ft) above Crag Lough, then through woods to Hotbank lane. Climb now past Milecastle 38 to crest Hotbank Crags (320 m/1,050 ft) and Cuddy's Crags, past Milecastle 37, through wood to ruins large Housesteads Fort. Wall crosses Knag Burn at wide saddle and ascends again above Broomlee Lough, continue to stile/gate through wall 100 m past 2nd top (King's Wicket 798694).
5.5 kms/3½ miles

(2) Pass N through Wall, yellow waymarks lead W over boggy ground to stile – poor path – and through new plantation to poorly defined green lane. After crossing Pennine Way near ruined kiln, lane becomes muddy tractor track. As track turns S to Hotbank, yellow waymark indicates poor path through gate leading W across fields, past two barns to farm track and lane N of Steel Rigg. Follow lane NW then SW past Melkridge Common to junction 726674. 100 m N faint RoW leads through gate over fields to farm track and Cawfields Farm, continue 500 m to Cawfields car-park/picnic site (713666: 180 m/590 ft).
11.5 kms/7 miles

(3) Take path (Pennine Way) leading W along N shore lake to Wall and Milecastle 42, continue along crest Cawfield Crags, pass Thorney Doors step, Turret 41A, descend to lane at Shield Gap. Climb steeply to ridge crest where Wall continues, up and down, to rocky summit and trig point WINDSHIELDS CRAGS (345 m/1,132 ft) descends to lane at Steel Rigg.
4 kms/2½ miles

Variation

(3.A) From Cawfields car-park continue W along line of Wall (Pennine Way) past inconspicuous ruins Great Chesters fort, over Cockmount Hill to impressive section of Wall along Walltown Crags to Point 249 m. (**5 kms/3 miles** thus far). To return Cawfields either descend S to small car-park at 675663 and follow farm lane E (not RoW) past Lowtown and along line Roman vallum, or descend N from Turret 44B (680666) and follow lanes and RoW N of Wall E via Low Tipalt. From Cawfields follow (3) as above.
Total extra 11 kms/6¾ miles

High Street and the Kentmere Horseshoe   Lake District

OS 1 : 50,000 Sheet 90

High Street, a strange name for a mountain! Were it more shapely it might well have been dubbed 'pride of the eastern fells' for it is the principal eminence in that rather formless area of Lakeland where the high tops start to fade away eastward. Its main claim to fame, of course, is the Roman Road which traverses its whaleback plateau, but it is also the apex of the longest high ridge in the Lake District, over 13 kilometres (8 miles) between Yoke and Loadpot Hill continuously above 600 metres (2,000 feet), and the hub of a complex valley system. Formed from

the erosion-resistant rocks of the Borrowdale Volcanic series – the same rocks that shape the rugged mountains of central Lakeland – High Street's dull crest belies the spectacular scenery of its boney flanks and its several handsome satellites. A horseshoe ridge-walk that samples much of the best encircles Kentmere, central of the three parallel valleys that probe towards High Street from the south. Kentmere once held two natural lakes and the relict mere in the lower valley, drained a century past and now unobtrusively excavated for diatomite, is all that

There is a fine view from Star Crag on the north ridge of Yoke into the head of the Kentmere valley: Ill Bell rises on the left and the route follows the skyline over distant High Street, Mardale Ill Bell, the Nan Bield Pass – above Kentmere Reservoir – and Harter Fell on the right.

remains. A second lake, silted up in ancient times, lay in the secretive upper valley, hidden beyond the rocky knolls and wooded moraines beneath which stand the church, a scattering of cottages and Kentmere Hall, a farmhouse incorporating a fine Pele Tower built against fourteenth-century Scottish raiders. Though the farm is private, a nearby right of way makes an alternative start to the walk, rejoining the scenic Garburn Pass packhorse road near the huge boulder known as Badger Rock, laced with climbing problems. At the head of the pass a gentle ascent commences towards Yoke, avoiding a fierce bog and giving glimpses into the pretty Troutbeck valley: higher up as the summit approaches, impressive views open up down the length of Windermere and over to Coniston and Langdale. Now the ridge takes on a most distinctive form, rising and falling northwards over the splendid conical summits of Ill Bell and Froswick above the dark waters of Kentmere Reservoir and the wild hollow of Hall Cove. From this side the three peaks appear as green triplets, the lower pair echoing their bell-shaped brother.

On Thornthwaite Crag, its domed grassy summit graced with a peculiar dry-stone column 14 feet high (4.25 metres) and often avoided, the ridge takes a sharp bend, dissolving into the gently swelling moorland of High Street. Climbing from Troutbeck over the shoulder of Thornthwaite, the straight scar of the Roman Road stretches ahead. Built in the first century AD to link the Roman forts at Ambleside and Brougham beside Penrith, it was a ridge-way before the Romans came and has been trodden continuously since they left, though there is now little to show for it but a well-used path and the hint of ditches. The summit of High Street some little way off is disappointing though the name 'Racecourse Hill' marked on the 1:25,000 map recalls the time when this wide expanse was the scene of an annual shepherd's meet, a great occasion when strayed sheep were exchanged, horses were raced, wrestlers fought, and ale was swilled. Mardale was the nearest community, just three kilometres away below the formidable eastern flank of High Street at the head of a remote valley with a bonny lake. 'The Inn at Mardale is full a mile from the water' observed the guide-

book writer Harriet Martineau in 1855. But in 1936 this pretty village was engulfed by the rising waters of Haweswater Reservoir, raised by no less than 29 metres (95 feet) to slake the thirst of Manchester.

On High Street the route turns back southward to Mardale Ill Bell, the first shoulder on the ridge, preferably following the lip above the fine eastern coombes. Though hardly a summit, Mardale Ill Bell appears suitably impressive from the shores of Haweswater, rearing over two wild and craggy coombes, each cradling a perfect glacial tarn: Blea Water and Small Water are among the grandest in Lakeland. Crossing the narrow ridge above Small Water is the old packhorse pass of Nan Bield linking Kentmere to Mardale, rocky and with a shelter wall on its desolate crest it has the distinct flavour of a

Himalayan pass. Steep zigzags now climb to the green dome of Harter Fell, crowned with its bizarre iron cairn and an excellent vantage point over Haweswater – a majestic lake when the water is high, an ugly travesty when it is low – before broad grassland undulates onwards towards Kentmere Pike. From this angle Froswick, Ill Bell and Yoke appear especially shapely and the latter is now seen to hold a huge cliff, Rainsborrow Crag, some 150 metres high (500 feet), its base riddled with abandoned quarries. Though its summit is unprepossessing, Kentmere Pike has an unusual 'edge-of-the-world' atmosphere and its descent, if easy, is not always simple to follow, but the latter stages especially, through the comely Kentmere valley, make a satisfying conclusion to an airy itinerary.

The Kentmere Horseshoe

Length: 19 kms/12 miles
Total ascent: 1150 m/3,770 ft
Difficulty: easy walking mostly on grassy paths, sometimes boggy, rarely steep. These are high and exposed mountains, appropriate equipment essential

Start: limited parking, with care, at 456041 by Kentmere church, popular at weekends. LDNP plans (1987) public car-park nearby (175 m/575 ft).

(1) Follow lane NW from church, keeping L, to yard entrance. Cart track goes R, signed 'Garburn Pass', bends L to gate, continues W, steady stony ascent to gate, crest Garburn Pass (447 m/1,467 ft). Continue 250 m to bend whence path strikes R over moorland, initially faint, boggy, cairns, then dryer track ascends gradually N to wall, follow to stile. Now path ascends steeply R to grassy plateau, cairn YOKE (706 m/2,315 ft). Continue along ridge to small rocky summit, 4 cairns ILL BELL (757 m/2,482 ft).
6 kms/3¾ miles

(2) Descend ½L, steepish, follow ridge over summit FROSWICK (720 m/2,361 ft) to long saddle, 700 m beyond low point take path ½L to plateau, wall, tall summit pillar, THORNTHWAITE CRAG (784 m/2,571). Strike ESE to join major path gradually ascending wide moorland NE, continue 800 m till near LH plateau edge, strike E 100 m to broken wall, OS pillar HIGH STREET closeby on E side -828 m/2,718 ft).
4 kms/2½ miles

(3) Go SE across moorland to scarp lip, poor path follows edge R to join larger path, shallow descent to vague top, twin cairns, MARDALE ILL BELL (761 m/2,496 ft). Stony path descends to outcrops, narrow Nan Bield Pass (625 m/2,050 ft) then ascends steeper rocky ridge to plateau lip and 200 m E to iron-post-cairn HARTER FELL (778 m/2,552 ft). Path, often boggy, follows fence then wall S along broad moorland crest to summit, OS pillar E side wall, KENTMERE PIKE (730 m/2,396 ft).
4.5 kms/2¾ miles

(4) Path descends S, forks where fence bends L, go R to stile, descend across W flank hill. At green spring, cairns, path zigzags steeply down, indistinct, to lower grooved path descending easily L across rougher slopes to lane at Hallowbank cottages. Go L through white gate 200 m, turn R signed 'Mardale Bridle Path' to muddy lane, go L to gate, stone steps through wall R, near huge boulder, lead to footbridge and lane beyond, go L to Kentmere church.
4.5 kms/2¾ miles

Nordic skiers run the crest of Kentmere Pike northwards towards Harter Fell. On the left, across the Kentmere valley, rise Ill Bell (left), Froswick and Thornthwaite Crag.

The Narrow Crests of Blencathra

Lake District

OS 1:50,000 Sheet 90

'. . . moorish Skiddaw and far-sweeping Saddle-back (are) the proper types of majestic form . . .' wrote John Ruskin in 1889. Indeed, the two major peaks of the Northern Fells both embody classic mountain shapes. But Skiddaw, its noble cone rising from a clutch of graceful satellites and one of the four Lakeland peaks topping 3,000 feet, on close aquaintance is a cragless hump, while Saddleback – or Blencathra as it is more generally known – an almost isolated and complex mountain, flaunts a succession of challenging features and despite its modest stature as only the thirteenth highest, is acknowledged to be one of the grandest peaks in the Lake District. Sculpted from the easily eroded Skiddaw slates, as are all the Northern Fells, the public face of Blencathra – its southern and eastern flanks – has somehow avoided the rounded outlines of its lumpy neighbours. Rearing above the broad pastoral Glenderamackin valley and the busy A66 trunk road, a trio of matching ridges bounded by a pair of boggy grassy buttresses and separated by four deep ravines form its symmetrical south face. These ridges are steep and narrow and rise to pointed tops on a scalloped crest, all three demand to be climbed but the central or Hall's Fell Ridge, its upper section known as Narrow Edge and finishing abruptly on the actual summit, provides the most entertaining route. Cradled on the eastern flank however, a secret and spectacular coombe holds moody Scales Tarn beneath a knife-edge rock arete, the finest such feature in England – Sharp Edge. Only to the north and west where the undulating moorland flanks are known as Mungrisdale Common, does Blencathra disappoint. When seen from the east the mountain assumes a distinctive saddle-shape, and the name Blaencathair, Celtic in origin, appears to reflect this.

Though comparatively short, this particular traverse of Blencathra is one of the best expeditions in the Lakes and can easily be varied to suit weather and inclination. It is described in an anticlockwise direction ascending over Sharp Edge and descending Hall's Fell, but should the prospect of scrambling along Sharp Edge sound too daunting, a clockwise circuit up Hall's Fell but descending through Foule Crag Cove to avoid Sharp Edge is an excellent alternative. Certainly the horizontal rock crest of Sharp Edge is the closest the Lake District comes to Snowdonia's Crib Goch arete, and although it is rather exposed and a steady head is useful, in good conditions the difficulties on the crest are minimal and hands are necessary in only a few places, the lower 'escape' paths seem only to complicate the route. The black tarn below adds atmosphere to the ascent. Surrounded by steep craggy slopes it seems rarely to catch the sun, indeed so dark are its waters that they were once believed to reflect the stars at midday! The famous Abraham brothers, pioneering rock

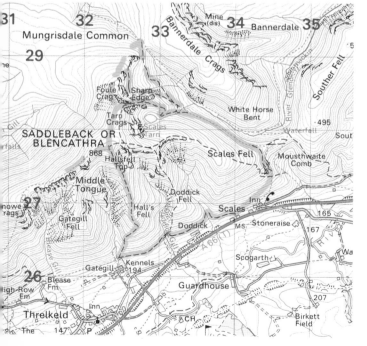

The path to Scales Tarn crosses the shoulder of Scales Fell into the upper Glenderamackin valley, ahead rears the imposing prow of Foule Crag with the black crest of Sharp Edge below it. The ascent is by no means as desperate as it looks here!

climbers and mountain photographers, first practiced their sport here in the 1890s but the rock is poor and it seems unlikely that anyone has climbed here since, except in winter when the broken crags are sealed tight in snow and ice.

Above the arete the east wing of Foule Crag rears up as a blunt headwall, appearing from a distance far steeper than it is: the easiest line is the direct one taking the obvious and well-scratched central weaknesses where the scrambling is slightly technical for a short way and the exposure as great but less apparent. On the close-cropped grass at the top you wonder what all the fuss was about, although descending in winter verglas is a very different proposition! A strange feature on the plateau-like northern ridge is a great cross of quartzite pebbles laid into the grass. Some five metres in length it was constructed in the late Forties by a local man as a memorial to a walker who had died nearby. Seen from here Blencathra's summit is a low-key affair but the southerly view across the deep green vale to the serried fells beyond, to High Street and Helvellyn, to Scafell Pike and the hills beyond Derwent Water, is superb. Especially imposing is the distant wall of the High Pennines lining the south-east horizon. A short diversion along the plateau lip to Gategill Fell Top, Blencathra's western summit, is rewarding and puts the topography of the south face into perspective. Narrow Edge drops straight below the small summit cairn, at first the path descends a steep grassy buttress and then, as the angle eases, the crest becomes narrow and rocky with the path winding between the outcrops and occasionally over short patches of glacis. In ascent, by selecting the rockiest sections of blunted crest, almost continual easy scrambling can be enjoyed for some 500 metres. But the path avoids all difficulties before again descending steeply through bracken and heather to the intake wall at the mouth of the deep Gategill ravine where remains of an old lead mine, worked from the seventeenth century until before the First World War, still stand beside the tumbling beck.

A Traverse of Blencathra

Length: 8 kms/5 miles
Total ascent: 685 m/2,250 ft
Difficulty: in good conditions Sharp Edge demands simple
though exposed scrambling. Foule Crag rather more awkward,
easier in ascent. Narrow Edge any difficulties avoidable, simple
scrambling if desired. Elsewhere easy grassy paths. This is a high
and exposed mountain, appropriate equipment essential
Note: route described anticlockwise, see text

Start: 340268 on A66 road 300 m W of Scales, park at layby
150 m W on N side road, or layby S side road at Scales
(215 m/705 ft).

(1) Between two roadside cottages track leads up to wall, gate,
signpost, diagonal path ascends R across hillside, up grassy spur,
round corner above broken slopes to wide saddle. Follow good
contour path L side narrow Glenderamackin valley, cross Scales
Beck and ascend L, steep stony path, to Scales Tarn, Sharp Edge
rises on RHS. Path leads up R to grassy shoulder, traverse narrow
horizontal rock arete, 150 m easy scrambling, exposed – or
narrow awkward paths RHS below crest – to little nick. Scramble
directly up more difficult rocky headwall above – Foule Crag – to
grassy plateau. Either follow path L round Tarn Crags edge or
take broad ridge above, past cross, tiny tarn, to small cliff-edge
cairn BLENCATHRA summit (868 m/2,847 ft).
4.5 kms/2¾ miles

(2) Immediately below cairn grassy Hall's Fell Ridge drops S,
initially steep. Path descends easily, avoiding simple scrambly
sections Narrow Edge as desired. Beyond flat shoulder path
descends R to meet Gate Gill above intake wall. Good contour
path leads L above wall, crossing two narrow valleys/streams, past
scattered conifers to gate, signpost, at start.
3.5 kms/2¼ miles

Alternative
(1.A) Avoiding Sharp Edge: at Scales Beck continue on traverse
path, after 150 m either taking L fork ascending into Foule Crag
Cove to gain grassy ridge on far side, or R fork leading to col at
head of Glenderamackin valley before striking L up same grassy
ridge. Then ascend slaty Blue Screes to summit Foule Crag.
Continue as for (1).
Extra 1 km/½ mile

On Blencathra's south face the narrow crest of Hall's Fell
Ridge leads directly to the summit providing a fine route for
ascent or descent.

The Newlands Fells – Grasmoor and the Coledale Ridges Lake District

OS 1 : 50,000 Sheet 89

> 'I shall return to Cumberland & settle at Keswick . . . I shall have a tendency to become a God, so sublime and beautiful with be series of my visual existence.'
>
> Samuel Taylor Coleridge: letter 1800

Walk down from Keswick to the shores of Derwent Water and gaze out across the tranquil lake, as Coleridge must have done, to the massed mountains ranked above the wooded western shore. One sharp peak stands out from the interlocking crests and domes of the Newlands Fells: obviously not the highest, it is certainly the most distinctive, and once identified Causey Pike will seem an omnipresent landmark hereabouts. Like Blencathra, carved from rocks of the Skiddaw Slate series, these north-western fells are typically rounded and genteel – in contrast to the scarred and rugged mountains like Scafell to the south, hewn from the harder Borrowdale Volcanics. Here the slates have formed a central clump of higher tops throwing out characteristic narrow ridges above long deep valleys, the gabled crests, seen end-on, sometimes assuming handsome shapes. There are crags among the innermost recesses but they are shattered and shaly while the many abandoned workings bear witness to a rich history of mining – for lead, copper, silver, even gold – going back before the Tudors. Today these fells are justly famed for superlative ridge-walks.

This walk starts near old workings above Farm Uzzicar in the heart of the Newlands Vale. Here once lay Uzzicar lake, drained in the thirteenth century to expose cultivatable land – hence the name 'New Lands'. As the elegant ridge curves over the shoulder of Rowling End towards Causey Pike the view unfolds across lush Newlands to Derwent Water and the majestic shapes of Skiddaw and Blencathra,

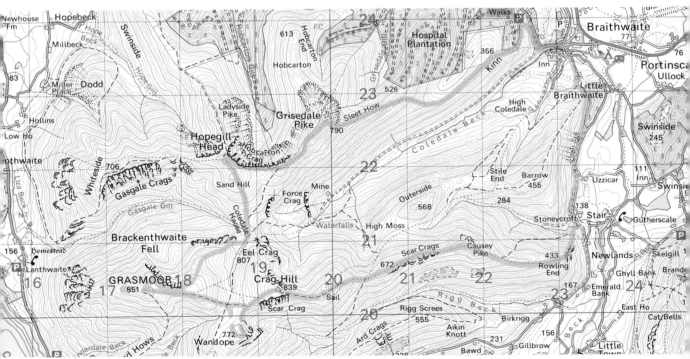

surely one of the most exquisite prospects in the Lakes. Most people will use their hands on the final rocky brow leading to Causey's popular summit which proves to be one of five knobs on an undulating ridge, disappointing perhaps but aloof enough to provide intriguing glimpses into the deep coombes of the neighbouring Derwent Fells. The pleasant ridge continues to Sail Hause, a useful pass with an abandoned cobalt mine below its northern flank, before a steep climb leads over Sail to Crag Hill, the highest 'Hill' in England, an unworthy appell-ation for such a massive crag-girt eminence which is the twenty-third highest point in the country! Perhaps this is why it is popularly known as Eel Crag, really the name of its rugged northern ridge-end above Coledale Hause. Tucked under the southern flank lies deep Addacomb Hole, the most perfect hanging valley in Lakeland, and a short diversion round

its lip to Wandope is recommended before pushing onwards to Grasmoor. Thrusting out its enormous bastion high above Crummock Water, isolated Grasmoor is the giant of the area and there are spectacular views from the outer lip of its turfy dome over the deep gulf of Buttermere to the knot of high mountains surrounding Scafell. Gras-moor derives from the Norse 'grise' – a wild-boar – and this is still tough hunting country where the fox is hunted on foot.

Coledale Hause, the wide green saddle be-tween Crag Hill and Hopegill Head, is the focal point of the Newlands Fells. Below its east flank the little Pudding Beck tumbles over a rocky wall, High Force cascade, into a hanging am-phitheatre from which it cascades again – as Low Force – over a taller crag into the head of long, narrow Coledale. It is a strange place but unfortunately a working barytes mine scars the

base of Force Crag. However, Hopegill Head is a pleasant surprise after the dull plod up from the Hause, its rocky peak – the meeting point of three narrow crests – jutting out above the imposing cirque of Hobcarton Crags, the prospect marred only by the angular geometry of the conifer plantation invading the lonely gill below. These dark cliffs, almost 150 metres (500 feet) high, are shaly and vegetatious and the only known English habitat of the red alpine catchfly, *Viscaria alpina*, a pretty herbaceous flower closely related to the famous Cheddar Pink. The crags continue below the cliff-edge path until it begins its stony climb to Grisedale Pike. Its sharp summit again the junction of sharp ridges and presenting a handsome end-on profile to the Derwent valley. The descent is long but pleasant and leads past the Royal Oak in attractive Braithwaite village – which may well ensure an easy return to Uzzicar.

Causey Pike, Grasmoor and the Coledale Horseshoe

Length: 18 kms/11 miles
Total ascent: 1450 m/4,750 ft
Difficulty: mostly easy going, grassy paths, occasional rocky sections: high and exposed mountains, appropriate equipment essential
Note: almost entire route National Trust property

Start: roadside parking at 233217 E side Stair-Braithwaite road. (125 m/410 ft).

(1) Go S 500 m, beyond Stonycroft bridge good path ascends directly up shoulder past wooden set, keeping L then following crest steeply to flatter ridge Rowling End whence path climbs steeply to 25 m (80 ft) semi-scramble and summit CAUSEY PIKE (637 m/2,090 ft). Path continues, grassy, hummocky, over flat summit SCAR CRAGS (672 m/2,205 ft) to shallow saddle Sail Hause, then more steeply, shaly, up rounded shoulder to dome, small cairn SAIL (773 m/2,536 ft). Narrow ridge, easy rocky steps, leads to broad slaty plateau, OS pillar, CRAG HILL (or Eel Crag, 839 m/2,753 ft).
5 kms/3 miles

(2) Descend easily WSW to wide grassy saddle, two tiny pools, climb W across broad shoulder, cairns, hints of path, to wide plateau, shelter cairn GRASMOOR (852 m/2,795 ft). Strike NE to cliff top, follow edge R then shoulder N to descend grassy slopes E, above waterfall, to broad saddle, many paths, Coledale Hause (600 m/1,970 ft). Ascend path N to round shoulder Sand Hill continuing to crag edge, go L to sharp rocky summit, slightly exposed, HOPEGILL HEAD (770 m/2,525 ft). Retrace path along crag lip continuing round cliff edge to follow broken wall to rocky minor top, now ascend steeper narrower stony crest to small summit GRISEDALE PIKE (791 m/2,594 ft).
7 kms/4½ miles

(3) Continue 70 m, drop R onto E ridge, stony path, initially steeper, then grassy. When ridge flattens path drops steeply R to long green spur, at stile by forest edge keep L till path cuts back R dropping to road. Go R to village, follow Stair road to bend 100 m past bridge whence take Braithwaite Lodge drive RHS, RoW passes through farmyard to contour L above wood rejoining road 600 m from start.
6 kms/3½ miles

Coleridge must have been familiar with this view across Derwent Water towards the sharp peak of Causey Pike and the Coledale summits beyond.

The Napes, Pillar and the Mosedale Crest

Lake District

OS 1 : 50,000 Sheet 89

'Mosedale . . . a noble amphitheatre of dark mountains . . .'

'. . . the famous Pillar Rock springs up vertically from the steep fellside, with a north face like a cathedral-front . . .'

Owen Glynne Jones: 1897

Wasdale, for its mountains and its mood, has no peer in all England. From its old and still-hospitable inn the great pioneers – men like O G Jones and the Abraham brothers, Haskett Smith, Cecil Slingsby and Norman Collie – would set off in their tweeds and hob-nailed boots at the turn of the century to explore the surrounding crags and fells. Indeed, rock-climbing was born here. Luckily Wasdale is a dead-end valley opening to the remote side of Lakeland, much of the dale and its enclosing mountains are protected by the National Trust so most of its visitors are mountain-orientated folk rather than mere tourists, and the atmosphere remains unspoilt. The sanctuary of Wasdale Head is a jigsaw of stone-walled pastures enclosed on one side by the Scafell massif, on the other by a chain of respectable peaks cradling the deep and deserted side valley of Mosedale. This itinerary almost traces the Mosedale crest – but not quite – for it extends the regular 'Mosedale Horseshoe' to include some of the most spectacular and adventurous walking in Lakeland.

The deep-walled lane leading past the tiny church of St Olaf – where victims of several early climbing accidents lie buried – provides intriguing glimpses into Mosedale before Burnthwaite, the last farm, where the sharp-angled pyramid of Great Gable seems to block the valley head. The fluted ribs jutting from Gable's flank are the 'glorious jumble of the Napes' and our route, climbing first across the bottom of the scree-covered hillside to the wide saddle of Sty Head, crosses back along a tenuous and rugged path known as the 'Climbers Traverse' to their base, before angling round the spur of Gavel Neese to the col of Beck Head visible on the left.

Named from the Norse 'stee' – a ladder – Sty Head was once a busy packhorse crossing and is still an important pedestrian pass, though as recently as 1934 there were moves to drive a road over it! Soon after the pass the little crag is Kern Knotts where the obvious crack was a celebrated test piece for many years, while at the base of the intricate Napes ridges, laced with

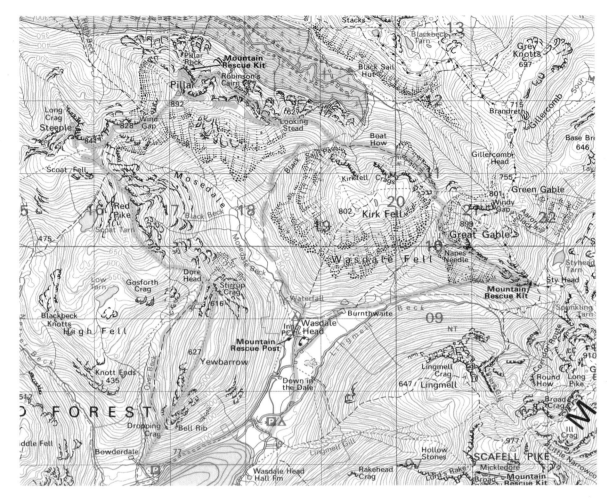

classic climbs, the famous Napes Needle is unmistakable. The first ascent of this remarkable 18-metre (60-foot) pinnacle by Walter Haskett Smith, solo, in 1886, is considered to mark the birth of rock-climbing as a distinct sport. Easier ground leads to Beck Head and pleasant walking continues round the desolate head of Ennerdale to join the usual Mosedale circuit at Black Sail Pass.

The ridge leads onwards to Pillar mountain via the fine vantage point of Looking Stead, but an ingenious traverse – the High Level Route – threads its way round the steep northern fellside to Pillar Rock, together with Scafell Crag and Dow Crag, one of the three greatest Lakeland

There is a fine view of Great Gable from the Styhead path just above Burnthwaite Farm: our route links Styhead Pass, seen on the right, to the saddle of Beck Head, seen on the left, across the south face of Gable below the Naples.

cliffs. At Robinson's Cairn, an impressively situated pile erected in 1906 to remember that notable early cragsman and principal discoverer of the High Level Route, Pillar Rock comes into sight, a dramatic view of this 'noblest of rock monuments' towering above moody Ennerdale. With its crenellated summit – High Man – rising nearly 230 metres (750 feet) above its lowest point, Pillar Rock is seamed with fine climbs ancient and modern, and though first ascended by an Ennerdale shepherd, John Atkinson, in 1826, and long considered an adventurous feat, its summit is still for climbers only. One character it attracted was the Reverend 'Steeple' Jackson, self-styled 'Patriach of the Pillarites', who climbed the Rock in 1875 at the age of 79. Unfortunately he was killed attempting his third ascent three years later. The narrow path winds sensationally up above Shamrock buttress ('sham–Rock', from some angles mistakable for the real thing) to the narrow neck linking the Rock to the main hillside before ascending the final 120 metres (400 feet) onto the rounded grassy plateau of Pillar mountain.

Wind Gap, a stony col, separates Pillar from Black Crag, a distinct and worthy summit until recently ignored on OS maps, whence the gentle sward of the northern edge provides spectacular views of shapely Steeple and Ennerdale Water as it is followed onwards to the old wall on which the summit cairn of Scoat Fell is actually perched. A short diversion will include Steeple's

tiny top before the route drops southwards to join the Red Pike ridge. Surprisingly the path avoids the sharp crest of this fine mountain, and with it the summit, its cairn seeming to overhang the broken precipices that plunge into the lonely head of Mosedale. On the long descent to Dore Head look out for a conspicuous outcrop where the Chair, a rudely constructed and ancient stone seat once famous but of unknown origin, sits among the boulders. Yew–barrow is the final summit and its northern bastion, Stirrup Crag, rears up imposingly above Dore Gap. Indeed a well-used path

descends the Over Beck valley to avoid the mountain, but the short scramble is far easier than it appears and the long aloof crest of Yewbarrow is the finest viewpoint in Wasdale, a fitting finale to a fascinating expedition. Now the path drops steeply towards Bowderdale farm, home of Joss Naylor the famous fell-runner, before cutting back to the Wastwater shore and the start.

Yewbarrow is the final summit of the Mosedale circuit: Stirrup Crag, the northern buttress of the mountain, rears above the saddle of Dore Head.

A Napes, Pillar Rock and Mosedale Circuit

Length: 21.5 kms/13¼ miles
Total ascent: 1460 m/4,800 ft
Difficulty: a strenuous expedition, high and exposed, much easy going but one short scramble pitch and several very rugged sections. Appropriate equipment essential

Start: NT car-park/camp site 182075, N end Wastwater (70 m/230 ft), limited parking out-of-season Wasdale Green 186085.

(1) Go NE through camp site, path fords river – or follow E bank to footbridge – continues to Wasdale Green. Walled lane passes church to Burnthwaite Farm whence main path ascends valley bottom before climbing diagonally across stony S face Gt Gable to Sty Head Pass, Rescue Box, path junction (480 m/1,575 ft).
5 kms/3 miles

(2) Avoid main Gt Gable (NW) path, cut back L (W) on tenuous traverse path below Kern Knotts cliff, huge boulders, then rising easily over rocky steps, across large scree shoot to broken ground below Napes crags where paths proliferate. Choose best traverse path passing well below conspicuous Napes Needle, across second scree shoot, descending slightly round SW corner of Mt to cross gentle scree flanks to grassy Beck Head saddle (620 m/2,040 ft). Good path descends NW from 2nd small tarn, traverses grassy N flanks Kirkfell, crosses rocky gully to ascend, indistinct, boggy, to cairn Black Sail Pass (550 m/1,800 ft).
4.5 kms/2½ miles

(3) Follow ancient fence posts NW over grassy summit LOOKING STEAD (627 m/2,058 ft) to saddle. Ascend rocky crest 150 m to platform, continue 20 m to low cairn whence narrow path drops R over edge, to traverse, undulating, round steep N flanks Pillar Fell to conspicuous Robinson's Cairn (1 km). Path continues towards unmistakable Pillar Rock, descending then climbing steeply L, scree, to slabby terrace leading R above crags to narrow neck behind Pillar Rock whence path zigzags steeply up rocky slopes to summit plateau, cairns, PILLAR (892 m/2,928 ft).
3 kms/2 miles

(4) Descend SW, grassy then shattered crest, to Wind Gap col (750 m/2,460 ft), ascend easily to stony crest, cairn BLACK CRAG (828 m/2,717 ft). Grassy slopes continue to wall, summit SCOAT FELL (841 m/2,760 ft). Path descends grassy slopes SE to wide saddle, ascend ridge crest to cliff-edge cairn RED PIKE (826 m/2,709 ft), descend long slopes, path, to Dore Head saddle (480 m/1,575 ft).
4 kms/2½ miles

(5) Ascend steep buttress ahead – Stirrup Crag – scramble, short, simple, to grassy top, cairns. Path along crest leads 900 m to cairn, summit YEWBARROW (628 m/2,058 ft), descend gradually onwards to narrow rocky neck – Great Door – drop R to shallow gully, easiest RHS, then descend steep grass L to good path crossing pasture to stile, descend far side wall until path strikes L to Wastwater shore, road returns to start.
5 kms/3 miles

Variation
(1.A + 2.A) Ascend direct to Black Sail Pass: behind Hotel take path E bank river 300 m to junction, fork L, good path leads into Mosedale then climbs R to Pass.
5 kms/3 miles

The Scafell Crest Lake District

OS 1 : 50,000 Sheet 89

'The summit is bare of everything that grows
. . . blocks and inclined planes of slate rock . . .
compose the peak . . . the greatest mountain
excursion in England.'

Harriet Martineau: 1855

'The Pikes of Scawfell are bold and
picturesque but their precipices are
slight . . .'

Owen Glynne Jones: 1897

Scafell Pike is indeed the highest point in
England, a noble rocky mountain that is not
unworthy of its crown. And the other descriptions, if not strictly accurate, do present a fair
impression of a summit that despite demanding a
long and rough walk to reach it, must be one of
the most popular in the country. However,
another summit standing one kilometre southward is only slightly lower, massive and moody
it is girt with formidable defences and far less
frequented: Scafell is unquestionably the finest

mountain in England. Centuries ago the rugged
high ground between the heads of Wasdale and
Eskdale was known as Scaw Fell, possibly
meaning 'Fell of the Rocky Tops', but as time
went by the name became appended to the
singular, most distinctive and seemingly highest
summit of the massif while the less imposing
tops became the Scawfell Pikes. To be strictly
accurate there are three 'Pikes' – the others are
Broad Crag and Ill Crag – before the undulating
stony plateau from which all three rise rears up
in the final bastion of Great End. In all the Lake
District the Scafell massif has no peer.

Rising at the hub of a series of radiating valleys
and accessible from many other starting points,
Scafell and its great crags frown down upon
Wasdale Head which has always been inexor-
ably linked with these mountains in word and

From Styhead the route ascends the northern flanks of Great
End by the shoulder known as The Band, seen here reflected
in the moody waters of Styhead Tarn.

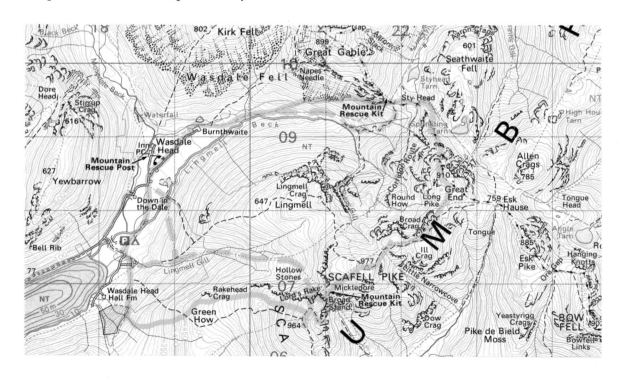

deed besides geography, and here our traverse starts, climbing first to the 'grand junction' of Sty Head. A more pleasant if slower alternative to the usual route previously described follows the old path up the valley bottom instead of mounting across the stony face of Great Gable. Then the northern ridge of Great End, its craggy butt dominating busy Sty Head and its dark tarn, is ascended by a less beaten path taking the easiest line, not always obvious, through fairly steep and rocky ground to the stony, mossy expanse of the summit plateau. From the lip one can peer down several impressive gullies, celebrated winter climbs, which slash deep into the large broken cliff that falls north-east above Sprinkling Tarn. The true summit, highest of three outcrops rising from the plateau, stands close to the cliff edge and provides impressive glimpses down Borrowdale to Derwent Water and unmistakable Skiddaw.

Now the tourist route from Esk Hause is joined, a highway almost, and followed to Scafell Pike, although connoisseurs will make short and obvious diversions to the bouldery summits of the other two Pikes, Ill Crag (927 m/3,040 ft) and Broad Crag (931 m/ 3,054 ft) both excellent vantage points – the latter summit is reputed to be the roughest in the Lakes. On a fine day folk will queue to ascend the monumental 3-metre (10-foot) cairn which crowns Scafell Pike: constructed in 1921, it dedicates the highest forty acres in England as a memorial to Lakeland men who died in the First World War. But the best views, especially down into Eskdale and Wasdale, are from the more lonely plateau edge while Scafell dominates the southern horizon as the path descends – or rather boulder-hops – to the narrow red crest of Mickledore and a nasty surprise. Formidable cliffs defend the entire northern flank of Scafell. Mickledore leads into the cliff and stops. Although a route does surmount the wall straight ahead, it is polished, exposed and often slimy if technically straightforward – a nasty place and scene of many accidents, Broad Stand is for climbers only. Two reasonable routes however outflank the precipice to left and right, the easier Fox's Tarn route involving a considerable descent. Classic and more interesting, the route via Lord's Rake contours down beneath the awesome cliffs of Scafell Crag, over 150

metres high (500 feet) and the greatest in England, passing the start of many famous climbs besides a discrete cross carved in the rock, a memorial to four climbers who died here in 1903 and are buried in the tiny Wasdale head churchyard. And so into the forbidding crag-walled gully of Lord's Rake itself. 'Reik' is old norse for a steep hillside path, and this one, floored with loose and now badly worn scree, is unwelcoming, but it is hardly scrambling and after a few minutes the gully is abandoned for the grassy shelf of the West Wall Traverse leading into the yawning portals of Deep Ghyll. The rocky scenery hereabouts is sensational, and the situation as atmospheric as any in Cumbria. A grassy plateau surrounds the wide gully mouth and the summit outcrops rises a short way southward from where the broad seaward prospect across

the shimmering sands of Morecombe Bay and the Duddon and Ravenglass estuaries is marred only by the incongruous towers of the Seascale nuclear facility rising against the distant shape of the Isle of Man. Nevertheless Scafell is a summit on which to linger. Despite its strong northern defences there are easy descents to south and west, the latter – the once popular Green How route dropping to the head of Wastwater – is swift and easy although the path fades low down and the final steep slopes should not be descended too soon.

The best views of the Scafell massif are from the summit crest of Yewbarrow. Here the entire route is visible from the green fields of Wasdale Head below, to Sty Head on the left, over Great End, Broad Crag and Scafell Pike to Mickledore and the great cliff-rimmed summit of Scafell itself; the descent lies down the slopes on the right.

A Traverse of the Scafell Crest

Length: 14.5 kms/9 miles
Total ascent: 1300 m/4,270 ft
Difficulty: a rugged and fairly strenuous expedition on high and exposed mountains, appropriate equipment essential: one major section of slightly exposed semi-scrambling could appear intimidating
Note: most of route National Trust property

Start: NT car-park/camp site 182075, N end Wastwater (70 m/230 ft), limited parking possible out-of-season Wasdale Green 186085.

(1) Ascend easily as described preceding walk (1) to Rescue Box, multi-path junction, summit Sty Head Pass (480 m/1,575 ft).
5 kms/3 miles

(2) Following major Esk Hause path, initially due E, 500 m, just before ford strike $\frac{1}{2}$R up grassy hillside, faint path, to conspicuous block on crest whence better path zigzags L up shoulder, through trough R of The Band peaklet, to steeper craggy slope. Ascend easiest line, hands useful, to scree patch, shallow gully. Steep boulder field leads to wide plateau, two cairns GREAT END, summit SE cairn/outcrop (910 m/2,984 ft). Descend SSW to saddle, join major path continuing SW, rough, bouldery, many cairns, over shoulders Ill Crag, Broad Crag, dropping to narrow red col before climbing steep eroded rocky slope to huge cairn SCAFELL PIKE (978 m/3,210 ft).
4 kms/2½ miles

(3) Rugged path descends SW, boulder slopes, to narrow rocky saddle Mickledore (c 790 m/2,600 ft). Ascend slightly to far end, descend steeply R skirting cliff, stony, path traverses easily below cliffs 250 m till above scree slope. Now narrow cleft-like gully – Lord's Rake – rises steeply to tiny col behind pointed buttress ahead. Ascend gully **with care**, steep, rocky, loose, until 10 m below col step L onto wide easy ledge – West Wall Traverse. Path, slightly exposed, ascends into chasm Deep Ghyll, climbs RHS gully into upper bowl then bears R up red cleft, shaly, loose, hands useful, zigzagging to grassy plateau. Go R 250 m to summit outcrop, cairn, SCAFELL (964 m/3,162 ft).
1.5 kms/1 mile

(4) Retrace steps 150 m to grass, small cairn, stony path descends L near cliff edge until above scree funnel Lord's Rake (see 3.A) whence line of cairns, faint path, leads L, descending grassy hillside W. Follow path, often near RH edge, to top prominent Rakehead Crag. **Avoid** very steep path dropping directly over edge soon after crag but continue SW, intermittent path, descending across slopes then following beck to bridle-path at intake wall above wood. Go R, follow path easily to start.
4 kms/2½ miles

Alternatives

(3.A) Continue Lord's Rake beyond West Wall Traverse over two tiny cols, stony hollow, joining (4), retrace to summit. Easier, less interesting.

(3.B) To avoid Lord's Rake: stony path drops steeply L from Mickledore skirting cliff 500 m until stream in small gully ascends R to tiny pool – Fox's Tarn – sometimes dry. Ascend steep slopes R to grassy plateau Scafell.
Extra ascent 100 m/300 ft

(3.C) Wasdale avoiding Scafell: continue to start Lord's Rake gully whence drop R down scree into Hollow Stones combe, good path leads down Brown Tongue, R bank stream, to Wasdale.
Scafell Pike to start: 4 kms/2½ miles

The Carneddau   Snowdonia

OS 1 50,000 Sheet 115

'The crests of solitude – the moaning wastes
of the Carnedds,
The frightened curlew flying from a wrecked
plane's wing.
The empty, perfect, peaceful top of
Llewellyn.'

'A Boy goes Blind' from Verses from My Country
Roger Redfern: 1975

The Carneddau is the secret range of Snowdonia, a complex tangle of great whale-back
crests cradling hidden cwms, remote tarns and
several sizable lakes. This is a very extensive
mountain group, a great block of country
stretching from the edge of the sea into the very
heart of Snowdonia. Here rises Carnedd Llewelyn, the second highest summit south of Scotland surrounded by more high ground above
3,000 feet than the rest of England and Wales put
together. The climate on these high tops is
virtually 'Arctic-Alpine' and in winter the range
offers the best skiing in the region: in many ways
the Carneddau resemble the Cairngorms.

Writing of his 1778 tour, unsuspecting
Thomas Pennant reported the Carneddau as '...

very disagreeable – dreary bottoms or moory
hills ...' and indeed at first sight the Carneddau
do appear unspectacular mountains. But an
apocryphal tale has it that early rock climbers
discovered the great crags of Craig yr Ysfa by
telescope from the summit of Scafell, and this
and the other imposing cliff of Ysgolion Duon –
the 'Black Ladders' – are the most famous
among several noteworthy crags. Though the
high tops are typically stony, the naked rock
remains characteristically secluded. An aura of
the unexpected, the unusual, seems to cloak the
Carneddau. Wild ponies breed, for instance, in
the sanctuary of Ffynnon Caseg – 'the Fountain
of the Mare' – and the feathery cataract of lovely
Aber Falls, plunging a clear 50 metres (170 feet),
is the highest waterfall in Wales. Dark Llyn
Colwyd is the most profound lake in Snowdonia
and it is said that smiling Llyn Crafnant is home

Yr Elen must be the prettiest Carneddau summit: it is seen
here from the west across the mouth of Cwm Llafar with the
Black Ladders – Ysgolion Duon – of Carnedd Dafydd just
visible on the right. A highly recommended route follows the
skyline crest.

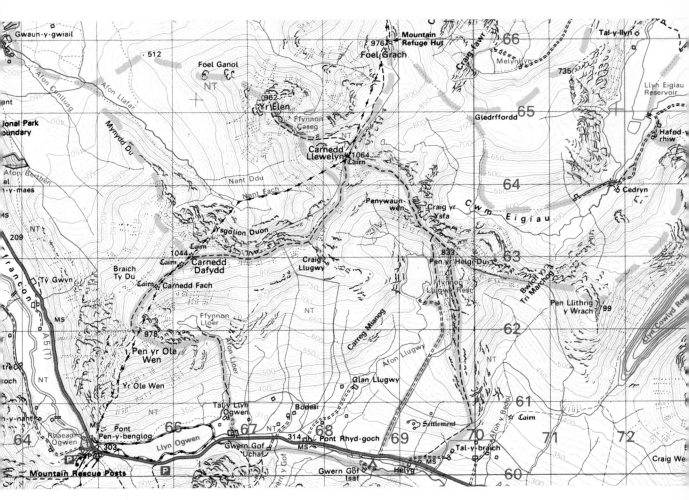

to the largest trout in the Principality. The sad ruins of a burst dam lie in Cwm Eigiau, there are aircraft wrecks scattered on the high tops while the seaward northern slopes are scattered with Bronze Age remains. Carnedd in fact means 'cairn' or 'burial mound' and the large burial cairn between Carnedd Llewelyn and Foelgrach is said to be the grave of Tristran, a Knight of the Round Table. Appropriately the two highest summits are named for Llywelyn Fawr – 'the Great' – who first united Wales and died in 1240, and for his son Dafydd who succeeded him. Certainly the Carneddau are special.

The circular walk described is not a difficult one, but the Carneddau are continuously high and lack the small-scale features so useful when navigating in all but perfect conditions: a mistake of a few degrees in mist can lead the walker miles from his destination so compass skills are a prerequisite. Admittedly the climb up onto Pen-yr-oleu-wen, (the 'Hill of the Moonlight') is a long pull, but the mountain occupies a strategic position in the Ogwen valley; there are superb views from its slopes to Cwm Idwal and the Glyders – and from the mouth of delightful little Cwm Lloer (the 'Cwm of the Moon') a most unusual view of *aiguille* Tryfan. Once up on the tops the going is never hard and the only strenuous ascent is that of the final scree-cloaked dome of Carnedd Llewelyn – just 75 feet (23 metres) lower than Yr Wyddfa. The extensive panorama, especially northward to the sea, to Puffin Island and the Great Orme, is a just reward. This peak is the hub of the Carneddau and three great ridges spring from its summit, while a fourth lesser arete leads to strangely isolated Yr Elen, (the 'Peak of the Fawn') a rugged and imposing mountain in its own right and most shapely of the Carneddau summits. Until recently the range was considered to contain six of Snowdonia's fourteen 3,000-foot tops but with the advent of metrification Carnedd Uchaf has been promoted to 926 metres – 3,038 feet – so there are now seven 'three-thousanders' in the range, making fifteen in all. Peak-baggers will be pleased to note that there are seventeen Carnedd 'tops' of which no less than twelve are classed as 'separate mountains'.

In so extensive a range there are many other good walking routes, and notable are the Cwm Eigiau Horseshoe and the classic Carneddau traverse, some 21 kilometres (13 miles) one-way from Ogwen to the coast at Aber. However, an excellent and rather shorter alternative (13 kilometres/8 miles) to the main circuit described is the round of Cwm Llafar – the 'Valley of Sound'. Starting from Bethesda where limited laneside parking is currently possible at 634662, the route enters the lower cwm before striking up the west and north-west ridges of Yr Elen

from where there are good views across to Dafydd and forbidding Ysgolion Duon – a favourite winter climbing cliff which dominates the wide head of the cwm. Llewelyn's summit can be ascended or avoided by a col-to-col traverse path across its south-west flank, and the main Carnedd crest followed to Dafydd from where, carefully avoiding the northern crags, easy slopes fall northwestward to the grassy Mynydd Du ridge dropping to the boggy path at the cwm mouth.

A Carnedd Circuit

Length: 18 kms/11 miles
Height ascended: 1030 m/3,380 ft
Difficulty: mostly easy going on grassy or rock-strewn terrain after long initial ascent, two short semi-scrambles in section (3)
Warning: this route is continuously high and can attract fierce weather. It passes close above major cliffs of Ysgolion Duon and Craig yr Ysfa where care is necessary in bad conditions. Appropriate equipment, including map and compass, essential.

Start: good roadside parking beside A5 at head Llyn Ogwen at 667605 (303 m/994 ft).

(1) Take lane N past Glan Dena hut almost to Tal y Llyn Ogwen farm were path crosses stile to ascend easy hillside beside Afon Lloer stream. Just before Ffynnon Lloer tarn strike up steeply left, rocky couloir giving best line, onto stony E ridge ascending to PEN YR OLEU WEN summit. (979 m/3,211 ft).
3.5 kms/2¼ miles

(2) Easy going along stony ridge curving NE above Cwm Lloer leads to CARNEDD DAFYDD (1044 m/3,424 ft). Ridge continues, rocky at first, descending gradually to flat section: just before, at 678632, safe escape is possible to SSE. Flat narrow ridge now curves to NE broadening to scree-covered slopes leading to wide top of CARNEDD LLEWELYN (1064 m/3,490 ft).
4.5 kms/3 miles

(3) Descend SE ridge, initially scree then grassy and narrowing, passing above cliffs of Craig yr Ysfa to easy rocky step down onto narrow saddle (Bwlch Eryl Farchog 750 m/2,460 ft). Slender ridge now climbs via a short slight scramble to spacious crest of PEN YR HELGI-DU (833 m/2,733 ft). Path descends long grassy S ridge, avoids Tal-y-Braich cottage on W and follows track to cross main road at Helyg (280 m/930 ft). Return to start by old road along S side of the valley bottom.
6 kms/3¾ miles

Variations

(1.A) At end of flat ridge at 683639 strike left on vestigal path across stony SW face of Carnedd Llewelyn to obvious col on its NW ridge, now follow narrow crest to YR ELEN summit (961 m/3,152 ft). Retrace steps to col and ascend NW ridge to Llewelyn summit.
3 kms/1¾ miles

(2.A) From Bwlch Eryl Farchog steep rocky path descends SW to Ffynnon Llugwy lake. Ugly CEGB jeep road leads down to A5.
3 kms/1¾ miles

The familiar view from the Glyders over the Ogwen valley towards the Carneddau peaks encircling Cwm Llugwy: our route drops off Carnedd Llewelyn on the left, crosses the Craig yr Ysfa saddle and climbs again over Pen yr Helgi-du on the right.

The Glyders   Snowdonia

OS 1: 50,000 Sheet 115

'. . . a fit place to inspire murderous thoughts, environed with horrible precipices . . .'

Thomas Pennant
describing Cwm Idwal, 1778

Although Pennant was not favourably impressed, many modern mountain lovers will consider the Glyder range their favourite Snowdonian mountain group and for the same reason. With imposing Cwm Idwal cradled at its very heart and Tryfan – the most striking peak south of the Scottish Highlands rearing closeby – the diverse scenery of these most craggy of Snowdonia's mountains is unsurpassed. Excellent and characterful walking, scrambling and climbing – the latter especially so in the easier grades – together with straightforward access to its summits, ensure the attraction of the Glyders.

The chain of the Glyders (strictly speaking – Y Glyderau) forms a north-east facing arc centred on Pen y Benglog at the end of Llyn Ogwen and holds no less than eight 'separate mountains' and two 'subsidiary tops'. Five summits rise above 3,000 feet. Although the lower south-west slopes of the range above the narrow Llanberis Pass are steep and rocky and hold a succession of popular climbing crags, the southern and western flanks of the Glyders are generally rounded and grassy and the spectacular scenery is concentrated above the Ogwen Valley. Here an array of twelve hanging cwms – some among the finest in Britain – are deeply gouged into the northern flanks of the chain. Jutting northwards from the main crest is Tryfan whose triple summits and craggy shark-fin outline are familiar to all who travel the A5 Holyhead road. Reputedly it is the only British mountain outside Skye where hands are required to reach the summit, twin monolithic blocks known as Adam and Eve. For scramblers Tryfan's long north ridge provides classic sport and if continued via the seemingly intimidating yet no more difficult Bristly Ridge to the main plateau, a superb alternative start to any route over the Glyders. Continuity, perfect rock, large holds, and an airy situation ensure the popularity of both ridges though the easiest lines are not always obvious and neither ascent should be underestimated by the inexperienced or in bad conditions. Descent from Tryfan to the Bwlch and Bristly's foot is by the shorter and easier south ridge, the regular ascent route and possibly just a scramble. But bleak and windy Bwlch Tryfan is more easily reached from Cwm Bochlwyd, its moody tarn overshadowed by the cliffs of Glyder Fach, and the usual route onwards by the diagonally ascending 'Miner's Track' crosses the scree-seamed headwall of Cwm Tryfan to the marshy plateau near pretty

The jagged peak of Tryfan rises steeply above the Ogwen valley: this distinctive mountain is seen from Helyg on a winter afternoon with the North Ridge on the right and Bwlch Tryfan and Bristly Ridge on the left.

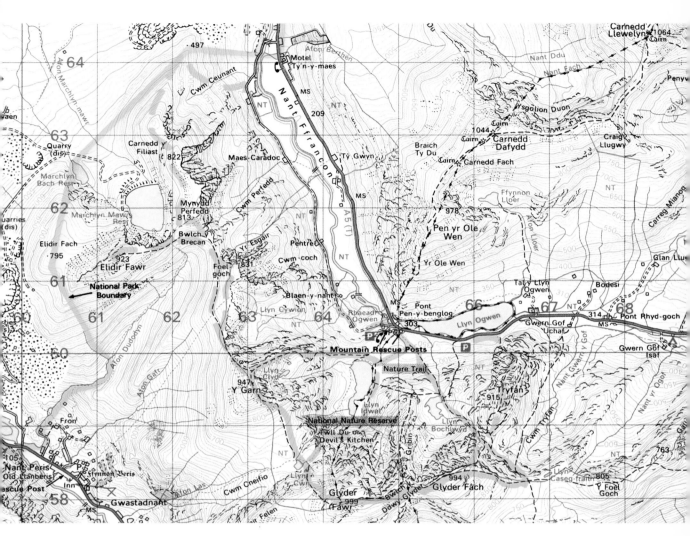

little Llyn Caseg-fraith – 'the Lake of the piebald Mare' – a superb viewpoint for Tryfan.

Charles Kingsley described the extensive stone and boulder covered summit plateau of Glyder Fach as an '... enormous desolation, the dead bones of the eldest born of time ...' An apt description for here are scattered strange configurations of rocks such as the famous Cantilever Stone and the bizarre bepinnacled Castell y Gwynt – the 'Castle of the Winds'. In poor visibility navigation can be difficult for the plateau extends over 2 kilometres ($1\frac{1}{4}$ miles) to Glyder Fawr, though it does narrow to a noticeable saddle between the two summits above the secret and delightful little Cwm Cneifion (known as the Nameless Cwm). The useful Gribin Ridge drops northward a little

further on. At the big Glyder the crest turns north-east and scree slopes fall steeply to the broad boggy saddle that holds tiny Llyn y Cwn, where once lived peculiar monocular fish, so they say. The vertical cliffs of Clogwyn y Geifr drop abruptly from the eastern edge of the saddle into Cwm Idwal and are rent by several great gullies: Hanging Garden, the Devil's Staircase, the Devil's Appendix, and others. Most famous of all is the Devil's Kitchen – Twll Du, 'the Black Hole' – into which tumbles a fine waterfall providing a splendid view for the cautious photographer. From the saddle a wide grassy ridge leads up to the triangular summit of Y Garn, a rocky top with fine seaward views – sometimes as far as Ireland. The north-east ridge, curving round shallow Cwm Clyd with its tiny tarns, provides an excellent descent to the north lip of Cwm Idwal with its celebrated view across the lake towards the dark wall of Clogwyn y Geifr: on a moody day the legend of Prince Idwal, how he was thrown by a giant from these cliffs into the lake over which – to this day – no bird will fly, may not seem so far fetched!

The ridge onwards from Y Garn, attractive and cwm-scalloped on its eastern flank, continues over Foel Goch and Mynydd Perfedd to Carnedd y Filiast before falling away as a great shoulder towards the abandoned Penrhyn slate quarries and the village of Bethesda.

The final Glyder top however is Elidir Fawr, a shapely rocky summit on a subsidiary spur jutting over the Llanberis valley. Sadly this worthy mountain has been despoiled over the years, first by the Dinorwic slatemen whose mighty quarries hacked to within 800 feet of the summit, and more recently by the CEGB who dammed the once perfect circular tarn of Marchlyn Mawr which lies in its beautiful crag-girt northern cwm.

The summit of Glyder Fach is crowned by the extraordinary Cantilever Stone.

A Glyder Circuit

Distance covered: 10.5 kms/6¾ miles
Height gained: 1040 m/3,400 ft
Difficulty: mostly rough stony ground, paths of sorts most of the way
Warning: serious and craggy mountain country, appropriate equipment essential: in bad weather navigation often difficult and mistakes serious. Cairns proliferate, do not navigate by them. Only easy escape routes on Ogwen flank between those described are scrambly Gribin Ridge and steep screes E side Bristly Ridge.

Start: Ogwen Cottage car-park W end of lake, 649603, phone, toilets, often refreshments. (303 m/994 ft).

(1) Subsidiary path forking L from well-used Cwm Idwal track leads over boggy ground to torrent, climbing steeply beside it to Llyn Bochlwyd and round E side of lake before climbing to Bwlch Tryfan saddle (709 m/2,325 ft). After slight descent E, path ascends diagonally to Glyder crest near tiny Llyn Caseg-fraith (762 m/2,500 ft) before turning W up broad rocky flank to stony plateau and OS cairn GLYDER FACH summit (994 m/3,262 ft). **4 kms/2½ miles**

(2) Cross plateau to W, pass round Castell y Gwynt, descend to shallow saddle. Avoiding path ascending N to head of Gribin Ridge, continue W to regain ridge crest and ascend gradually to GLYDER FAWR summit (999 m/3,279 ft). Descend screes NW to wide saddle near Llyn y Cwn (637585: 715 m/2,350 ft). Grassy slopes ahead lead to rocky summit Y GARN (946 m/3,104 ft).
3.5 kms/2¼ miles

(3) Just N of summit, path descends grassy NE Ridge into Cwm Clyd before dropping steeply to Llyn Idwal N shore and main Ogwen Cottage path.
3 kms/2 miles

Variations

(2.A) Direct descent to Cwm Idwal: From Llyn y Cwn cairned path leads NE into mini canyon near abrupt cliff edge before descending ramp across cliff face, scrambly near bottom, to foot Devil's Kitchen cleft. Descend boulders to well-marked path leading beneath famous Idwal Slabs to Llyn Idwal E shore and onwards to Ogwen Cottage.
Distance from Glyder Fawr: **3.5 kms/2¼ miles**

(3.A) Continuation to Elidir Fawr: From Y Garn grassy crest leads N over FOEL GOCH (831 m/2,727 ft) to MYNYDD PERFEDD (812 m/2,665 ft). Descend SW to col above Cwm Marchlyn. Lower contour path on W avoids these summits. Ascend easily to rocky summit ELIDIR FAWR (924 m/3,030 ft).
Distance from Y Garn: **4 kms/2½ miles**
Ascent: 370 m/1,210 ft

(4) Descent from Elidir Fawr: Contour Mynydd Perfedd N to CARNEDD Y FILIAST (821 m/2,695 ft). Go 500 m NNW then descend pronounced ridge NNE for 1 km until faint path crosses SE into Cwm Ceunant, descend to old Nant Francon road at bottom of slope. Pleasant walk up valley to Ogwen Cottage.
Distance: 9 kms/5½ miles
Ascent: 200 m/650 ft

It is said that no bird will fly over the moody waters of Llyn Idwal! In this view of the lake from its outfall, the Idwal Slabs are seen on the flank of Glyder Fawr on the left, ahead is the cleft of the Devil's Kitchen while the slopes of Y Garn rise on the right.

Snowdon and the Horseshoe

OS 1:50,000 Sheet 115

'We've the stones of Snowdon
and the lamps of heaven'

Charles Kingsley 1856

Rising to the highest summit and the most magnificent peaks south of the Highlands, the compact Snowdon massif contains surprisingly little high ground and only four tops above 3,000 feet (900 metres). But the summits are characteristically sharp and rocky and the five great ridges that spring from the central peak of Yr Wyddfa are rugged, often narrow and cradle several fine craggy cwms – two of which hold the greatest cliffs in Wales, Clogwyn D'ur Arddu and Y Lliwedd.

Snowdon predictably is a major tourist attraction and since 1896 Yr Wyddfa has suffered the indignity of a mountain rack–railway to its summit. While the antiquated little steam trains themselves are quaint and perhaps even sufferable, the obtrusive ugliness of the sordid summit terminus is unacceptable. Luckily the tourist season is short and for six months of the year visitors must walk to the summit, many by the

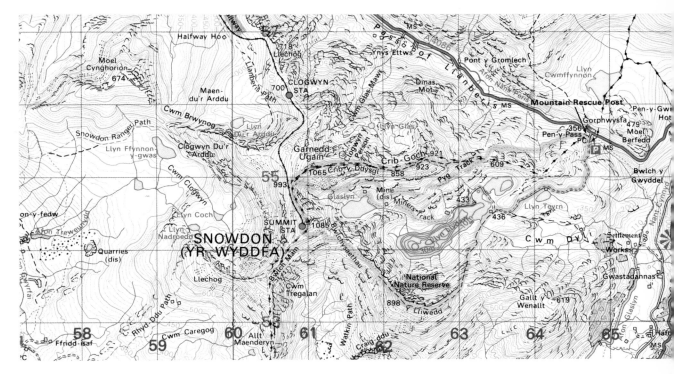

easiest way, a trudge of nearly 8 kilometres (5 miles) up the gentle flanks of the NW or 'Railway Ridge' from Llanberis. In poor conditions however the route can be deceptive, weather up high can be very different from that in the valley and when icy the track above Cwm Du'r Arddu can be very treacherous.

The classic route on Snowdon is the 'Horseshoe' – the most famous ridge scramble in Britain though many folk may find it rather challenging. The Crib Goch arete falls precipitously on the northern side and although in good conditions surefooted climbers will delight in skipping along the actual crest, most ordinary mortals will scramble along on the southern side using the crest for handholds. Under snow it can provide a superb expedition for experienced alpinists. The view from Yr Wyddfa – 'the Tomb' where they say Rhita Gawr, an early prince was buried – is justly celebrated. On the best days the panorama extends to the Preseli Hills, the Wicklows of Ireland and the Lake

District; on hazier days the silhouette of Harlech Castle beyond the glinting Glaslyn Estuary is sufficient inspiration. Five hundred metres (1,600 ft) below lies tiny Glaslyn in whose fathomless green waters is sometimes glimpsed the legendary Afanc, a monster supposedly banished from the Conway in days gone by.

Most convenient of the easier routes to the summit, the Miners Track also provides a swift descent. Derelict workings of the Britannia copper mines, abandoned before the Great War, are scattered above the northern shores of Llydaw and in Cwm Glas, the copper accounting for the colour of the 'green lake': a massive iron casting from the crushing mill still lies near the path. The miners lived in barracks beside Llyn Teyrn below. Other straightforward routes to the summit follow the south ridge, the Watkin and the Snowdon Ranger paths. The latter, ascending from Llyn Cwellyn to the west ridge of Crib-y-ddysgl and passing above the finest climbing crag in Snowdonia, awesome Clogwyn Du'r Arddu – the 'black cliff of darkness' – was the regular route before the coming of the railway. It was tramped by Thomas Pennant in the 1770s and named from the professional guide who once frequented it.

All the four summits traversed by the Snowdon Horseshoe circuit are seen in this classic view over Llynau Mymbyr: left to right they are Y Lliwedd, Yr Wyddfa, Crib Goch and Crib-y-ddysgl.

The Watkin Path, rising from lovely Nant Gwynant, was constructed in 1892 by the local proprietor to facilitate the ascent on horseback. Today only feasible on foot above Bwlch y Saethau, the final steep rubbly traverse below the top can demand caution, especially in snowy conditions. Rather more rugged, the south ridge can be reached from Rhyd-Ddu, Beddgelert or Nant Gwynant but ideally over the pretty little peak of Yr Aran. The almost scrambly ridge crosses Llechog (605537) often now considered –

since metric maps show a ten-metre contour ring round its 931 m (3,054 ft) spot-height – to be a 'three-thousander'.

Bwlch y Saethau itself – the 'Pass of the Arrows' – was the site of King Arthur's last battle and it was into Llyn Llydaw that Sir Bedivere hurled the sword Excalibur. His knights lie sleeping to this day in a cave, accessible only to skilled climbers, in Slanting Gully on Y Lliwedd, awaiting again the call to arms. Surely Lliwedd is one of the great moun-

tains of Wales, through it just fails to reach 3,000 feet, and the traverse of its splendid twin summits above thousand-foot crags laced with vintage rock climbs and frowning down upon Llyn Llydaw completes the Horseshoe. Llydaw is second deepest of the Welsh lakes – nearly 200 feet – and the black pipes feed the old Cwm Dyli HEP scheme in the valley below. There is one final view of Yr Wyddfa from the shore before the Miner's Track is joined and followed round the corner towards Pen y Pass.

A Snowdon Circuit – The Horseshoe

Distance covered: 11 kms/7 miles
Height gained: 1060 m/3,500 ft
Difficulty: committing, rugged, high and susceptible to bad weather, long section of exposed but straightforward scrambling potentially hazardous in bad conditions. Route always well defined. Serious undertaking for the inexperienced

Start: Pen y Pass car-park (charge) on A4086 at 647556 (356 m/1,170 ft), telephone, refreshment and YHA.

(1) From W corner car-park take Pyg Track, rough path contouring hillside above Llanberis Pass up to grassy saddle Bwlch y Moch 633553 (570 m/1,870 ft).
1.5 kms/1 mile

(2) Ascend E ridge Crib Goch ahead, initially path then rocky steps on steep narrowing ridge leading to narrow rocky shoulder – false summit. **Easy scrambling: beware loose stones** often dislodged from above. Exposed knife-edge rock arete continues horizontally 150 m to tiny summit CRIB GOCH (922 m/3,026 ft), not difficult but **caution required**. Narrow crest continues to Pinnacles – short, steep, airy scramble avoidable unpleasantly on L. Steep descent to grassy saddle, Bwlch Coch. Narrow rocky ridge continues over short scrambly steps to widening summit plateau CRIB-Y-DDYSGL (1065 m/3,496 ft). Slopes descend to finger stone at broad saddle Bwlch Glas near railway track. Virtual highway ascends to summit YR WYDDFA (1095 m/3,560 ft).
3.5 kms/2¼ miles

(3) Descend SW ridge 150 m to finger stone, drop L onto Watkin Path descending across steep rubbly SE Face **caution loose rock** to wide saddle Bwlch y Saethau, continue easily over bluffs to Bwlch Ciliau where Watkin Path descends R. Ascend instead narrowing stony ridge ahead to sharp twin summits Y LLIWEDD (898 m/2,947 ft), exposed on L. From E summit path descends along crest 750 m before dropping L down steep grassy slopes to Llyn Llydaw (436 m/1,430 ft). Here join stony Miners Track contouring hillside to Pen y Pass.
6.25 kms/4 miles

Variation
(3.A) From Yr Wyddfa retrace route to Bwlch Glas fingerstone and drop E down steep stony zigzags into Cwm Glas. **Caution** in winter conditions. Path forks above Glaslyn, upper route – Pyg Track – contours via Bwlch y Moch to Pen y Pass **5 kms/3 miles**, or lower Miners Track returns via Llyn Llydaw **6.5 kms/4 miles**

Yr Wyddfa – Snowdon itself – appears almost Himalayan in this spring view from the east over the ice of Llyn Llydaw: the northern cliffs of Y Lliwedd rise on the left.

The Pennant Hills Snowdonia

OS 1:50,000 Sheet 115

'Oh god, why didst Thou make Cwm Pennant so beautiful and the life of an old shepherd so short?'

Eifion Wyn

Cwm Pennant is indeed a delectable valley. A narrow lane winds some four miles into its secret depths, crossing and recrossing the tranquil Afon Dwyfor, passing scattered farms and a chapel or two. Sweeping hillsides, once the home of Bronze and Iron Age folk, hold but few conspicuous traces of the miners who toiled here after slate and copper, and are today deserted save for shepherds, their dogs and their ubiquitous sheep. Surrounded by a horseshoe of fine mountains, the valley opens southwards to the sea and the sun, and seems always to hold a smile.

The so-called Nantlle Ridge that walls Cwm Pennant on the north is notable for its series of craggy north-facing cwms, especially Cwm Silyn with its twin tarns and Great Slab which climbers hold in high regard. Deeply scarred by abandoned workings and containing the celebrated 180 metres (600 feet) deep Dorothea slate pit, the Nantlle valley strongly contrasts Cwm Pennant. Holding three shapely mountains and three subsidiary peaks, the Nantlle Ridge provides an excellent traverse, though not an obvious circuit. A single public path climbs from the Nantlle valley, and after a period of local antagonism to mountain goers, access to and along the crest is now by Courtesy Footpaths carefully negotiated by the National Park. An ancient right of way does however run to the head of Cwm Pennant, past the sad ruins of the

Ogof Owain Glyndwr – the lonely cave on Moel y Ogof, a satellite of Moel Hebog, in which the Welsh leader is supposed to have hidden for six months.

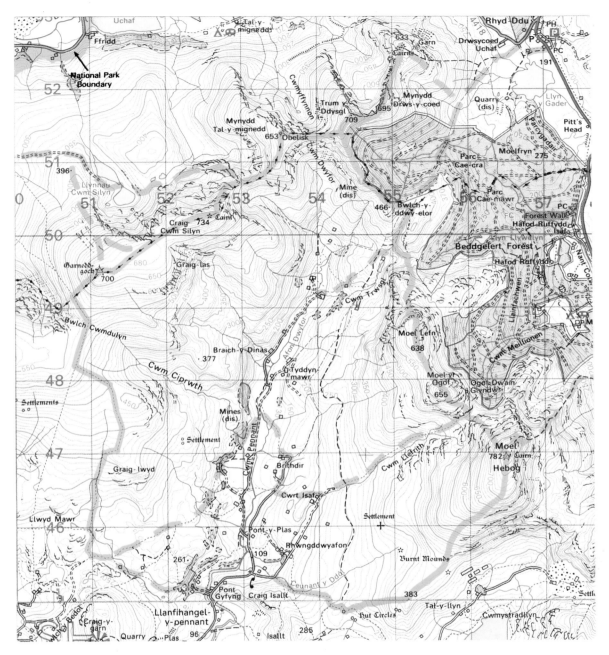

Prince of Wales Quarry abandoned in 1882 after a fatal rockfall, to narrow Bwlch-y-ddwy-elor, the 'Pass of the Two Biers', which leads over to Nant Colwyn and Rhyd-Ddu. The Bwlch is a useful and legal access point to the mountains on either side. Local tradition claims that the hillsides between the Bwlch and Rhyd-Ddu were those most frequented by fairies in all Wales, indeed the last claimed sighting was in 1899, but now alas, regiments of sombre pine have conquered fairyland.

The regular route up the big dome of Moel Hebog, the 'Hill of the Falcon', is by its northeastern slopes above Beddgelert but it can just as well be climbed from Cwm Pennant over the shoulder of Braich y Cornel with its scattering of prehistoric hut circles. The summit is smooth and round and a rewarding viewpoint, both out to sea and over Nant Colwyn to the Snowdon massif. Moel Hebog is associated with Owain Glyndwr. Surprised in Nantmor by his enemies he is said to have swam the then tidal Glaslyn estuary and fled up the eastern slopes of the mountain, the pursuit hard on his heels, finally escaping up a deep chimney in the summit crags. But he was still not safe and for six months he is supposed to have laid up in a cave under the rocky lip of Moel yr Ogof – the 'Hill of the Cave': hardly a comfortable retreat, it can be located at 559478. The route north from Hebog descends to strange Bwlch Meillionen – the 'Clover Pass' – lush with ferns and mossy outcrops, before climbing to the reedy pools and jumbled rocks of Moel y Ogof and onwards to Moel Lefn, Bwlch Sais – the 'Englishman's Pass' which leads nowhere – and through a complex area of bogs and bluffs to Bwlch Cwm Trwsgl and Bwlch-y-ddwy-elor.

One can return now into Cwm Pennant or continue onwards to join the Nantlle Ridge at the grassy table-mountain summit of Trum y Ddysgl – the 'Dished Ridge'. It is worth diverting here to visit little Y Garn, via the airy arete of Mynydd Drws-y-Coed – 'Mountain of the Wooded Gap' – before returning to descend over a narrow green *mauvais pas* and climbing to the conspicuous Victorian Jubilee pillar, over 6 metres high (20 feet), which crowns Mynydd Tal-y-mignedd – the 'Mountain at the Bog's End'. A useful but ill-defined path strikes down from the deep saddle beyond and contours to the twin lakes of Cwm Silyn. However the rocky arete above rises to the major peak of the Nantlle Ridge – Craig Cwm Silyn – and although intimidating at first sight, a straightforward scrambly route a little right of the actual crest leads to the stony summit plateau.

Easy ground leads onward to the final top, Garnedd Goch, from where the normal route descends northward to the Cwm Silyn road-head. To return to Cwm Pennant however, without trespassing, one must join an ancient right of way on the next saddle and follow its gradual convoluted descent back into the peaceful valley below Moel Hebog.

A steep climb leads up from Bwlch Dros-bern towards the top of Mynydd Tal-y-mignedd. The rocky ridge leading to Craig Cwm Silyn is seen across the saddle.

The Pennant Horseshoe

Distance: 24 kms/15 miles
Height gained: 1540 m/5,050 ft
Difficulty: rough mountain country, several rocky, airy – but avoidable – sections: a long and strenuous route that can be conveniently shortened. Appropriate equipment essential
Warning: concealed workings Bwlch Cwm Trwsgl vicinity
Note: visitors to Cwm Pennant should be tactful and especially considerate of local sensibilities
Parking: approved only at lanehead 540492 (140 m/460 ft) – parking fee payable Braich y Dinas farm, customary but unapproved at Dwyfor bridge 532476 (110 m/360 ft) and 504512 below Cwm Silyn (330 m/1,100 ft)

Start: because parking is difficult and varied, route is described from lane junction at 531454 (chapel + phone box).

(1) From junction follow narrow lane 400 m E, then footpath branching R towards Cwm Ystradllyn. After 1200 m at highest point, strike NE on faint path up spur to Braich y Cornel shoulder, continue to MOEL HEBOG summit (782 m/2,567 ft).
5 kms/3¼ miles

(2) Path descends NW flank R of wall to deep col, climbs through rocky outcrops to MOEL Y OGOF (655 m/2,150 ft), over MOEL LEFN (638 m/2,094 ft) and descends N to Bwlch Trwsgl, before bearing W to join Pennant track by old workings (340 m/1,115 ft).
4 kms/2½ miles

(3) Ascend track R to Bwlch-y-ddwy-elor crest then climb boggy path past forestry edge to grassy ridge leading to summit plateau TRUM Y DDYSGL (710 m/2,329 ft). (To include MYNYDD DRWS-Y-COED (695 m/2,280 ft) and Y GARN (634 m/2,080 ft) follow path contouring NE 150 m (500 ft) before summit plateau. **Extra 3 kms/2 miles; 175 m/575 ft return.**) Descend ridge E from S end plateau, over narrows, ascend to stone pillar on MYNYDD TAL-Y-MIGNEDD (655 m/2,148 ft). Follow boggy path S to steep shaly ridge, descend to Bwlch Dros-bern (500 m/1,640 ft) and climb steep rocky ridge beyond to summit plateau CRAIG CWM SILYN (734 m/2,408 ft). Continue SW, via walls and stiles, to GARNEDD GOCH (700 m/2,297 ft).
6.5 kms/4 miles

(4) Descend bouldery slopes SSW to Bwlch Cwmdulyn. Legal route follows grassy/heathery rights of way S and E to reach Cwm Pennant N or S of starting point.
7.5 kms/4¾ miles

The Moelwyn – Melancholic Mountains

Snowdonia

OS 1:50,000 Sheets 115, 124

'By fair Festiniog, mid the Northern Hills,
The vales are full of beauty, and the heights,
Thin set with mountain sheep, show statelier far
Than the tamer South.'

Sir Lewis Morris *Songs of Britain*: 1887

These are strange hills. They rise at the southern extremity of a wild inland region of high rumpled moors, rocks and lakes of which Moel Siabod is the northern bastion, yet with their faces to the sea they seem almost maritime mountains. Indeed, until William Madocks MP built his embankment across the Glaslyn estuary in 1811 and reclaimed the sands of the Traeth Mawr, the Moelwyn peaks stood literally with their feet in the waves.

Standing beyond the high central massifs of Snowdonia, the Moelwyn lack somehow the assertiveness of the real peripheral ranges such as the Rhinogs or the Arans and exude an aura of melancholy. Probably this is the fault of history, of the once frenzied slate industry that hacked and tunnelled into their flanks, changing the face of the landscape and then disappearing to leave a plaintive legacy of deserted quarries, roofless barracks and grey mountains of spoil. The Moelwyn are certainly intriguing, their mood is compelling and they offer excellent hill walking.

Seen from the heights of Snowdonia to the north-west, the Moelwyn appear as a knot of distinctively shaped peaks. This itinerary traverses the three major summits by their most challenging routes, sampling meanwhile much of the unique atmosphere of the area. It starts at the tiny village of Croesor, perched at its lane-

Cnicht has been dubbed 'the Matterhorn of Wales' – a title which belies its gentle nature. The south-west ridge of this graceful mountain, our line of ascent, is seen straight-on from the water-meadows beside the Afon Glaslyn. Cwm Croesor on the right.

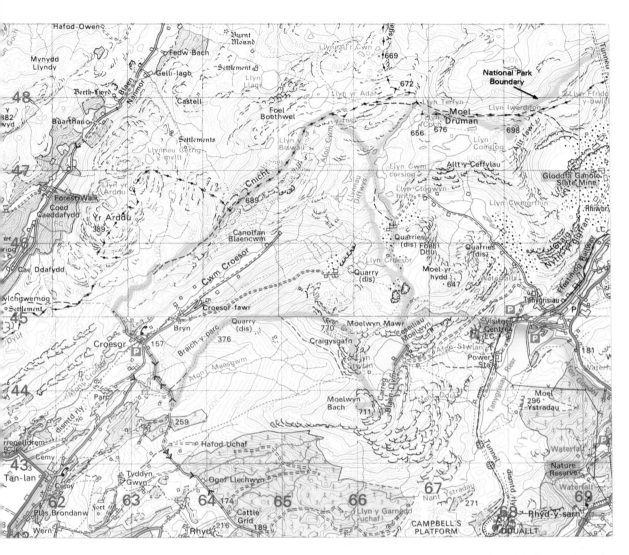

end high above the lush meadows of the Traeth Mawr and dominated by the beautiful cone of little Cnicht rising steeply behind. Perversely Cnicht is not a Welsh name, it is middle-English for knight and to medieval sailors in Tremadog Bay its sharp outline no doubt recalled the pointed bascinet helmet then worn by chivalry. After ascending its shapely little south-west ridge, Cnicht is discovered to be a sham – merely the butt end of a long moorland crest – but none the worse for that as the moorland is wild and lonely and scattered with small tarns, its rumpled surface broken by rocky outcrops and unfrequented tops. Beside Llyn yr Adur – the 'Lake of the Birds' – the old path from Nant Gwynant to Ffestiniog is joined, leading boggily around the

head of Cwm Croesor. The old slate mines in this deep defile between the flanks of Cnicht and Moelwyn Mawr, though never a commercial success, are notable for being the repository of the treasures of the National Gallery during the Second World War. The path leads onward to the wide saddle of Bwlch y Rhosydd.

When the extensive quarries closed in the 1920s the Bwlch must have been an industrial wasteland but time and weather have mellowed the acres of grey slate, and surrounded by grey-green hillsides the incongruity is now one only of shape rather than colour. Among the eyeless buildings the inquisitive will discover rusted machinery, old shafts and dripping adits: one adit is said to emerge eventually at workings a mile away beyond the mountain but exploring it is for experienced cavers only. Old and carefully constructed quarryman's paths lead onwards into Cwm Stwlan. One can continue thus directly to the rather sinister looking defile of Bwlch Stwlan between Moelwyn Bach and Moelwyn Mawr, but it is better to descend to the dam, visiting *en route* the rocky knob immediately above its northern end from which an interesting view unfolds. Although it extends to Cader Idris, the Rhinogs and the nuclear power station at Trawsfynydd, probably most intriguing is the valley immediately below with its hydroelectric station, the tiny steam trains of the quaint Festiniog Railway puffing along the shores of Tanygrisiau Lake, and the magnificent Vale of Ffestiniog beyond. The wild goats that roam these hillsides might also be glimpsed. Though purists will resent its existence, the engineering of the dam is impressive and it does impart some scale to the cwm: it was commissioned as the upper reservoir of the Ffestiniog Pumped Storage system in 1963, though there was a small lake here before.

Moelwyn Bach, the 'Little White Mountain', ringed with crags and with its well-known nose appearing actually to overhang Bwlch Stwlan, appears almost inaccessible from this eastern flank until close scrutiny reveals several steep grassy breaches in its defences. Whatever route through the cliffs is chosen, it is wise to take careful note of the diagonal descent path that leads from the shoulder left of the summit down to the Bwlch. The panorama from the summit itself, reached across a peculiar plateau of narrow pools between rocky spines, is superb, particularly down the twisting Dwyryd estuary and over the glinting sea to St Tudwal's Islands off the distant Lleyn.

The well-trodden route up Moelwyn Mawr rears steeply up from Bwlch Stwlan. There are two short rocky steps – almost scrambling – before the angle eases along the shoulder of Craig ys Gafn: soon the path steepens again and zigzags up the final rounded grassy ridge to the airy summit and OS cairn. Now the circle is almost closed and the north-east ridge, at first quite narrow in places and then becoming a sweeping grassy crest, provides a delightful descent back to Croesor.

A Moelwyn Horseshoe

Length: 16 kms/10 miles
Height gained: 1150 m/3,770 ft
Difficulty: a fairly committing circuit with some strenuous walking – mostly grassy, often boggy, sometimes rocky. Both Moelwyn peaks involve a little easy scrambling
Warning: some confusing ground where detailed navigation is not easy in poor conditions, beware old quarry workings with loose rock, pits and shafts. Craggy Moelwyn Bach deserves respect

Start: good SNP car-park in Croesor village at 631447. (157 m/515 ft)

(1) Follow main lane leading NW through village, over brow, through gate/stile and woods to 628451 where path leads off right just before next stile, occasional waymarks. Fork right at ruin to stile/wall on crest 150 m above and continue on L flank of ridge over scree section to grassy saddle. Now follow good path – scrambly alternatives to left and right – up prominent SW ridge to CNICHT summit (690 m/2,265 ft).
3 kms/2 miles

(2) Continue NE descending gradually on vague path to wide boggy plateau by Llyn yr Adur. At path junction (657478, flattened cairn) turn SE onto twisting undulating path leading to large abandoned quarry on wide saddle. Behind prominent ruin ascend inclines, etc to moor where well-marked path leads off left of first large quarry pit towards lowest point of skyline ahead – shallow col at 666449 with path junction and cairn. Follow good path contouring R round hillside for 400 m, then descend left on vague path to N end of Llyn Stwlan Dam. (500 m/1,640 ft).
5.5 kms/3½ miles

(3) Cross dam and ascend broken ground into shallow cwm below NE flank of Moelwyn Bach. Pass left below prominent quarried cave to reach stepped rib on left skyline. Scramble up steep grass and rocky bluffs to summit plateau, cross glacis and pools NW to cairn of MOELWYN BACH (711 m/2,334 ft). If unsure of route or ability on scrambly ground, use (3.A).
1.5 kms/1 mile

(4) Strike ENE 150 m to vague marshy saddle above cliffs where narrow path descends diagonally left across steep scree-covered hillside to Bwlch Stwlan (585 m/1,920 ft). Ascend rocky ridge N, two easy scrambly steps, to shoulder then zigzag up steep grass ridge to MOELWYN MAWR summit (770 m/2,527 ft). Descend grassy ENE ridge to metalled lane 0.5 km S of Croesor village.
5.5 kms/3½ miles

Variations
(3.A) Do not descend to dam from the contour path but follow it easily to Bwlch Stwlan. From here ascend MOELWYN BACH by descent route used in (3) above. Bwlch also reached easily from S end dam.
1.5 kms/1 mile

(4.A) From MOELWYN BACH summit descend easy grassy W ridge to metalled lane 1.5 kms S of Croesor, thus avoiding Moelwyn Mawr.
4.5 kms/2¾ miles

Moelwyn Bach displays its most rugged flank to Llyn Stwlan, dammed in 1963 by the CEGB. Our route crosses the dam and ascends the mountain virtually by the left skyline and the descent route to Bwlch Stwlan is also well seen.

Cader Idris Gwynedd, North Wales

OS 1:50,000 Sheet 124

... the polished peak ... where God had sat on the seventh day to bless the Mawddach estuary.

Huw Jones 'Cadir Idris' from the Anglo Welsh Review

The walls of Dolgellau, they say, are a mile high. Actually Cader Idris with its 6 miles (10 kilometres) of northern crags rears to only 2,928 feet, a little over half a mile above this little grey town at the head of the beautiful Mawddach estuary. But certainly this magnificent mountain dominates the country hereabouts and is a magnet for mountain lovers of all kinds.

Cader Idris: the name means Chair – or Stronghold – of Idris. Was Idris a giant, or a poet, or a prince? Was he Arthur himself? No one is sure. Certainly Cwm y Gadair (Gadair = Cader), scooped deep from the northern crags below the summit, appears like a gigantic chair when seen from afar. More of a range than a single mountain, the 20-kilometre (12-mile) ridge of Cader springs from the sea above Tywyn and rises to nine tops above 2,000 feet (600 metres). The valleys to north and south, the one a striking tidal estuary with wooded shores, the other a deep and tranquil lake-floored glen, are among the finest in Wales. The glory of Cader however is its succession of sculpted cwms, five holding tarns, of which Cwm Cau –

Mynedd Pencoed rises a thousand feet over the tranquil waters of Llyn Cau on the southern flanks of Cader Idris.

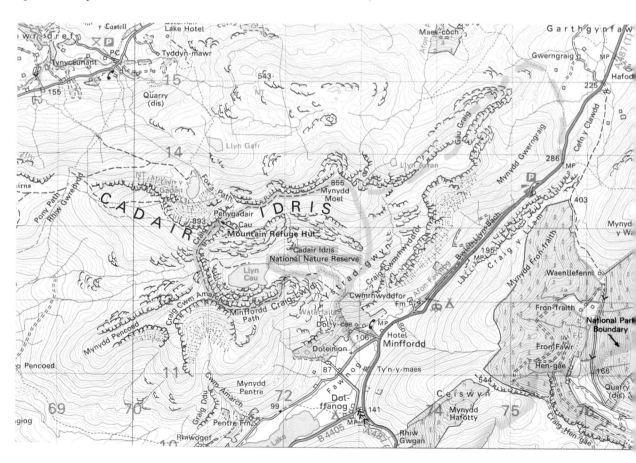

the 'Hollow Cwm' – is famed as a perfect example and with its surrounding hillsides and the ferny woods below is now an NCC Nature Reserve.

The regular tourist route to Penygadair, Cader's summit, ascends the broad easterly ridge – the so-called Pony Path – but discerning hill walkers will find the more rugged route suggested here far more interesting. Craggy bluffs and enchanted woods through which the Nant Cader plunges in a series of continuous cascades and waterfalls, provide a truly dramatic start and before beginning the steep ascent from the cwm above to the ridge of Craig Lwyd it is worth diverting along the grassy floor to visit the shores of Llyn Cau, invisible until almost the last moment. The view across the dark waters to the brooding crags is an imposing one. Owen Glyn Jones, greatest of Victorian rock climbers, cut his teeth amid this great phlanx of gullies and buttresses but such climbs are not really fashionable today. Meanwhile, the main path traverses the ridge above and shortly before reaching Mynydd Pencoed the now sparkling waters of the lake can be glimpsed through the forbidding cleft of Great Gully, one of Jones's classic climbs, and for years a major *tour de force* until rock fall destroyed its crux.

From Bwlch Cau beyond connoisseurs can continue their ascent around the steeply rising cwm lip rather than remaining on the main trail. Crouching among the boulders below the rocky summit is a little refuge shelter, successor to a refreshment hut erected here by Dolgellau guide Richard Pugh in the 1830s, when the ancient belief that those who spent the night alone on the summit would descend either blind, mad or a poet already seemed rather far fetched! The wide summit panorama includes most of the mountains of North and Central Wales but the proximity of the sea is part of Cader's charm and the closeby lip of Cwm y Gadair provides a foreground for the splendid vista down the glinting estuary to the sands of Barmouth and the wide ocean. This cwm is hardly less impressive than Cwm Cau and the jagged rib rising above its little lake is the celebrated Cyfrwy Arete on which Sir Arnold Lunn met with the near fatal accident that switched his destiny from rock climbing to be the founding father of modern skiing.

Descending eastward the route passes the head of the Fox's Path, the steep, shaly and unpleasant descent into Cwm y Gadair, and continues along the precipitous northern edge of a wide ridge carpeted with short wiry grass and patterns of small stones. On this airy upland the rise to Mynydd Moel is hardly noticed. Just past the cairn the ridge drops steeply round the secluded little Cwm Aran with its tiny tarn then runs flat to the more broken ground of Gau Graig. One can continue along this ridge and return beneath the unfrequented south-east face of the mountain – in summer a maze of rocky ribs and bilberry-hung hollows – but it is a long route and the recommended descent strikes directly down steepening grassy slopes to rejoin eventually the ascent route at the lip of Cwm Cau.

From Mynydd Moel the main crest of Cader Idris stretches westward towards Pen y Gadair, the main summit of the mountain. Little Llyn y Gafr is visible on the right.

Over Cader Idris

Length: 10 kms/6¼ miles
Height gained: 960 m/3,150 ft
Difficulty: in ascent good paths, often on steep and rocky ground, on descent rough ground and steep slopes
Warning: this is a large isolated mountain taking full force of bad weather, paths often run close to cliff edge, take care in windy or bad conditions
Note: much of route lies within NCC Reserve, respect by-laws posted at Idris Gate. Path erosion is occurring, avoid such areas especially on descent route
Start: good SNP car-park at 732115 just W of junction A487 with B4405, or on foot 300 m along B4405 at iron gates bearing legend Idris Ltd. Good campsites and B&B at two cottages closeby. (94 m/308 ft).

(1) From either start follow signed paths that cross river and meet at edge of woods. Badly eroded path now climbs steeply through woods above stream gorge to open ground and cwm lip. Easy path continues to shore of Llyn Cau. (470 m/1,540 ft).
2.5 kms/1½ miles

(2) At prominent shark-fin rock 400 m before lake (720123) fork left to rocky path zigzaging steeply up to ridge and continuing over minor top of CRAIG LWYD (686 m/2,251 ft) to fence and stile at summit of MYNYDD PENCOED (798 m/2,617 ft).

Short descent N near cliff edge leads easily to Bwlch Cau before path zigzags up wide final slopes to meet 'Pony Path' among rocky tors close to PENYGADAIR (893 m/2,928 ft) – summit of CADER IDRIS, refuge shelter 30 m N of trig point.
2.5 kms/1½ miles

(3) Descend NE over boulders to long wide ridge, follow easily edge of N cliffs over slight eminence of Twr Du to stile in fence and MYNYDD MOEL summit and shelter-circle (863 m/2,831 ft). Descend SE down wide steepening slopes keeping well L of NCC fence and eroded traces of path closeby. Avoid scree band to L or R. When angle eases stile leads W into shallow re-entrant, cross brackeny hillside to NW corner of fir plantation where stream is easily crossed and ascent route rejoined.
5 kms/3¼ miles

Alternative
(3.A) From Mynydd Moel path descends steeply SE round head of Cwm Aran, continuing NE over several stiles along wide undulating ridge to Gau Graig. Descend long spur, indistinct traces of path and some broken ground, to strike bridle path at 754152 continue S to A487 road. At 751133, 300 m SW of large layby, fork right to follow old road descending to valley floor eventually rejoining highway 500 m NE of Minffordd junction close to start.
10.5 kms/6½ miles from Penygadair

The Llangollen Escarpments

Clwyd, North Wales

OS 1 : 50,000 Sheets 116, 117, 125

The first real landmark in Wales as the traveller hurries up the A5 road towards Snowdonia is the sharp cone of Dinas Bran above the Dee, and with luck the arched ruins on its summit might even be silhouetted against the evening sun. Before reaching the outskirts of Llangollen our traveller might briefly glimpse behind Dinas Bran a broad hillside topped by tiers of white limestone. This is Eglwyseg Mountain, not really a mountain, more the impressive edge of a domed moorland plateau.

Eglwyseg – a name likely to confuse an English tongue – is pronounced 'egloosik' and means 'church lands': presumably it once belonged to the medieval Cistercian Abbey of Valle Crucis whose ruins still stand beside the Eglwyseg river a couple of kilometres below the mountain. The Llangollen area has a long history, the town with its fourteenth-century bridge is one of the traditional 'Seven Wonders of Wales' while the fortress of Dinas Bran itself – 'Castle of the Crows' and surely all but impregnable – was the eighth-century stronghold of the Princes of Powys. Until the designation of the 270-kilometre (168-mile) Offa's Dyke Long Distance Footpath in 1971, Eglwyseg Mountain and the beautiful valleys beneath it were comparatively unknown to other than local walkers. The definitive route however follows an old and narrow path – known locally as the Cow's Path – below the western scarp of the hill, and the magnificent sequence of gleaming limestone crags that line the edge above has become more familiar. Indeed this is probably the most

The great limestone prow of Craig Arthur rears over the Eglwyseg valley in this view northward along the Cow's Path that runs along the base of the escarpment.

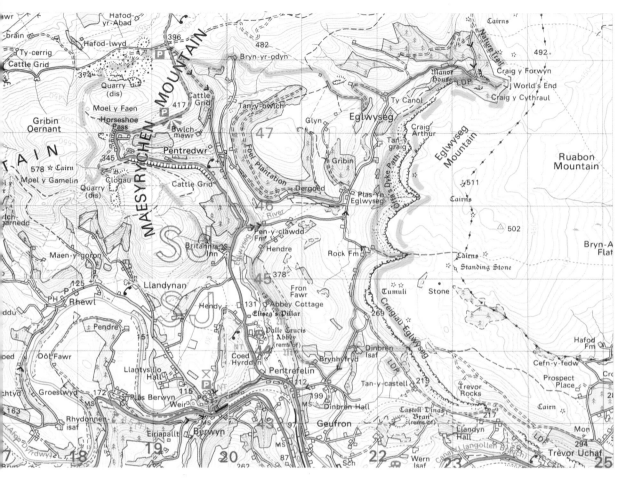

spectacular section of the Offa's Dyke route.

The delightful valleys of the Eglwyseg River and its tributaries are deep and narrow and H-shaped – and must have been even more beautiful before the modern road, the A542, was constructed over the Horseshoe Pass. The first leg of our suggested route follows an ancient grassy track around the lonely head of the western valley, over a gorsey saddle, down a green lane from where there are excellent views of the imposing cliffs ahead, and into the secluded eastern valley – Eglwyseg Valley itself. This tranquil glen beneath the white cliffs is traversed only by the burbling kingcup-bordered stream and a narrow lane set about with hazel, hawthorn and tall ash trees. At the head of the valley stands an ancient manor house, a jewel in a sylvan setting. Perfect proportions and geometrically patterned black and white timbering emphasize the date above the door

–1563. The furthermost recess of the valley is appropriately known as World's End and here our route joins the 'Cow's Path' traversing the base of the cliffs. At first a low line, the cliffs rapidly grow higher and the first major bastion is Craig Arthur, over 30 metres high (100 feet), the largest unbroken crag on Eglwyseg Mountain and a favourite modern rock-climbing area. The path crosses patches of scree and areas of greens-ward dotted with stunted thorn bushes and passes the remains of old lime-kilns: ravens circle overhead.

Above Rock Farm a rugged and popular path leads up beside waterfalls through a cleft in the cliffs to the wide moorland above. Strictly speaking – and inexplicably – this path is not a right of way although 500 metres beyond the cliff top it becomes one and continues for some 5 kilometres (3 miles) over the managed grouse moor of Ruabon Mountain to the village of

Penycae. Neither is our suggested route south-ward along the cliff top a right of way – though traversing 'common land' and well frequented – and the entire cliff-top section can be bypassed by continuing the Offa's Dyke route along the lane at the cliff bottom. Nevertheless the limestone moorland along the cliff edge is delightful walking country and the splendid views from it over the smiling vale extend to the distant mountains of Snowdonia.

It is important, but not too difficult, to locate the correct gulch down which the next path – this time a right of way – drops through the crags to the lane below. A rural section follows, descending to what was once an old tramway running through the woods along the edge of the Valle Crucis meadows – a meander of the Dee abandoned long ago – where a path leads off

to the closeby abbey, a worthwhile diversion. The final section of the route takes the quiet lane up the western branch of the Eglwyseg Valley, once the old road over the Horseshoe Pass, and climbs steeply back to the start. A more strenuous alternative crosses the steep moorland of Maesyrchen Mountain with its grey slate quarries, a very different landscape from the limestone mountain across the valley.

Eglwyseg Scarps

Length : 18 kms/11¼ miles
Total height gained: 800 m/2,650 ft
Difficulty : a fairly gentle route on good paths or metalled lanes, but moorland section trackless though easy going with rocky outcrops to negotiate and two steep sections
Warning : exercise caution when walking above vertical cliffs especially in windy or bad conditions

Start : large parking area at 192482 opposite Ponderosa Café. 396 m/1300 ft.

(1) 100 m SE down Pentredwr Lane grassy track forks L descending gradually to Bryn-yr-odyn, path continues behind cottage, contouring hillside to broad saddle and green lane, descend taking L fork at each junction via hairpin into deep valley on L, past cottage to lane at Ty Canol. Follow lane L to woods and ford at World's End (233477, 310 m/1,017 ft).
5.5 kms/3½ miles

(2) Follow 'Offa's Dyke' waymark S from lane through wood onto open stony hillside, good path contours steep slopes below cliffs to stream at 220454 (255 m/837 ft) in cwm above Rock Farm. Rough path leads steeply E up narrowing valley through cliffs to re-entrant in moorland above, at c 225453 strike up R off path and contour above upper tier of cliffs (425 m/1,395 ft) W then S to 2nd deep re-entrant in cliff top (225441) with straggly pine trees beyond. Descend steep rocky path N side stream, dropping diagonally R below cliffs to lane.
6 kms/3¾ miles

(3) RoW opposite descends W across fields to Dinbren Isaf farm and lane beyond, descend lane steeply L, bridle path leads off R 60 m before junction, passes Brynhyfryd and contours N through woods to Hendre. Neglected RoW cuts corner across fields L to lane by bridge (204457: 140 m/460 ft). Ascend lane L 60 m, RoW cuts corner R to lane leading N up valley through Pentredwr hamlet and steeply up to start.
7 kms/4¼ miles

Variations
(2.A) From World's End ford follow FC Nature Trail up valley E through woods to open moorland, strike R along forest boundary until above cliffs and contour S above Craig Arthur to join (2) at path above Rock Farm in 3rd large re-entrant in cliff top (225453).
Extra 1 kms/½ mile

(3.A) From bridge 204457 continue W to main road then S 250 m to Britannia Inn. RoW leads NW behind Inn up steep moorland slopes Maesyrychen Mountain. Pass abandoned quarry on first summit (450 m/1,475 ft) to saddle and follow contour path N above Horseshoe Pass to extensive abandoned quarry on Moel y Fan, descend NE to start.
Extra 2.5 kms/1½ miles, +65 m/213 ft

Seen westward from Offa's Dyke footpath, the cliffs of the Eglwyseg escarpment look down on the strange peak of Dinas Bran, crowned with its ruined fortress.

The Gentle Berwyn Clwyd, North Wales

OS 1:50,000 Sheet 125

'Aren't they the hills on the left just past Llangollen?' The Berwyn is just a name to the majority of English mountain lovers heading up the A5 to Snowdonia, and its northern flanks seem unremarkable as they rise steeply above the twisting Dee. But the Berwyn is actually quite an extensive range, its single ridge stretches almost 30 kilometres (18 miles) from Llangollen to Lake Vyrnwy and twenty-one summits top 2,000 feet (600 metres), no less than ten of them classed as separate mountains. And only one road crosses the Berwyn near its southern end at the little pass of Milltir Gerrig. The Berwyn range is not the border but it is the final crest of Wales before the Shropshire plain. Holding its beauty close to its heart, the Berwyn is best approached from the head of the Rhaeadr valley above the sleepy little 'waterfall village in the pig glen' –

Llanrhaeadr-ym-Mochnant. Rhaeadr – 'water-fall' – is the operative word here for at the very end of the valley stands Pistyll Rhaeadr, the 'spout waterfall' and one of the traditional Severn Wonders of Wales: it is certainly the jewel in the heart of the Berwyn with the highest summits and most impressive scenery standing to the north within easy reach. Any walk on the Berwyn should include a close look at the waterfall, the ferny spray-damp wood at its foot easily reached by a short public path from the lane end. The water pours over the lip of a dark crag 75 metres (240 feet) above and falls as a white column to explode behind a natural arch

Little Llyn Lluncaws lies below the eastern flank of Moel Sych. The most interesting ascent route is by the east ridge from where Cadair Berwyn is seen rising across the cwm.

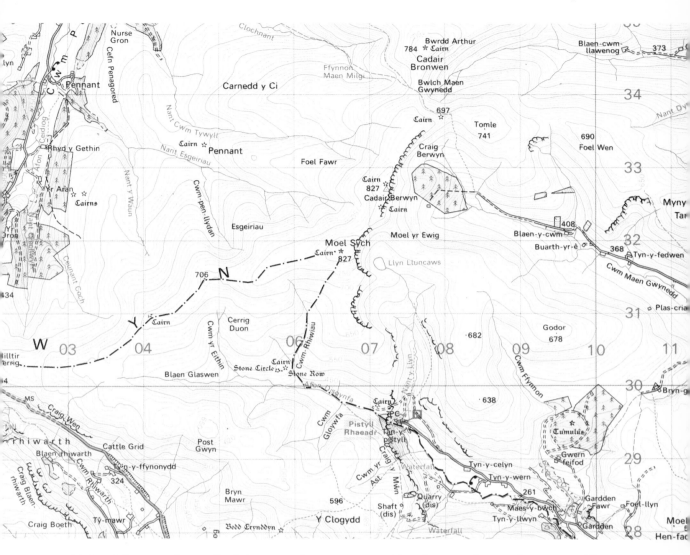

two-thirds of the way down. Dr Johnson, who came here in the 1770s, is said to have remarked 'Very high and in rainy weather very copious', an accurate if unimaginative comment on such a beautiful and imposing place.

A path zigzags steeply up above the falls where the Afon Dysgynfa flows down a wide green valley and into a series of pretty cataracts among the trees before surging over the cliff edge. Another path leads northwards from Tan-y-pistyll up an attractively desolate little glen towards Llyn Lluncaws, the only tarn in the range, lying beneath the craggy eastern face of Moel Sych. At 827 metres (2,713 feet) Moel Sych is the highest point in the Berwyn and its east ridge, cradling the little lake like a protective

arm, provides the most interesting ascent. Unfortunately the top is flat and unprepossessing, just a small cairn and a junction of fences, though the view is excellent round a full 340 degrees encompassing both Snowdon and the Long Myndd, while the little lake below, its surface covered with strange wide-leaved pondweed, appears in certain lights as a pool of molten metal. Cadair Berwyn, a kilometre northward along the grassy but sharp-lipped ridge, is a more imposing summit with a rocky edge, a stone shelter-circle and an OS pillar, unfortunately in real terms it is just one foot lower than Moel Sych although its metric height is the same. The ridge continues easily northward to Bwlch Maen Gwynedd, crossed by an ancient trackway

Pistyll Rhaeadr, where the Afon Dysgynfa plunges over a high crag on the southern flanks of the Moel Sych massif, is the highest waterfall in Wales.

and presumably once the border of the medieval principality of Gwynedd, before rising again to Cadair Bronwen whose actual summit is known as Bwrdd Arthur – 'Arthur's Table' (78 metres/2,572 feet). Northward now the ridge flattens out and loses height, and though still good wide walking country its character becomes one of moorland hills rather than gentle mountains. However the long eastern spur that springs from the ridge just south of Bwlch Maen Gwynedd is not without interest and its second top, Foel Wen (690 metres/2,265 feet), is a worthy little summit and a 'separate mountain'. From Moel Sych the main crest falls away southward as a flat heathery ridge to Milltir Gerrig but its grassy southern shoulder drops to the Disgynfa valley not far above the falls and provides a convenient descent route.

Like so much of Wales, the Berwyn is sheep country, in fact no fewer than ninety-six farmers hold grazing rights on the western flanks alone, a large part of which has recently been acquired by the Nature Conservancy Council. Although the range is well frequented by walkers, well-used paths traverse the main ridges, and a major fell-running event is held every September, the only rights of way to approach the high tops are the ancient track over Bwlch Maen Gwynedd and the path leading almost to Llyn Lluncaws, all other access is only on the acquiescence of the local landowners though happily an atmosphere of tolerance and reason currently prevails. Starting from Tan-y-pistyll several circular routes will suggest themselves, both short and long, and nothing that has been described is in any way difficult or dangerous – except the abrupt lip of the Falls: the discerning hill-walker can safely make his own arrangements to explore these airy ridges and lonely recesses.

Berwyn Circuits

Difficulty: easy going on all obvious routes, grassy hillsides, bracken, several short steepish sections. This is high isolated ground, appropriate equipment should be worn/carried

Start: good parking beside the narrow lane at Tan-y-pistyll 076293, or beside the swiss chalet-style tea room beyond for modest charge

Routes: as legal access here is not formalized, walkers should seek permission locally, as necessary, for the itineraries they plan

Pumlumon – the Heart of Wales

OS 1 : 50,000 Sheet 135

Raising its crest almost to 2,500 feet, just above the level of the great upland plateau of Central Wales and the convoluted valleys that cut into it, Pumlumon has been variously described as a 'sodden weariness' by one ancient traveller and as 'probably the best viewpoint in Wales' by the great geologist Sir Arthur Trueman. To most people the truth will lie somewhere in between.

Pumlumon, often Anglicized to Plynlimon, would appear to mean 'five tops' – there are certainly four definite summits above 2,000 feet (600 metres) besides several points which might appear as tops. Pumlumon is in fact a series of smooth hills covered in peat bog and poor grassy pasture strung out along a broad flat-topped ridge that runs south-west to north-east for some 14 kilometres (9 miles). A dozen cwms cut into the ridge flanks, most of them shallow and

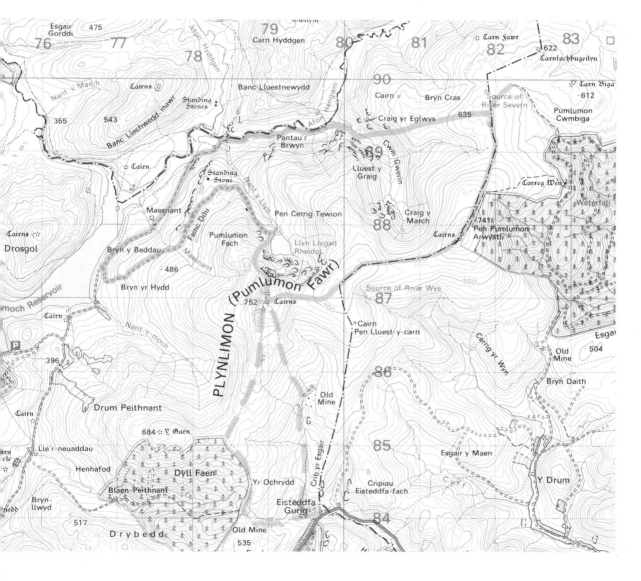

unremarkable, although three – which our route will visit – are of interest.

Pumlumon's claim to fame, besides being virtually in the centre of Wales and the highest point between Cader Idris and the Brecon Beacons, is that two of Britain's great rivers – the Wye and Severn – rise high on its eastern flanks; for several centuries travellers have ascended the mountain to visit their sources. Most of the upland area of the mountain is common land belonging to the Crown Estates to which free public access was granted in 1932, thus with its generally gentle contours and straightforward terrain Pumlumon offers excellent walking in a fine situation under a wide sky. Both the two popular routes approach the summit from the south: the shorter waymarked route leaves the A44 road at Eisteddfa Gurig farmhouse café (a small charge is made for parking) on the pass between Wye and Rheidol, the main watershed of Wales at 415 metres/1,360 feet. A longer ridge route leaves the A44 at Dyffryn Castell hotel 4 kilometres (2½ miles) further west. Linking the two routes gives a circuit of about 7.5 kilometres (4¾ miles).

A more interesting circuit leaves the narrow road above the twisting and not unattractive Nant-y-Moch Reservoir to follow a good access track up to what might be termed the 'Jewel of Pumlumon', the little Llyn Llygad Rheidol – the Source of the Rheidol – nestling beneath steep craggy hillsides in a dark and moody cwm, surely the most spectacular scenery on Pumlumon. The lake itself is the Aberystwyth water supply and is guarded by a notice threatening dire penalties to '. . . persons found poluting same by bathing or other acts . . .' Steep but easy green hillsides lead to the small grassy summit plateau, where ugly fences intrude on the magnificent view which on a good day encompasses the entire width of Wales from the Marches to Cardigan Bay and the two headlands of Lleyn and Preseli. The route onwards is well scarred by passing feet and leads easily to Pumlumon's third highest and least inspiring summit, Pen Pumlumon Llygad-bychan – 'Pumlumon of the Little Source' – where adjacent to the small cairn

Llyn Llygad Rheidol has been called the 'Jewel of Pumlumon' and lies under the mountain's northern face surrounded by the most rugged scenery in the area.

is a slate slab bearing the enigmatic inscription '1865 ↑'. Blaen Gwy – the Source of the Wye – lies some 400 yards due east across a plateau of short windblown grass, over a fence and down into a green gulch which frames the distant hills of the Radnor Forest. It is a worthy birthplace for a famous river.

Stretches of easy path and a fence line continue along the ridge, over a scabby saddle where the overlying peat has eroded away, to the two cairns and several stone shelter-rings that mark Pen Pumlumon Arwystli, Pumlumon's second summit. Now a long gradual descent and a little compass work leads to a wilderness of peat hags, colourful in summer with heather, moss, and bog plants. Here, amid a maze of small trickles, a white post proclaiming 'Tarddiad Afon Hafren' stands over a reedy pool. Here rises the mighty Severn. It is pretty perhaps but hardly impres-

sive. The route now descends into Cwm Gwerin, a fine rugged little valley which leads down to a sizable river, the Afon Hengwyn, where the melancholy ruins of a remote farmstead – Pantau'r Brwyn, the 'Hollow of the Rushes' – stand in an overgrown crag-girt dell beside the water. The boggy track along the river makes heavy going as far as a second ruined cottage below Llyn Llygad Rheidol. Just across the river in the wide green flats beside the Afon Hyddgen two standing stones mark the site of Owain Glyndwr's decisive victory over a small English army in 1401. Gathering his men in this secluded valley, Glyndwr had found himself suprised by superior forces, and the ensuing bloody engagement initiated his bid to create a united and independent Wales.

A dryer and easier trail leads past Maesnant, now an outdoor pursuits base, back to the start.

A Pumlumon Crossing

Length as described: 16 kms/10 miles
Height gained: 600 m/1,970 ft
Difficulty: easy walking on high bleak plateau mostly above 650 m/2,000 ft

Start: careful parking is possible on the roadside verge near track junction at 768874. (350 m/1,150 ft).

(1) Follow jeep-track up and across hillside to Llyn Llygad Rheidol at 510 m/1,675 ft. Ascend steep hillside above W end of dam to obvious shallow re-entrant leading to small grassy col. Climb knobly shoulder to summit plateau where trig point and stone shelter-ring mark summit of PEN PUMLUMON FAWR (752 m/2,468 ft).
4.5 kms/2¾ miles

(2) Descend E on peaty path alongside fence to broad saddle then ascend to plateau and small cairn of PEN PUMLUMON LLYGAD-BYCHAN (723 m/2,372 ft) (not named on map). Source of Wye lies at head of gulch 400 m. E. Follow wide flat ridge parallel to fence, occasional path, over saddle and stony areas to large cairns at summit of PEN PUMLUMON ARWYSTLI (741 m/2,427 ft). Descend N past two small tarns to white cairn at 819899, post 400 yds E over stile marks source of Severn. Retrace steps to first tarn then strike E down wide ridge before descending steeply past rocky outcrops into lower Cwm Gwerin, follow stream to ruin at 797893. (375 m/1,230 ft).
8.5 kms/5¼ miles

(3) Very boggy riverside track leads W to second ruin at 783891 then good path leads to Maesnant. Continue 1 km along road to start.
4 kms/2½ miles

High on the eastern flank of Pumlumon Fawr lies this little hollow – Blaen Gwy – the source of the River Wye.

Lonely Elanydd
Central Wales

OS 1 50,000 Sheet 147

> 'All that I heard him say of (Wales) was, that instead of bleak and barren mountains, there were green and fertile ones;'
>
> James Boswell *Life of Samuel Johnson*: 1774

The Elan Valley cuts deeply into the wide plateau of empty moorland and smoothly domed hills that separates Pumlumon from the Brecon Beacons, an upland so remote that it has been called 'the desert of Wales' but is more generously known as Elanydd. This ancient name was applied loosely by Giraldus Cambrensis, the twelfth-century Welsh traveller, to the southern mountains of Wales, as opposed to Eryri, the northern mountains, but in context he clearly meant this particular upland. Elanydd was the final refuge of the Romano-British chieftain Vortigern from the Saxon invader and from here Owain Glyndwr waged his guerilla campaigns in the fifteenth century. Here too were early Celtic monastic communities and later an important satellite of the famous Cistercian abbey of Strata Florida whose ruins stand on Elanydd's western extremity. Lead, silver and copper were mined here from Roman times until the late nineteenth century and important hordes of bronze axes and Romano-Celtic gold jewellery have been discovered locally. Even the extensive bogs were once celebrated and the local peat was said to be '... scarce inferior to coal ...' But sheep were the real wealth of Elanydd – as they are today – and in the Middle Ages wool was exported to the Continent after a special customs exemption had been granted by King John. Eventually communications improved the romantic appeal of the rugged hillsides and green valleys attracted tourists, among them

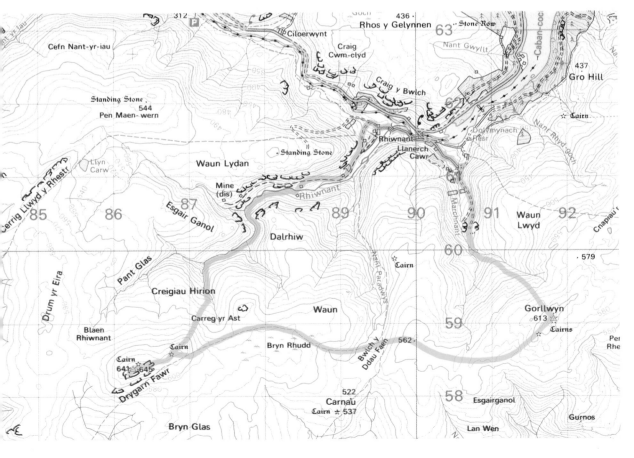

the poet Shelley who stayed near Llanerch in 1810.

However, these hills suffer an annual rainfall of seventy inches and in 1904 three great dams were completed for Birmingham Corporation Waterworks, flooding over 1,200 acres of valley bottom. Parliament meanwhile guaranteed free public access to the entire seventy-one square miles of the Elan watershed. A fourth dam was commissioned in 1952 and today good roads lead along the wooded shores of a chain of attractive lakes into the heart of Elanydd. The Welsh Water Authority, landlord to some hundred people and 45,000 sheep, has created in a low-key but farsighted way, the nearest approach to an American-style National Park – one actually owned by its administrator – in Britain. Recreational use such as walking, pony trekking, fishing and bird watching is en-

couraged, an interesting visitor centre is open daily below the Caban-coch dam and there is a helpful ranger service.

The freedom to roam at will – always respecting the rights of the farmers – gives scope for several interesting walks especially among the rugged lesser hills either side of the Elan 'portal' or over the ancient drove roads which once linked the Wye Valley to West Wales. The circuit described traverses the two highest of Elanydd's three 2,000-foot tops. It has a rare lonely quality, for these unfrequented hills are virtually trackless and rippling purple moor grass sets off distant vistas of cloud-shadowed plateaux shared only with semi-wild ponies, the circling red kites and the ubiquitous sheep. Easy going leads to the summit of Gorllwyn from where there are superb views southward over the patchwork field of the Irfon valley to all four

massifs of the Brecon Beacons National Park. At first Drygarn Fawr with its twin tower-like cairns is visible in the distance only to disappear as the wide swell of the ridge becomes broken by peat hags eventually fading into a desert of awkward tussocky moorland slashed by the deep gash of Bwlch y Ddau Faen. Then a cairn reappears and firmer ground leads to the rocky bilberry-hung crest that links these two strange drum-shaped constructions. Typically the mountain streams of Elanydd are deep-cut and the Afon Rhiwnant followed on the descent route is no exception. Presenting an unexpected contrast to the moorlands above with its twisting tumbling course and pretty cataracts, some intriguing pot-holes, a fine waterfall and isolated rowans overhanging deep pools, it makes a fitting final stage to an unusual excursion.

A Circuit in Elanydd

Length: 16 kms/10 miles
Total ascent: 490 m/1,600 ft
Difficulty: strenuous but never steep walking over trackless, rough and often boggy country. Confusing terrain, map and compass useful even in good conditions
Note: patrolling rangers ask that walkers out overnight leave a note on their car or more securely on the ranger ansaphone tel: 810-880

Start: small car-park at 901616 near Llanerch Cawr phone box. (260 m/850 ft).

(1) Cross bridge opposite car-park following lane SE past bungalow, across stream bridge then up R on wide track to old quarry (907603). On trackless hillside ahead ascend rib between streams, continuing upwards to summit GORLLWYN, low rocky cairns, OS pillar (613 m/2,009 ft).
4 kms/2½ miles

(2) From tall cairn 300 m SW, follow high ground W to Pt 562, peaty hollows, ponds, stony patches. Irregular sequence small concrete posts, useful for navigation, marks watershed. Descend to marshy Bwlch y Ddau Faen (510 m/1,670 ft), cross stream to gain boggy plateau Bryn Rhudd. Continue WSW, featureless, rough, to conspicuous 6 m (20 ft) white-topped cairn on rocky outcrop DRYGARN FAWR and on to OS pillar at second cairn (641 m/2,104 ft).
6 kms/3¾ miles

(3) Descend easily NE from first cairn into stream defile, continue past cataracts, join Pant Glas stream at 872598. Take S bank path through narrow glen, past major confluence, waterfall, into Rhiwnant valley. When possible cross stream to join mine track descending N side valley, best place probably near old workings 884607, track continues to Rhiwnant Farm then E to start.
6 kms/3¾ miles

The strange tower-like cairn that crowns the remote summit of Drygarn Fawr. This is desolate, open moorland dotted with peat bogs.

The Beacons Crest South Wales

OS 1: 50,000 Sheet 160

The Brecon Beacons National Park, a group of
four adjoining upland massifs, contains the most
splendid mountain country in southern Wales.
There are other mountains more lonely or
perhaps more atmospheric but these are the
steepest, the highest and the least gentle mountains south of Cader Idris. Each flaunts, in one
form or another, great north-facing scarps of
Old Red Sandstone, yet their character is very
different. Pen y Fan, the culminating point of the
massif that lends its name to the entire Park, is
just 94 feet short of the magic 3,000-foot mark,
its name aptly meaning 'Top of the Beacon' for
its summit can be seen from deep into Central
Wales, from the Cotswolds and even from the
Exmoor coast. Five other tops rise above 2,500
feet (760 metres) and this large area of high
ground displays the harsh environmental conditions that go with it.

From the Vale of Usk these seem shapely
peaks but from elsewhere the truncated main
tops appear disappointing and suggest a comic
strip rendering of minor lunar volcanoes. Their
glory however is in their northern flanks where
steep and sometimes craggy faces rise from a
sequence of deep and attractive glacier-carved
cwms. Elegant ridges divide the cwms and their
meeting points with the scalloped northern scarp
form the main summits – a classic formula. With
their typically grassy yet steep enough slopes and
real mountain ambience the Beacons offer many

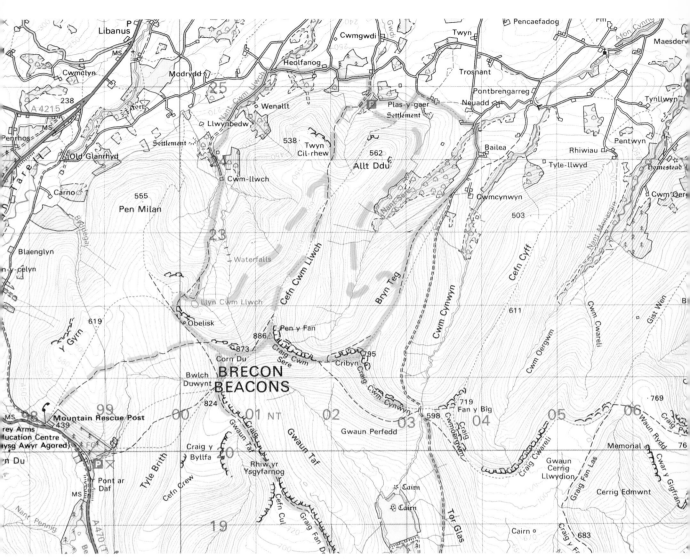

worthy and challenging routes to discerning hill walkers. The heavily eroded 'tourist' route to Pen y Fan ascending the western flank from the Storey Arms – once an inn now an outdoor centre – on the A470 Brecon to Merthyr Tydfil road, is of little interest.

Routes from the south, typically gentler than those from the north, start from the mountain road traversing the attractive Taf Fechan valley. A pleasant 'horseshoe' of about 12 kilometres ($7\frac{1}{2}$

From the steep final section of the north-east ridge of Y Cribyn there are impressive views across Cwm Sere to the precipitous face of Pen y Fan – the Beacon's highest peak. The truncated summit of Corn Du can be seen in the distance.

miles) round the upper reaches of the valley follows the ancient trackway – thought to be a Roman Road – to Gap, the lowest pass over the Beacons at 032205, before crossing the four highest tops: a convenient start is the forest car-park at 036171 below the Neuadd – or 'Zulu' – reservoir. The challenging east-west traverse of the entire range crosses the peaty plateau east of Gap keeping to the easier terrain of the scarp edge, to link Talybont to Libanus, a distance of some 18 kilometres (11 miles). But the classic circuit from the north detailed here includes the highest summits and the most spectacular scenery.

The grassy Bryn Teg ridge – which translates

as 'beautiful hill' – surely lives up to its name, for it sweeps up steeper and steeper in smooth symmetry to the seemingly sharp summit of Cribyn. But there is a good path and the final rocky step is hardly a scramble. Across Cwm Sere rears the 400-foot (120-metre) north-east face of Pen y Fan – the largest crag in the Beacons. The red rock layered with green ledges and slashed by dark gullies is an impressive sight but the sandstone is poor and too broken to hold any attraction for the climber; although under a plastering of ice and snow it has provided some excellent sport. From Cribyn the path onwards is fairly steep but badly eroded and carves an ugly scar up a grassy and otherwise graceful hillside. Cliffs fall abruptly from the flat summit of Pen y Fan but the view northward over them, across the Vale of Usk with its little fields and white cottages to the Mynydd Eppynt and the distant blue tablelands of Elanydd must be the best in the Beacons. On the summit, and also on the tops of Cribyn and Corn Du, there are traces

of Bronze Age cairns: the latter summit was excavated in 1978 and contained pottery cremation urns dated to around 1,800 BC. It is believed that folk then lived high on the Beacons – hunting, herding and subsistance farming – until overgrazing, soil exhaustion and deforestation interacted with a wetter colder climate and forced them to lower ground. Steep at first, the descent from Corn Du leads down to the dark little lake cradled in the green head of Cwm Llwch past the Tommy Jones obelisk marking the spot where in the summer of 1900 after a month long search the body of a missing child was found, dead from exposure. Tommy was a miner's son who disappeared while visiting his grandfather at Cwm-llwch farm. A tragic tale.

Some good waterfalls on the eastern fork of the Nat Llwch are worth visiting on the way down the valley and before reaching the hill fence. Then pleasant lanes and pretty woods complete the circuit back to the starting point.

A Brecon Beacons Circuit

Distance covered: 15 + kms/9½ miles
Height gained: 825 m/2,700 ft
Difficulty: grassy hillsides and well-marked paths except for steep semi-scramble below summit of Y Cribyn
Caution: on steep edges in high winds or winter. Conditions often unexpectedly severe high up, appropriate equipment essential

Start: car-park on lane 500 m S of Blaen-Gwdi farm at 024248. (297 m/975 ft).

(1) Strike E above fence line 600 m to track, 200 m N lane leads E via Plasgaer to minor road and Nant Sere bridge at 039244, continue S to lane end at 036235.
3 kms/2 miles

(2) Good footpath ascends open hillside SW onto Bryn-teg ridge. Path steepens to final short very steep stony section leading directly to small summit Y CRIBYN (795 m/2,608 ft). Easy path leads W along scarp edge, via pronounced saddle, to broad summit PEN Y FAN (886 m/2,906 ft).
4 kms/2½ miles

(3) Continue easily to wide summit CORN DU (873 m/2,863 ft). Descend NW Ridge, initial steep section avoidable by traverse path from next saddle to S (extra 0.5 km). At Tommy Jones obelisk (000218) strike N along scarp then descend to Llyn-cwm-llwch (580 m/1,900 ft). Easy path descends valley to lane head at Cwm-llwch farm (005238) (305 m/1,000 ft). Follow RoW then lane N to crossroads at 012252, minor road leads E to Blaen-Gwdi.
7.5 kms/4¾ miles

Alternatives
(1.A) From car-park contour above hill fence into Cwm Sere ascending to join Bryn Teg ridge at c 600 m/2,000 ft. Occasional path.

(3.A) From Pen y Fan summit good path descends initially steep N Ridge directly to start, white markers indicate lower section.
3.5 kms/2¼ miles

(3.B) From Llyn-cwm-llwch contour hillside to NE to join (3.A) path near 019238, rough going, no path.
Saves 2 kms/1¼ miles

Steep crags fall northward from close beside the summit of Pen y Fan, the highest point in Wales south of Snowdonia. In the distance lies the Vale of Usk.

Carmarthen Fan – The Black Mountain

South Wales

OS 1: 50,000 Sheet 160

Mynydd Ddu the Welsh call it – the Black Mountain. And so it seems even on balmy summer days when skylarks sing and cloud shadows dapple its swelling moorland plinth, while this long wedge of mountain broods dark and mysterious against the sky.

The most westerly massif in the Brecon Beacons National Park and also the most desolate, for the connoisseur the Mynydd Ddu is also the most beautiful. The spectacular scenery is concentrated at the north-eastern corner of the 20 kilometres (12 miles) of upland which bears the name Black Mountain on the map, while the extensive limestone moors to south and west are windswept, almost featureless and very remote – and of considerable ecological interest. Under the moors lies the famous Dan-yr-Ogof cave system. Yet at night from the summit the whole southern sky glows orange from the lights of the villages and towns that cluster along the valleys beyond the moors, a reminder that Swansea is but twenty miles away.

There are five 2,000-foot summits on the Black Mountain, the two highest standing either side of the distinctive north ridge of the main massif and either side of the boundary between the ancient counties of Brecknockshire and Carmarthenshire – now Powys and Dyfed – hence their names. Bannau Brycheiniog – the 'Horns of Brecon' – is also known perversely as the Carmarthen Fan and is the higher at 802 metres (2,632 feet) while Bannau Sir Gaer, at 750 metres (2,460 feet), is the 'Horns of Carmarthenshire'. Indeed, from the north the mountain does appear to rise to two shallow but sharp tops. Perhaps the Mynydd Ddu is best seen from the western heights of the Fforest Fawr, from above the rumpled moorland that from lower altitudes tends to mask its stature. The precipitous eastern

The view westward at dawn from near the summit of Bannau Brycheiniog along the lip of the 'scarp to Bannau Sir Gaer: the dark waters of Llyn y Fan fach can be seen beyond.

scarp, narrow at first, rises gradually higher and higher to run unbroken for more than 5 kilometres (3 miles) over the slight hump of Bannau Brycheiniog with its graceful prow to the more gentle sweep of the north ridge. Below the prow lies dark Llyn y Fan fawr, lonely and unspoilt, the source of the River Tawe and – through 200 yards of moraine – also a major feeder of the infant Usk. In the valley below and close beside the Tawe and the bleak mountain road that links Trecastle to Glyntawe, stands Cerrig Duon, a strange stone circle, one of the few in Wales. Many would claim that Cerrig Duon reflects the mood of Bannau Brycheiniog – and of Brycheiniog himself who is said to have been a local prince during the Dark Ages.

At the head of the north ridge the escarpment turns sharply back on itself and runs south for a short way before curving round westward and rising again to Bannau Sir Gaer. The impressive scarp edge runs for another 3½ kilometres (2 miles) and holds a few narrow bands of red cliff before fading out in the green ridge that cradles Llyn y Fan fach – the source of the Sawdde which feeds the Tywi. There is a change of

mood on this side of the mountain and this little lake is a smiling one: possibly it holds the sun longer. Anyway, the lake is known to be the home of the Tylwyth Teg – the fairy folk. Tradition has it that a local herd boy met one of the fair ladies of the lake, wooed her and won her. Her fairy father, though he provided a rich dowry of sheep and oxen, cast an unfair spell on the marriage that should he strike her thrice she would vanish. Eventually he did, gently to be sure and under great provocation, and she disappeared again into the lake. But not before they had raised three remarkable sons who, steeped in the secrets of fairy medicine, became the celebrated Physicians of Myddfai, the first of a line of medical men that seems to have lasted in that village until the mid-nineteenth century!

Fairies or not, it is fitting perhaps that the National Park has defined the Black Mountain as a special 'remote area' and has decided not to promote and formalize access to it. The only

Here Bannau Brycheiniog is seen to the west from the heights of the Fforest Fawr some four miles distant. The ascent route climbs to the shallow saddle left of centre past Llyn y Fan fawr which is hidden in a fold of the moor below the scarp.

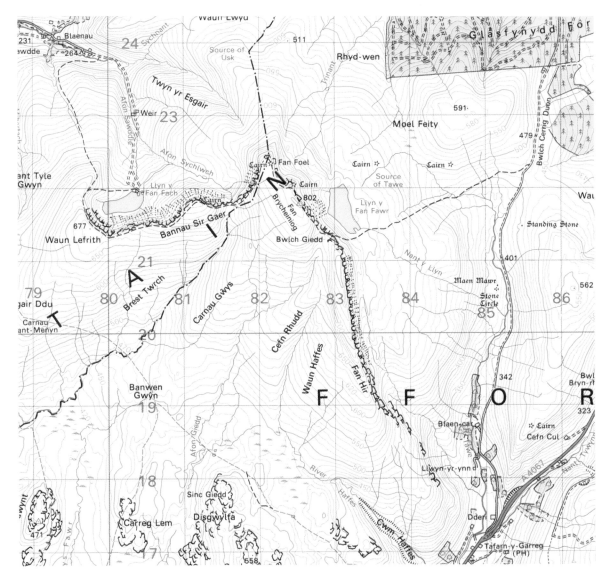

rights of way are ancient long distance routes that cross the moors from south to north linking the Swansea Valley to the Vale of Llandovery: all other access is, strictly speaking, 'unofficial' yet apparently tolerated. Thus we suggest no routes on Mynydd Ddu and leave it to the enthusiastic hill walker to taste the delights of exploration and find his or her own way over the mountain armed only with a map, compass and experience.

The Eildon Hills

Scottish Borders

OS 1: 25,000 NT 53 (OS 1: 50,000 Sheet 73)

> 'I was but three days in Scotland, and
> was glad to get back to my own dull flat
> country, though I did worship the . . . Eildon
> Hills, more for their Associations than
> themselves. They are not big enough for that.'

Edward Fitzgerald, letter 1874

Small the Eildon Hills may be but they are as
shapely as any in Britain besides being, as one
might say, 'well-connected'. These are
connoisseur's hills, and though hardly justifying
more than an afternoon's expedition their ascent
can be conveniently combined with other at-
tractions in the area or used to break a journey to
the real mountains of the north. The Eildons are
indeed a *bonne-bouche*. Dominating the entire
Melrose region, the triple summits of these hills
are unequalled as viewpoints in all the Border
Country. Their steep cones, cloaked with
heather and wiry grass, are the remnants of a
great dome – or laccolith – of volcanic lava
which invaded the bedding planes of the surrounding old red sandstone some 300 million
years ago to be exposed only when the softer
sandstone was eventually worn away.

The Eildon Hills are closely associated with
Sir Walter Scott who lived at Abbotsford, the
Tweedside mansion in their shadow, from 1811
until his death twenty-one years later, writing
meanwhile a stream of celebrated works, several
of them stimulated by his 'delectable mountains'
as he dubbed them. 'Scott's View' – his favourite
– looks to the Eildons over the horseshoe bend of
the Tweed below Bemersyde Hill a short way
from Dryburgh Abbey where he is buried
among the ruins. Another ruined abbey lies
immediately at the foot of the Eildons: the
Gothic splendour of Melrose Abbey, where
traditionally the heart of Robert the Bruce is
interred. It is one of the treasures of the Borders
and is an important Cistercian foundation dating
to 1136 whose great wealth was based largely
upon wool: standing upon the ancient route into
Scotland it was regularly sacked by the English.
But the Hills were important as early as the Iron

The Eildon Hills belie their modest height in this dusk view
from the east near Dryburgh Abbey.

Age when a large fortified village some forty acres in extent on the summit of North Hill was the chief settlement of the Selgovae, a Pictish tribe: archaeologists have located over 300 houses within the still clearly discernible twin ramparts. But in AD 79 the Roman legions marched up the Tweed and the inhabitants fled. The invaders built a fort called Trimontium on the river bank at Newstead below the hill's north-eastern flank and established a signal station on its summit.

These are popular hills and the route sugges-ted is a short and simple one which traverses all three summits and the less-frequented eastern flanks of the range. Though all summit paths are steep and stony, only on the southern flanks of Wester Hill and Mid Hill is the going at all 'sporting' – and easily avoidable by other paths obvious on the map. But climbing these short steep slopes of scree, heather, grass and rock does add a flavour of the mountains to the expedition.

It is easy to see how the Eildon Hills command the surrounding countryside in this view from the summit of North Hill towards Mid Hill (right) and Wester Hill.

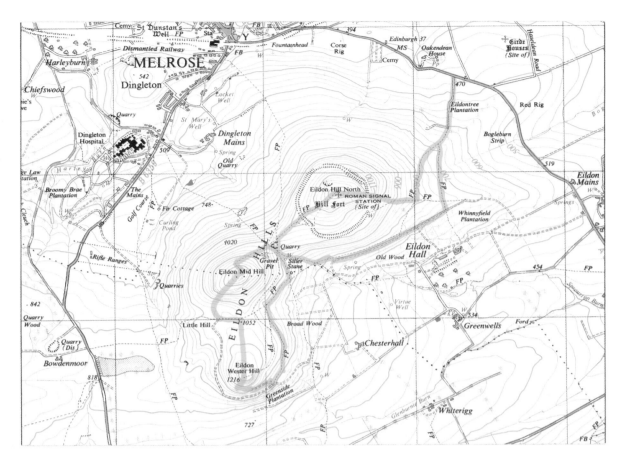

An Eildon Traverse

Length: 7.5 kms/4½ miles
Total ascent: 510 m/1,675 ft
Difficulty: lower paths perhaps muddy, upper paths steep, rough; trackless sections scree, heather, avoidable by paths elsewhere
Note: military firing ranges below W flank Mid Hill, red flags and sentries guard limited dangerous area: territorials may shoot on Sundays

Start: limited vergeside parking on N side A6091 at 563336. (140 m/460 ft).

(1) Across road farm track ascends S, becoming narrow stony trail to gate on to open hillside, several paths diverge. Go L 150 m along plantation edge, find well-defined path leading steeply up vague ESE ridge of EILDON HILL NORTH (404 m/1,325 ft).
1.5 kms/1 mile

(2) Obvious path descends SW to wide northern saddle, descend track L 300 m then strike R on narrow contour path along forest edge, keep L at junctions until below SE flank Wester hill where vehicle track emerges from Greenside Plantation, follow to end at 'quarry', vague contour path continues on scree 100 m across S face. Strike straight up very steep hillside to small cairn EILDON WESTER HILL (371 m/1,217 ft).
3.5 kms/2 miles

(3) Path descends gently N to wide southern saddle, go L to join steep scree path ascending vague SW ridge to topograph and summit trig pt EILDON MID HILL (422 m/1,385 ft). Descend steep path to northern saddle, drop SE again to forest edge, follow wide track just inside wood 800 m ENE until path strikes L across open moorland to rejoin ascent route near gate.
2.5 kms/1½ miles

White Coomb and the Grey Mare's Tail

Southern Uplands

OS 1:50,000 Sheets 78, 79

The traveller driving up Moffatdale towards Birkhill Pass and Selkirk is unlikely to enthuse at the scenery, though he may recognize the dale as an excellent example of a U-shaped glacial valley perhaps more Welsh in character than Scottish. Where the dale starts to steepen he will pass a National Trust for Scotland parking area set between two humped bridges and here he should stop and take a short stroll, for folded into the hillside closeby and unseen from the road is one of the most spectacular sights in southern Scotland – the Grey Mare's Tail – one of Britain's finest waterfalls.

A few minutes ascent up a steep path provides a good and very popular glimpse of this graceful plume of water pouring from the lip of a hanging valley some 60 metres (200 feet) clear into a narrow defile. A series of smaller cataracts above more than double the total fall. The best view is seen from the depths of the defile itself but the approach path has collapsed and the route is now accessible only to those with mountaineering skills: the waterfall itself has been climbed in winter when frozen solid. However, few panting tourists will follow the regular path above the falls and the jewel of the Tweedsmuir Hills is fortunately left for those prepared to continue upwards, albeit easily, for a further 1½ kilometres (1 mile). It appears quite suddenly almost at your feet as you emerge from a maze of heather-cloaked moraines. Loch Skeen, shimmering water and rippling reeds: a blue boulder-ringed lake studded with tiny islets and cradled by handsome swelling hills. It is a new unsuspected world up here, a gentle world of heather, grass and sky and the valley below is invisible and forgotten. The craggy spur of Mid Craig rises above the western shore of the loch and offers a stylish route to the crest of the hills behind, though not necessarily a shorter circuit. Meanwhile the suggested route follows a fading path along the opposite shore before climbing to the tight windswept grass of the Lochcraig Head plateau. A few metres north of the cairn the junction of stone wall and wire fence marks the meeting point of three counties, Dumfries, Peebles and Selkirk, as well as the boundary of the National Trust's 2,800-acre Loch Skeen property.

After the initial steepish descent to the saddle above Lochcraig – the raven-haunted rocky slopes that arc round the head of Loch Skeen – the boundary wall leads onwards over undulating dun-coloured moorland to Firthhope Rig. A lower boggy alternative avoids the crest of the hills but visits the impressive little gorge of the Midlaw Burn cutting deep in the plateau lip below Donald's Cleuch Head. From Firthhope Rig the most scenic route to White Coomb now follows the southern edge of the plateau

Flowing from Loch Skeen, the Tail Burn plunges over the famous waterfall of the Grey Mare's Tail: the route to Loch Skeen crosses the hillside at upper right.

with good views along the craggy edge of Raven Craig as it sweeps round above the deep trench of the Carrifran glen to shapely Saddle Yoke. Providing transport can be arranged, a traverse round this edge with a descent down the beautiful south ridge of Saddle Yoke, or even onwards over Hart Fell and round the imposing Blackhope glen, provides a fine continuation for energetic walkers.

In bad conditions however the tumbledown wall offers the surest route to White Coomb the small and as yet invisible summit cairn standing just one hundred metres south-west of the first sharp bend in the wall. White Coomb is the fourth highest summit in southern Scotland and like its neighbours it is an unexciting yet strangely tranquil hill surrounded by an ocean of rounded tops and flowing ridges reaching, like waves, an almost consistent height. A second cairn rises a hundred metres further across the broad plateau and provides better views, though the best viewpoint is the plateau edge where the spur of Coomb Craig makes a impressive foreground. The wide pano-

rama extends from the Eildon Hills and the Cheviots to the Lake District and the Mull and hills of Galloway while a keen eye can discern Annan power station fifty miles distant.

The easiest return route follows the wall down through an area of broken slopes to the lowest reasonable crossing point of the Tail Burn above the cascades. A more elegant but rougher descent continues down the long east ridge before dropping through a craggy band to the steep hillside south of the Grey Mare's Tail and the final return to Moffatdale.

White Coomb from Loch Skeen

Length: 13 kms/8 miles
Total ascent: 770 m/2,530 ft
Difficulty: high trackless moorland, usually easy going but occasionally boggy, rarely steep. Map and compass skills desirable

Start: large NTS car-park below Grey Mare's Tail at 187145. (230 m/755 ft)

(1) Ascend steep tourist path above true L bank of Tail Burn, above Falls muddy path continues beside burn to Loch Skeen (510 m/1,675 ft) take rough E shore to small tributary burn near NE corner of loch. Now follow burn to steepening grassy slopes near wall, bearing L ascend to wide summit plateau, near junction wall and fence is small cairn LOCHCRAIG HEAD (800 m/2,625 ft).
4.5 kms/2¾ miles

(2) Descend wide ridge steeply SW and follow broad crest curving to S, usually marked with wall over FIRTHYBRIG HEAD (783 m/2,569 ft) and DONALD'S CLEUCH HEAD (775 m/2,543 ft) to FIRTHHOPE RIG (801 m/2,621 ft). Descend E to shallow saddle then contour S and W round White Coomb plateau to steep shoulder of Coomb Craig, strike N to 2nd small cairn WHITE COOMB summit (822 m/2,695 ft).
5 kms/3½ miles

(3) Go NE 100 m to wall, follow it R, occasional faint path, descending 500 m to rocky step, wall continues down to crossing place Tail Burn above highest cataract, now return down ascent route.
3 kms/2 miles

Variations

(1.A) For shorter circuits: from S end Loch Skeen poor path ascends prominent nose of Mid Craig then traverses ridge above W shore Loch to plateau near Firthybrig Head; continue L to White Coomb or R to Lochcraig Head

(3.A) Descend E ridge White Coomb over rough ground to poor path descending steep shoulder S of waterfall.
2.5 kms/1¼ miles

(3.B) From White Coomb descend SW past Coomb Craig to saddle, ascend broad ridge S to CARRIFAN GANS (748 m/2,454 ft). Descend well-defined E ridge breaking L when convenient to join burn at forestry edge, path follows burn to road 1 km below start.
5.5 kms/3½ miles

White Coomb is surrounded by an ocean of rounded tops and flowing ridges. This is the view south-west over the spur of Coomb Craig towards Carrifan Gans with Loch Fell, in the Ettrick group, in the left distance.

Loch Enoch and the Merrick   Galloway

OS 1:50,000 Sheet 77

Galloway is the far south-western corner of Scotland, a region of grand upland country all too often ignored by the hurrying Sassenach to whom Glencoe is the next stop beyond the Lakes. This was the Stewartry from which sprung the Scots royal line, the home of Black Douglas and Young Lochinvar, here was the setting for John Buchan's *Thirty Nine Steps* and the wilderness sanctuary of Robert the Bruce. Beyond a periphery of pastoral prosperity the rugged Galloway heartland is an area of rolling mountains, tangled granite hills, scattered lakes and spreading forests far more akin to the Highlands than to the rest of the genteel Southern Uplands, more the haunt of red deer and wild goats than of people. Jutting into the Gulf Stream, Galloway is wetter and lusher than the adjoining Borders and boasts the most extensive peat bogs outside Sutherland and Caithness. Here rises the Merrick, at 843 metres (2,764 feet) the highest point on the Scottish mainland below the Highland Line.

Enjoying easy access, beautiful surroundings and a top-of-the-world summit, the Merrick is a walker's mountain among the best, though it is not a particularly conspicuous one. The culminating point of the range known as the Awful Hand – a fanciful description of its western outline – its feet stand among the serried conifers of Glentrool Forest, part of the huge Galloway Forest Park, the second largest woodland in Scotland, which dates only to the 1920s: the

From Glen Trool our route passes Loch Neldricken on its way to the Merrick, seen here rising to the north beyond the Loch.

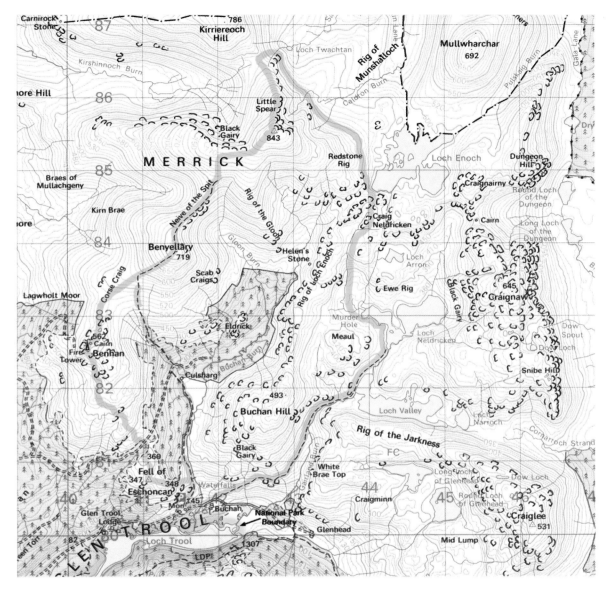

natural oak, birch and pine of the ancient Forest of Buchan are mostly long gone. The waymarked regular Merrick ascent strikes north from a popular car-park above the attractive shores of sinuous Loch Trool where a memorial stone nearby commemorates Bruce's 1307 victory over English troops on the opposite bank: Trool derives from the Gaellic *t'struthail* meaning 'river-like'. Our recommended route however continues eastwards to a good path climbing away towards Loch Valley and the spectacular granite wilderness at Galloway's heart.

In primeval times three great granite batho-

liths intruded into the shaly sedimentary rocks of present-day Galloway. The original rocks forced up to roof this massive intrusion have worn away, remaining only as crests of higher ground – the Rhinns of Kells ridge to the east and the Awful Hand chain to the west – overlooking the now exposed central granite mass where heavy glaciation has left a line of peculiar, low and extremely rocky granite hills with strange names such as Mullwharchar (692 metres/2,270 feet), Rigg of the Jarkness and Dungeon Hill. On one flank of these hills lies the dangerous quaking bog of the Silver Flowe, on the other the chain of enchanting lochs linked by our route, first the shallow lily-scattered Loch Valley and then Loch Neldricken where the so-called Murder Hole at its western extremity is a curious circular rush-fringed lagoon featured in *The Raiders*, a tale by the Edwardian novelist SR Crockett. Surrounded by glacis slabs and erratic boulders and fringed with beaches of silver sand, the third loch, Loch Enoch, fills an ice-scooped rock basin, deep, remote and splendid. It is said that originally its name was 'Loch-in-Loch' because one of its three small islands holds a tiny tarn of its own, equally implausible is the legend that because of the rough rock of the lake bed Loch Enoch trout have evolved without ventral or lower tail fins.

Fairly strenuous going over tussocky slopes leads now below the sculpted eastern flank of the Merrick to a well-defined col between it and Kirriereoch Hill, the adjoining summit, and the ridge rising to the Merrick, with its final narrow shaly crest, is surely the most elegant line on the mountain. The summit, a domed plateau of sparse stony grass and the occasional granite erratic dumped by the departing ice, is the meeting place of no fewer than six ridges, Merrick aptly meaning 'branched finger'. A fascinating viewpoint, the wide prospect stretches not only from nearby Loch Doon – Burns' 'Bonnie Doon' – to spiky Arran, bizarre Ailsa Craig and the long finger of the Mull of Galloway; but on a clear day even to Ben Lomond, the Mountains of Mourne, the Isle of Man and the Cumbrian summits: a shadow on the southern horizon could be Snowdonia. It is worth looking over Black Gairy into the lonely north-western corrie before joining the faint path and following the dry stone dyke which crosses the gentle grassy saddle known as Neive of the Spit, to Benyellery – 'Eagle Hill'. Those anxious to stay high to the bitter end will continue over Bennan and some extraordinarily rough ground to the rocky lip of Eschoncan Fell, a superb eyrie perched over Loch Trool and a worthy final top.

The granite heartland of Galloway: in this view below the eastern flank of the Merrick, Mullwharchar (left) and Dungeon Hill rise beyond remote Loch Enoch.

A Circuit over the Merrick

Length: 17 kms/10½ miles
Total ascent: 840 m/2,750 ft
Difficulty: part easy going on paths, part trackless and a little extremely rough. No scrambling, exposure or crag dangers

Start: good FC car-park at Glen Trool roadhead 416804. (145 m/476 ft).

(1) Follow track E, at bend 200 m past bridge fork L up good path leading to Loch Valley, continue N along shore and onwards to Loch Neldricken. Faint path follows W shore, past Murder Hole, to 2nd burn descending from crags to N, strike N up L bank continuing over rocky ground to SW corner Loch Enoch (494 m/1,620 ft).
6 kms/3¾ miles

(2) Traverse glacis terraces then boggy ground below steeper hillsides, round NE face Merrick to 3rd burn – that descending to tiny Loch Twachtan. Steep ascent leads to hummocky saddle below Kirriereoch Hill. Climb steep grassy slopes S over shoulder Little Spear to narrow shaly crest and summit plateau, OS cairn, THE MERRICK (843 m/2,764 ft).
3.5 kms/2¼ miles

(3) Descend W to follow corrie lip to southernmost point, strike S down wide grassy slopes to join path descending SW to wall, follow wall along grassy ridge and up to large cairn summit BENYELLARY (719 m/2,360 ft). Follow path 500 m SW until it strikes down S, thence continue SW along wall then S to cairn, radio mast, THE BENNAN (562 m/1,844 ft). Rocky shoulder, rough ground leads S then SE to trees, cross forestry track, continue across flat shoulder to OS cairn FELL OF ESCHONCAN (347 m/1,138 ft) and viewpoint SW 200 m. Descend steep craggy slopes to start, easiest to E.
7.5 kms/4½ miles

Alternative
(3.A) From Benyellary continue on regular path dropping S into forestry, then down beside Buchan Burn to start.
4 kms/2½ miles

The Arrochar Alps Southern Highlands

OS 1 : 50,000 Sheet 56

This is virtually sacred country for Glaswegians. Easily reached by both rail and road from the great conurbations of the Clyde, the group of rugged mountains clustered around the head of Loch Long has been a mecca for hill-lovers since the end of the last century. Here was rugged, scenic and relatively wild mountain country – containing the best rock-climbing in the Southern Highlands besides several real Munros – accessible on a day trip from Glasgow long before car ownership became universal. In the Depression years of the Thirties many folk escaped to Loch Lomondside and the Arrochar Alps from industrial squalor and the dole queue and discovered for the first time the freedom and

camaraderie of the mountains. *Always a Little Further*, Alastair Borthwick's delightful anecdotal account of those days, recently republished, is essential reading for those interested in Scottish mountaineering.

Chief among these mountains is the extraordinary three-peaked Cobbler. By no means the highest of the Arrochar Alps – and not even a Munro – it must be considered one of Scotland's more important mountains if only because of its character and steep crags. 'This terrific rock forms the bare summit of a huge mountain' wrote John Stoddart nearly two centuries ago, 'and its nodding top so overhangs the base as to assume the appearance of a cobbler sitting at

work . . .' (*Local Scenery and Manners in Scotland* 1799.) Contrary therefore to most modern descriptions the frowning North Peak is traditionally the Cobbler himself, the highest point, Centre Peak, is the Cobbler's Last while the sharp South Peak is the Cobbler's Wife. More prosaically the mountain is known also as Ben Arthur and its bizarre shape is due to the highly contorted and folded mica-schist rocks of which it is composed.

The route up from Loch Long beside the cascading Allt a'Bhalachain – the 'Buttermilk Burn' – is an old favourite and in dry conditions it is fun to pad up the smooth water-worn slabs of the stream bed. Little now remains of the celebrated 'doss' – or bivouac – of the Thirties beneath one of the twin Narnain Boulders from where the view of the Cobbler cirque really opens up. Several paths diverge nearby, one directly onwards to the Bealach a'Mhaim and another, the Cobbler regular route, winding

upwards to the ridge just left of North Peak: our more sporting route however takes us round the base of the formidable South Peak and along the narrow connecting ridge to the summit. In good conditions intrepid scramblers may decide to traverse the crest of both via the south-east ridge, a fairly long and exposed scramble graded 'moderate', followed by a steep and slightly awkward thirty-metre descent to the ridge beyond whose actual crest, falling sheer into the coire on the right, is straightforward except for one short crack at about 'difficult' standard.

Though it looks intimidating, the little rock tower that is the Cobbler's true summit is easily ascended by crawling through the window from left to right and scrambling up a short gangway on the western face: but it stands on the edge of all things and is not for the faint hearted particularly when the rock is damp. Thereafter the route onwards to North Peak is a walk and the slabby summit above the huge overhanging

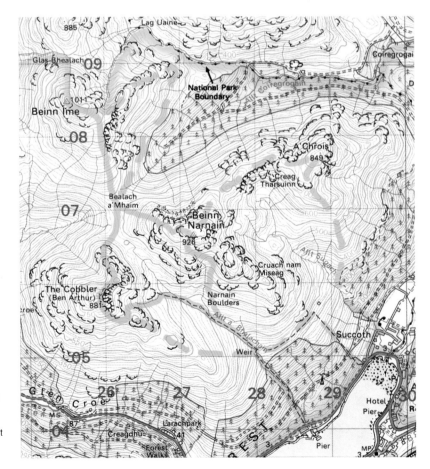

Beinn Ime commands a wide view: beyond its East Top the summit of A'Chrois is just visible with Loch Lomond and Loch Arklett in the left distance and the cone of Ben Lomond on the right skyline.

prow is worth gaining for its spectacular views across the cirque.

Beinn Ime – the 'Butter Mountain' – the highest of the Arrochar Alps, is shapely and quite rugged though unfortunately it presents its least interesting flank to the Cobbler. However, the well-defined north-east ridge, quite narrow near its rocky top, offers an excellent if considerably longer route to the summit, itself an excellent viewpoint especially eastwards over Loch Lom-

ond to Loch Katrine and the Trossachs. Now easily reached, the craggy final mountain, Beinn Narnain, is perhaps the most prominent of the Arrochar Alps with a characteristic flat top looking right down the bright fjord of Loch Long towards the distant ocean. Winding down between rocky outcrops and past waterfalls the intricate descent provides an entertaining finish to the day but if energy permits a delightful return may be made northwards over A'Chrois:

the superb panorama from this fine little peak – 'The Cross' – standing at the heart of the Arrochar Alps includes the most impressive flanks of the surrounding mountains rising over the deep recess of wild Coire Grogain.

Though it appears intimidating, the summit tower of the Cobbler proves a simple scramble; it looks down, past the South Peak, to Ardgartan at the mouth of Glen Croe and Loch Long.

The Arrochar Alps

Length: 14.5 kms/9 miles
Total ascent: 1600 m/5,300 ft
Difficulty: rough hillsides, some steep and rocky ground: all difficulties avoidable but serious scrambling is involved, rope useful, if Cobbler taken direct. These are real mountains, area known for poor weather, rock slimy when damp. Easy escape possible after each stage
Note: route lies within Argyll National Forest Park

Start: limited vergeside parking 284039 at S junction of MoD Torpedo Station road with A83, larger laybys at 294049. (10 m/30 ft).

(1) 300 m E along A83 at W side bridge narrow path enters scrubby woods, follow path uphill near burn, unrelenting, muddy, after climbing 300 m (1,000 ft) emerge from pines onto open hillside. Muddy path crosses small dam to N side burn continuing to conspicuous Narnain Boulders. After 300 m cross burn L soon striking off path aiming L across upper coire for obvious skyline saddle L of Cobbler S Peak. Go R at saddle then L of big buttress on scrambly well-defined path below vertical cliffs, zigzag up R to gain narrow ridge between S and Centre Peaks. Path leads just L of arete to small plateau below 5-metre summit tower COBBLER (884 m/2,900 ft).
4 kms/2¹⁄₂ miles

(2) Path descends near ridge to broad saddle whence easy scrambly slabs gain exposed summit NORTH PEAK (c 850 m/2,790 ft). Path, steep in parts, drops N to wide boggy saddle Bealach a'Mhaim (620 m/2,040 ft). Ascend broad grassy S slopes Beinn Ime ahead, path soon disappears, keep R near coire edge for best terrain, to trig pt summit BEINN IME (1011 m/3,318 ft★) on 2nd rocky kopje.
3.5 kms/2¹⁄₄ miles

(3) Retrace route to Bealach a'Mhaim, ascend steeper grass slopes NW ridge Beinn Narnain, best going far LHS, rocky near top, to summit plateau BEINN NARNAIN (926 m/3,036 ft★).
3 kms/2 miles

(4) Immediately S summit cairn grassy couloir leads down through craggy band to wide grassy ramp descending L into shallow corrie above Narnain Boulders. Retrace (1) to start.
3.5 kms/2¹⁄₄ miles

Variations
(1.A) Do not cross dam on burn above forest but bear W up boggy slopes to join long SE ridge of South Peak, rejoining (1) at skyline saddle.

(2.A) From Bealach a'Mhaim make descending traverse N into Coire Grogain, contouring L between crags and forestry to base NE ridge Beinn Ime below Ben Vane. Ascend ridge easily round rocky steps, potential scrambling if desired, to Beinn Ime plateau.
Extra 2.5 kms/1¹⁄₂ miles, plus 460 m/530 ft

(3.A) Descend easily NNE from summit to traverse knobbly NE ridge to sharp peaklet A'CHROIS (849 m/2,785 ft). Descend initially 400 m SSW to reach grassy SE ridge descending to path beside Allt Sugach burn and Succoth Farm.
Extra 3 kms/2 miles, plus 170 m/560 ft

Ben Cruachan Argyll

OS 1:50,000 Sheet 50

'We thought it the grandest mountain we had seen . . .'

Dorothy Wordsworth:
Journal 1803

'For elevation, magnitude and magnificence, I do not recollect a mountain superior to Cruachan.'

Thomas Wilkinson:
Tours to the British Mountains 1824

In medieval times it was thought to be the highest point in Britain. Ever since then this isolated and distinctive mountain rising above the shaggy and accessible shores of lovely Loch Awe has been celebrated in story and song, as a tourist 'sight' and even as one of the few Scottish mountains whose name appears on large-scale maps. Its correct gaelic name – Cruachan Beann – meaning 'Mountain of Peaks', is an apt description, for its main ridge, fashioned from coarse diorite and granite, extends some 8 kilometres (5 miles) and with its two subsidiary southern spurs rises to no less than seven summits about 3,000 feet (900 metres), two of them Munros.

Charmingly dubbed 'Queen of the Southern Highlands' and certainly a commanding height, Cruachan dominates Mid Lorne and its characteristic shape is easily recognized from a distance: from the north, from the Blackmount peaks or Ben Nevis, the sharp twin western summits linked by a gracefully arcing ridge rise above the horizon, while from Mull or out at sea

High and stately, Ben Cruachan dominates the Firth of Lorne in this view from Duart on the north-eastern point of Mull.

the 'Red Peak' – Stob Dearg – also known as Taynuilt Peak because it rises behind that village, is seen as a single shapely pyramid. Because of their form, their character, certain mountains demand attention and Cruachan is one of them. The traverse of its ridges is a classic expedition along an exhilarating and predominantly rocky crest that in winter can be truly alpine. While fairly long and strenuous the route is uncommitting and escape southwards is possible from many places: it is most aesthetically taken from east to west to enjoy to the full the splendid and ever changing vistas over Loch Awe and Loch Etive that almost moat the mountain, and beyond to Loch Linnhe, the Firth of Lorne and the hills of Mull.

Loch Awe points its 35-kilometre (22-mile) finger through the rolling landscape of Nether Lorne to the foot of Ben Cruachan, where our route starts above a wooded shore with rich historical associations. Kilchurn Castle, an imposing fifteenth-century pile at the eastern corner of the loch, is perhaps the finest castle ruin in Scotland and on small islands stand the remains of another castle, a nunnery and a crannog – an ancient Celtic lake dwelling. The rugged slopes of Meall Cuanail drop steeply into the gloomy defile of the Pass of Brander where the loch becomes the River Awe, a strategic narrows where Wallace defeated the men of Lorne in 1298 and Bruce overcame the MacDougalls in a bloody engagement ten years later.

Technically our route is a straightforward one. The first summit, Beinn a'Bhuiridh – 'ben a vuie', the 'Peak of the Bellowing' (… stags) – gives a good overview of the circuit and the

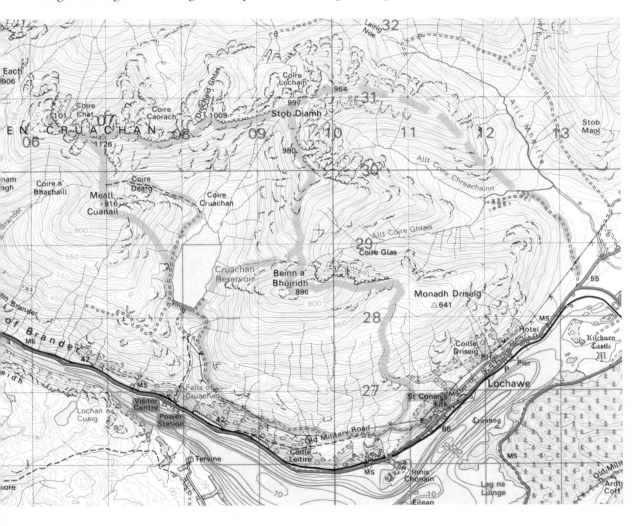

return route. A steep descent over broken slopes, sometimes awkward in winter, leads to the Larig Torran, the col over which Wallace is said to have outflanked the enemies challenging him in the Pass of Brander. Stob Garbh, the 'Rough Peak', leads on to Stob Diamh – pronounced 'daff' – the 'Peak of the Stags', from where the ridge starts to become narrow and stony. Drochaid Glas the next summit – in English the 'Grey Bridge' – demands careful navigation in mist for it stands a little proud of the ridge northwards at the head of a steep spur dropping towards remote Glen Noe, a flank of the mountain rarely frequented. The main summit is quite small and sharp and the final approach over steep and slightly exposed blocks can be another winter 'crux', while Stob Dearg, the far western peak ringed by redish granite slabs to north and west, is a viewpoint not to be missed.

A swift descent into Coire Cruachan leads to

the lake. Pent up behind a hideous 45-metre high (150-foot) concrete dam, the lake, entirely artificial, is almost the only outward manifestation of the huge 400 megawatt pumped-storage power station buried in a massive cavern hewn into the heart of the mountain nearly 300 metres (900 feet) below. Completed in 1965, it was a great feat of engineering, but surely the corrie could have been less brutally raped? The dam access road, though hard on the feet, gives an easy four-kilometre descent to the starting point.

A Traverse of Ben Cruachan

Length: 18 kms/11½ miles
Total ascent: 1700 m/5,500 ft
Difficulty: a long rugged route on a continuously high mountain chain, careful navigation essential in mist. A serious though not difficult expedition in winter conditions
Note: the hydro-scheme Visitor Centre at 077268 is interesting

Start: limited parking at 112266 beside Hydro Board access road immediately before locked gate.

(1) From access road climb N up long steep hillside to wide grassy crest Monadh Driseig, ascend W to summit BEINN A'BHUIRIDH (896 m/2,940 ft). Descend steep rocky flank N – hints of poor path – continuing across grassy saddle to easy ridge leading N to summit cairn STOB GARBH (980 m/3,215 ft) whence short descent, ascent leads to summit cairn STOB DIAMH (997 m/3,272 ft★).
7 kms/4½ miles

(2) Descend easily W to saddle and follow narrow rocky ridge upwards fading out into wider bouldery slopes, ascend R to narrow rocky summit DROCHAID GLAS (1009 m/3,312 ft). Go SW to regain main ridge, descend to low point whence long narrow rocky crest ascends, finally steepening to summit BEN CRUACHAN (1126 m/3,695 ft★) trig pt. Descend W to col and short steep ascent to summit STOB DEARG (1101 m/3,611 ft).
3.5 kms/2¼ miles

(3) Retrace route E until easy ground leads R across SW face Ben Cruachan to S ridge, descend to narrow col whence short ascent reaches summit MEALL CUANAIL (916 m/3,004 ft). Grassy slopes lead to lakeside, cross dam and follow undulating road back to start.
7.5 kms/4¾ miles

Variations
(1.A) Avoid Beinn a'Bhuiridh by following access road to hydro dam and ascending grassy slopes NE to saddle below N flank.
Extra 1 km/½ mile: saves 180 m/600 ft ascent

(1.B) Reach Stob Diamh by E ridge: start at 132284 junction on A85 road, follow old rail track 1.5 kms NW until bridge crosses Coire Chreachainn burn, ascend NE onto long curving ridge leading over SRON AN ISEAN (964 m/3,163 ft) and Stob Diamh 500 m beyond.
6 kms/3¾ miles from road: 1010 m/3,300 ft

(3.A) Avoid Meall Cuanail by descending easily E from narrow col N of peak to lake.
Extra 1 km/½ mile: saves 80 m/250 ft ascent

February snow swirls from the main ridge below the summit of Ben Cruachan. The westernmost top, Stob Dearg, is seen beyond with the Isle of Mull in the far distance.

Bidean nam Bian Glencoe, Central Highlands

OS 1:50,000 Sheet 41

'Glencoe itself is perfectly terrible . . .
It is shut in on each side by enormous rocks from which great torrents come rushing down . . . On one side of the pass there are scores of glens, high up, which form such haunts as you might imagine yourself wandering in, in the very height and madness of a fever.'

Charles Dickens: letter 1841.

'The Anvil of the Mist' the MacDonalds called it, a flat-topped rock beside the old road at the head of the glen. In dialect it became the Stiddie – the Anvil – and later 'The Study', a famous viewpoint where Queen Victoria herself enjoyed a picnic in 1873 and surely gazed down the desolate defile of Glencoe. As a connoisseur of Highland scenery she would have appreciated the wild view rather better than did Mr Dickens

and a multitude of other hyperbolic Sassenach travellers. She would have noted the triple great glowering buttresses – the Three Sisters of Glencoe – that wall the glen on its southern side but she could not have glimpsed the great mountain from which they spring, Bidean nam Bian. The monarch of Glencoe stands well back from the glen, a reclusive ruler, withdrawn, aloof and accessible only to those who work hard for an audience.

Bidean nam Bian is a shapely summit, the highest in Argyll, and best seen rising above its satellites from a distance, from the Mamores, the Blackmount or from Loch Linnhe. Bill Murray suggests that its name in English means 'Sharp Peak of the Bens' for it is the culminating point of four great ridges which rise to nine separate summits and cradle three deep and distinctive hanging corries. Uncompromisingly precipi-

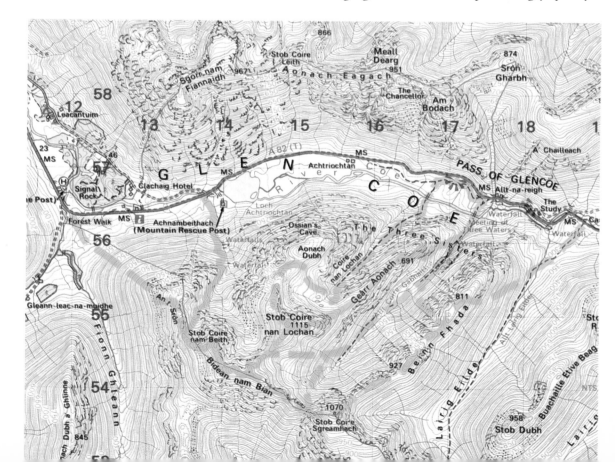

The celebrated view down Glen Coe from The Study: on the left rise the Three Sisters. The zigzag route up the great buttress of Gearr Aonach is clearly seen with the sharp summit of Stob Coire nan Lochan just visible to the left above: the 'Lost Valley' is the first valley on the left.

tous, its rock largely firm volcanic rhyolite and with plenty of accommodation at its foot served by excellent communications, the massif is among the most frequented in Scotland in both summer and winter. Glencoe itself, scene of the treacherous massacre of some forty of its MacDonald inhabitants by soldiers under Campbell command in February 1692 – it was not so much the doing as the manner of the doing that earned disapprobation in those more savage times – attracts a steady stream of tourists in its own right.

Such a complex massif offers a wide variety of routes. The one described here gives a good introduction within a reasonable day's expedition though it involves some straightforward scrambling which may appear from its imposing location a little intimidating to some. Several alternatives are suggested, both as easier variations or as itineraries for a subsequent visit.

Before commencing the climb up Gearr Aonach it might be interesting to divert a short distance into the Lost Valley – the popular name for Coire Gabhail. Pronounced 'Gyle', the name means 'Capture' for this was the sanctuary where the MacDonalds, notorious cattle rustlers, would hide their plundered herds. The inner corrie, a flat and charming fairy meadow, is sealed off by a great maze of gigantic boulders that fell from the cliffs above in the distant past and is thus invisible and unsuspected from the glen below.

Route-finding on the zigzag ledges of Gearr

Aonach – the 'Short Height' – is fairly obvious and the scrambly section just a few easy moves: the flat ridge top, a different and airy world from the forbidding glen below, proves easy going. Soon there are good views of the elegant cliffs of Coire nan Lochan above three tiny tarns, a renowned winter climbing venue, before the ridge climbs entertainingly round the lip of the corrie to the Stob above. The next steep ascent is to Bidean itself and passes above the famous crags of the Diamond and Churchdoor Buttresses, the latter immediately under the small summit. The junction of three sharp ridges, this can be the perfect mountain top in ideal winter conditions.

Stob Coire Sgreamhach – the 'Peak of the Rocky Corrie' – and twin topped Beinn Fhada – the 'Long Mountain' – can provide an excellent continuation around the head of the Lost Valley besides returning to the road nearer the start, but the westward descent will take us over the great cliffs of Stob Coire nam Beith – the 'Peak of the Birch Tree corrie' – and provide fine views back to Bidean over the craggy depths. The final path

down the lower corrie leads past conspicuous waterfalls, below the beetling and strangely symmetrical western cliffs of Aonach Dubh – the 'Black Height' – and finally down to the road at about the one place in the glen from which the summit of Bidean is visible: and the Clachaig Inn is but a stone's throw distant.

Two climbers follow our route up the north-east ridge of Stob Coire nan Lochan, the flat crest of Gearr Aonach is seen below. The twin summits of Beinn Fhada on their long ridge rise from the shadow across Coire Gabhail.

Bidean nam Bian

Length: 9.5 kms/6 miles + 3.5 kms/2¼ miles on road
Total ascent: 1200 m/4,000 ft + 55 m/180 ft
Difficulty: well-frequented rocky slopes and ridges, several short scrambly sections, route finding, map/compass skills essential: this is serious, high, very rugged mountain country demanding respect, particularly in winter, snow often remains into summer
Note: interesting Visitor Centre at 127564 operated by NTS who own area

Start: large car-park S side A82 at 171568 (170 m/560 ft): several similar car-parks nearby. Useful parking Achtriochtan junction 138567

(1) Paths descend S to cross 'Meeting of the Three Waters' footbridge (145 m/480 ft) whence muddy path ascends towards Coire Gabhail. Soon strike up R, faint path climbs hillside to base lowest crag L of Gearr Aonach. Better path leads up diagonally L along bottom of cliff to small cairn where obvious rake, initially steep, rocky, then grassier, leads back R below vertical blank wall to exposed nose where obvious route leads steeply back up L, short easy scramble before angle eases. At easier ground ascend rocky heather direct, or ledge ascending R, to flat ridge top. Go L to small cairn GEARR AONACH (691 m/2,267 ft).
2.5 kms/1½ miles

(2) Grassy undulating ridge continues SW eventually rearing into rocky ridge, easy scrambling over blocks, short walls, small gap, exposed on RHS, leads to steep final slopes STOB COIRE NAN LOCHAN (1115 m/3,657 ft).
2 kms/1¼ miles

(3) Descend easily SW to col whence steep easy rocky ridge, exposed on RHS leads via 2 false tops to summit BIDEAN NAM BIAN (1148 m/3,766 ft★).
1 km/¾ mile

(4) Descend W over subsidiary top then rocky ridge curves N to small cairn STOB COIRE NAM BEITH (1104 m/3,621 ft), drop steeply W to rejoin ridge. At lowest point before An t-Sron descend R down steep scree, trending R to join rough path near stream junction, path descends near W bank stream, steep in places, to road just W of Achtriochtan bridge.
4 kms/2½ miles

Variations
(1.A + 2.A) From footbridge below Coire an Lochan ascend well-used path to small lochans upper corrie. Strike NW to ascend easy N ridge, exposed on LHS, to Stob Coire nan Lochan.
4.5 kms/2¾ miles

(3.A) From col S of Stob Coire nan Lochan descend easy slopes directly L into Coire Gabhail trending L to avoid upper stream gorge. Follow path to Meeting of the Three Waters.
5 kms/3¼ miles

(4.A) From Bidean take easy stony ridge SE to STOB COIRE SGREAMHACH (1066 m/3,497 ft). Straightforward descent possible to Coire Gabhail from col immediately W of peak, Bealach Dearg – avoid gorge (3.A). Or continue ridge NE, several short, narrow, scrambles, over two summits of BEINN FHADA (S top 951 m/3,120 ft). Descend steeply L from col beyond N top into Coire Gabhail or more easily R into Lairig Eilde from pt 811 m beyond.
6 kms/3¾ miles to road + 2.0 kms/1¼ miles on road to start

Ben Nevis and the Carn Mor Dearg Arete

Lochaber, Central Highlands

OS 1: 50,000 Sheet 41

'Read me a lesson, muse, and speak it loud
Upon the top of Nevis blind in Mist
I look into the Chasms and a Shroud
Vaprous doth hide them:'

John Keats
On Ben Nevis, 1818

Climbers know it affectionately as the 'Ben'. It's a mountain of superlatives, the highest summit with the biggest cliffs and the worst weather in Britain: mean, ugly and treacherous into the bargain. But these very qualities make it a very special mountain attracting not only top-class international alpinists intent on the superb winter climbing found on its northern flank, but also hundreds of ordinary summer tourists to whom the zigzag path on the whale-back western flank leads to the high point of their holiday. Every September fell-runners race up and down – their record stands at less than eighty-seven minutes – sometimes the mountain is descended by hang-glider or balloon and occasionally even insulted by such stunts as 'ascents' by car, motorcycle or even bedstead. Nevertheless to the keen hill-walker this magnificent mountain and its only close neighbour Carn Mor Dearg provide one of

The cliffs of Ben Nevis rear above the hanging corrie of the Allt a'Mhuilinn in this view over the Great Glen from the north-west. Our route is seen almost in its entirety, with the Carn Mor Dearg Arete as the skyline crest linking Carn Mor Dearg on the left to the Ben itself.

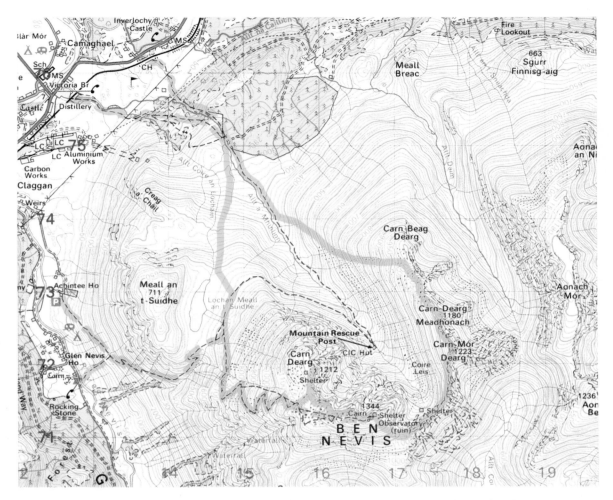

the finest – and certainly the highest – ridge traverses in these islands.

Nevis is an appropriate name, apparently derived from a Gaelic compound word which translates loosely as 'head in the clouds', indeed on less than one day in three is the top clear. Readings taken at the old summit observatory record an average of 261 gales a year and wind speeds often rising well over one hundred miles per hour besides a mean temperature of **minus** 0.3°C! (31.4°F). Snow may fall in high summer and neve patches usually persist until winter at the base of the great northern cliffs. Rising straight from the sea and almost alone, the Ben stands right in the storm track of North Atlantic hurricanes and it has been calculated that were the mountain but a couple of hundred feet higher it would hold a small glacier.

Although the vast southern flanks above Glen Nevis are not devoid of interest, they encompass for instance the longest hillside in Britain rising over 1200 metres (4,000 feet), attention naturally focuses on the precipitous northern flanks which rear above the rugged glen of the Allt a' Mhuilinn (pronounced 'voolin') – the 'Mill Stream'. Our route climbs viciously boggy yet attractively sylvan slopes to the lip of the glen where suddenly the serried spurs of the Ben come into view facing the smoothly curving slopes of Carn Mor Dearg across the defile ahead. The ascent of the latter – a full 600 metres (2,000 feet) – will appear daunting, but starting from the first scattering of boulders to approach the path, and following a series of shallow depressions and hollows in the hillside, the going is surprisingly good and the granite screes of the rounded summit ridge are gained without undue effort.

A handsome mountain in its own right with craggy eastern slopes and an arcing snow-wreathed crest, the twin summits of Carn Mor Dearg – the 'Big Red Cairn' (pronounced 'Jerrak') – provide good walking with everchanging views of Britain's grandest mountain wall. The north-east faces of the Ben and Carn Dearg extends over 6 kilometres (4½ miles), rising in places over 600 metres (2,000 feet) and sculpted into an array of famous ridges and buttresses, walls, gullies and corries carved from smooth dark porphyry, an igneous rock which provides excellent rock-climbing when dry.

The arete that sweeps round the secluded hollow of Coire Leis towards Ben Nevis is the crux of the expedition; in summer merely a sustained but easy and entertaining scramble, though quite exposed and potentially a serious undertaking in winter conditions. The southern end of the arete merges into the slopes of the Ben and can be approached from the corrie but route finding through the broken cliffs is awkward, hence the abseil posts which facilitate a safe winter descent. A rough path now winds up steep rocky slopes to the summit plateau giving the corrie edge a wide berth: the series of marker posts was placed here after a multiple accident in winter some years ago to indicate a safe descent line.

The summit, a flat stony desert surrounded on three sides by the abrupt cliff edge, is graced by several cairns, an emergency bivouac shelter and the ruins of the Observatory occupied between 1883 and 1904. One of the famous climbing gullies ending closeby is named Gardyloo after the old street cry 'Gardez l'eau!' because this was where Observatory staff used to tip their rubbish. In clear weather the panorama is incredible, ridge after rumpled ridge of hills stretching

away in every direction: from Torridon to Ben Lomond; from Barra to the Cairngorms; the sparkling pathway of Loch Linnhe leads south-west to Mull, the Paps of Jura – and on the very clearest of days even to Ireland.

Even in early summer the snow cornices can be immense and several profound gullies cut deeply into the plateau edge so the route onwards towards Carn Dearg along the plateau rim must be followed warily, but there are continuous exciting views of the jagged ridges and buttresses beneath against a backdrop of Loch Eil and the Great Glen. The stony desert rises to Carn Dearg where another emergency bivouac is sited and where the cliff edge just south – supposedly rising just above 4,000 feet and thus higher than the official summit – provides a last glimpse of Ben Nevis. The Tourist Track zigzags steeply down beside the cleft of the Red Burn and back to the valley.

Ben Nevis and Carn Mor Dearg

Length: 19.5 kms/12 miles
Total ascent: 1620 m/5,300 ft
Difficulty: rough ground with section of sustained but easy scrambling, straightforward in good weather: these are very high serious mountains with notorious weather, in bad conditions accurate navigation crucial. Route is quite committing especially in winter when it can become a serious alpine expedition

Start: Ben Nevis Distillery at 126757, parking currently on adjoining lot W. Golf Club currently threaten £3 fee at 137762 but convenient layby 500 m W on A82

(1) Cross Distillery yard then railway following streamside path E over old tramway, go L then steep, boggy path leads to stream dam at lip of Mhuilinn glen. Or from Golf Club take track under railway, arc L round greens to fence/stile across old tramway, boggy path leads steeply to dam. Easy path continues above R bank of burn until glen starts to narrow after 1 km/½ mile, now strike L up hillside, easiest line working R via series of shallow hollows to broad shoulder, ascend easily S to rocky summit CARN DEARG MEADHONACH (1180 m/3,870 ft), after short descent, ascend along edge to CARN MOR DEARG (1223 m/4,012 ft★).
7.5 kms/4½ miles

(2) Descend ridge S, angle eases to almost level narrow rocky arete, easy scrambling, difficulties avoidable. After 500 m ridge steepens, pass metal abseil sign (for descent into Coire Leis) and ascend wide broken slopes NW, stony path passing marker posts to summit plateau, emergency shelter, etc. BEN NEVIS (1344 m/4,406 ft★).
2 kms/1¼ miles

(3) Follow well-used Tourist Route path S then W along deeply indented cliff edge, **beware cornices**, bear R off path, descend slightly to low point and ascend gradually NW following edge past No 4 Gully marker post and across stony plateau to summit cairn CARN DEARG (1221 m/3,961 ft).
1.5 kms/1 mile

(4) From nearby shelter strike S 500 m to rejoin main stony path, descend steep zigzags to flatter ground above Lochan Meall an t-Suidhe, at path junction continue R 500 m until small cairn indicates poor path dropping L over moorland towards lake outlet, nearby drop N over lip descending grassy slopes to Allt a' Mhuilinn dam. Retrace route to start.
8.5 kms/5¼ miles

Alternative
(4.A) At path junction above Lochan Meall an t-Suidhe fork L to descend regular Tourist Route direct to Achintee roadhead.
6.5 kms/4 miles from Ben Nevis summit.

The summit of Carn Mor Dearg is seen from its northern top, Carn Dearg Meadhonach. On the right the great spur of North-East Buttress falls from the flat summit of Ben Nevis. Tower Ridge is the buttress on the far right beyond snow-filled Tower Gully with Gardyloo Gully at its head. Part of the Carn Mor Dearg Arete is seen above the figures.

Ben Macdhui and the Cairngorm Plateau

Eastern Highlands

OS 1:50,000 Sheet 36

'Nothing could be grander and wilder; the rocks are so grand and precipitous, and the snow on Ben Macdhui had such a fine effect'

Queen Victoria:
describing Loch Avon, 1861

The Cairngorms are unique. Once known as the Monadh Ruadh – the 'Red Hills' – from their bare red granite, they rise between the broad straths of the Dee and the Spey. These are mountains carved from a high plateau, the largest area of high ground in Britain with no less than 150 square miles (240 square kilometres) above the 2,000-foot contour (610 metres) and nine of the dozens of tops reaching about 4,000 feet (1220 metres). They appear featureless from a distance except in early summer when snow wreaths hint at the outline of cliff and corrie, but an actual visit reveals a wealth of unsuspected grandeur: great deep-gouged corries, a graceful rolling plateau, jagged crags and remote lochs, tumbling torrents and stately pines, all forming a rare 'arctic-alpine' environment nurturing a specialized flora and fauna. The climate is extreme, here are the largest snowfields in the country with snow patches lingering throughout the summer: this is the home of Scottish skiing.

In such a wide area it would be invidious to specify a particular itinerary to introduce the Cairngorms, rather one should select a sequence of the best characteristic features and link them by an interesting route. Any expedition among these mountains is governed by the weather, in winter it may be impossible to venture onto the plateau where experienced well-equipped mountaineers, unable to escape, have died of exposure: a week later it may be shirt-sleeves on the summits. Not even in high summer should the Cairngorm traveller let down his guard – he must be prepared for anything.

The unsightly ski-road zigzagging up from Loch Morlich gives a head start to any expedition from the north and some entertaining but easy scrambling up the blocky arete of the Fiacaill Ridge of Coire an t'Sneachda – pronounced Corrie an'Treck and meaning 'Toothed Ridge of the Snow Corrie' – ascends to the open plateau. So easily accessible, this is possibly the most popular winter climbing venue in Scotland. A slight ascent along the plateau edge leads to the vague summit of Cairn Lochan perched close to the lip of Coire an Lochain. Typically the impressive vertical cliffs here hold harder climbs while the long slopes beneath are famous for the thick nevé that accumulates on them in winter; forming a miniature glacier which avalanches in spring to obliterate the tiny lochan below with huge ice blocks.

Across the wide plateau of broken rock and gravel the going is easy all the way to the large cairn on the rounded summit of Ben Macdhui, the second summit in Britain and held to be the highest until the 1847 survey. The name has been variously translated as 'MacDuff's Mountain', 'Dark Hill' or even 'Hill of the Black Pig'. Ferlas Mhor, the Great Grey Man, is reputed to haunt Macdhui in misty conditions and Professor Collie, the noted mountaineer and distinguished scientist, was just one of several reputable people to have encountered this frightening apparition. Nevertheless in clear weather the views extend from Caithness to the Southern Uplands and across the great defile of the Lairig Ghru into the Garbh Coire where glaciologists claim relict glacial ice still lingered well into the eighteenth-century.

A descent past the moody waters of Loch Etchachan leads down over an edge to the head of Loch Avon. Surrounded by a horseshoe of beetling cliffs over which white cataracts foam down from the plateau, this is one of the most

There is a fine view up Loch Avon from the path below the Nethy Saddle. The hanging valley followed by our route from Loch Etchachan can be seen on the left between the dark crags of Beinn Mheadhoin and the twin cliffs of Carn Etchachan and Shelter Stone Crag which rise above the head of the loch.

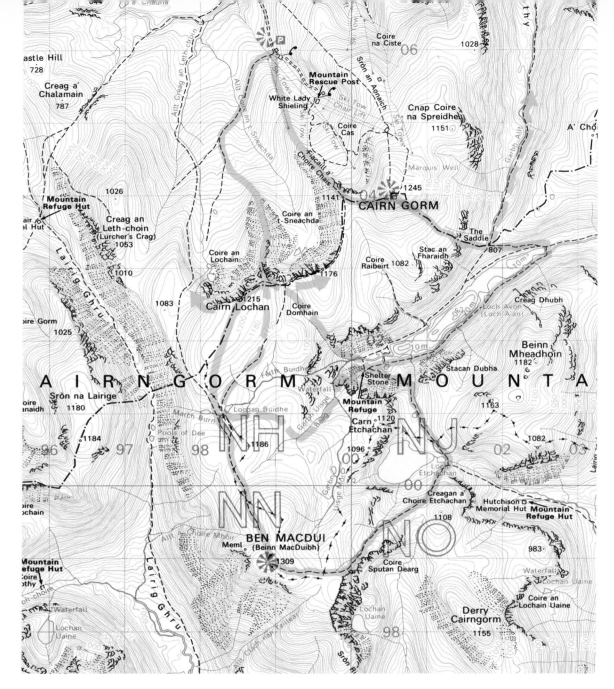

inaccessible places in the Cairngorms, indeed one of the most awesome sanctuaries in Britain. The path passes below the peaked crags of Carn Etchachan to the renowned Shelter Stone, a huge boulder lying beneath the 250-metre (800-foot) monolith of Shelter Stone Crag. Under it is a large and excellent cave, described in 1794 as 'accommodating eighteen armed men' – though only half that number can sleep in modern comfort. It probably weighs some 1,400 tons.

They say that a milk-white water-horse ranges brooding Loch Avon, a hardy beast for the waters are usually frozen half the year. A considerable river flows from the loch and crossing it near the outflow may be hazardous or even impossible, in which case one must decide whether to try again at Fords of Avon nearly two kilometres downstream or to play safe and retrace ones steps round the head of the loch.

Above the eastern end of the loch is the Nethy

Saddle, a low col at the head of Strath Nethy, a long valley which leads 13 tedious kilometres (8 miles), to Glen More via the pretty Pass of Ryvoan: this is the easiest low-altitude escape from Loch Avon. Gravelly slopes above the Saddle lead onto the great dome of Cairngorm, – the 'Blue Mountain' – its summit crowned by a little radio controlled met' station and easily

accessible to tourists from the top station of the chairlift. The prospect northwards over Loch Morlich and the Glen More forests to miles of low hills sweeping towards distant Ben Wyvis and the Moray coast is an exciting one marred only by the ugly developments in Coire Cas – the 'Steep Corrie', the heart of the ski area – around which leads the final descent.

A Cairngorm Circuit

Length: 22 kms/14 miles
Total ascent: 1340 m/4,400 ft
Difficulty: very committing route but generally good going on straightforward ground, one easy scrambly section. In dubious weather less experienced parties should take shorter alternatives. Proficient compass navigation, stamina, essential. In winter a major expedition
Note: 100 sq miles of highest terrain is National Nature Reserve, Loch Avon basin belongs to RSPB, access not affected

Start: huge chair-lift car-park at head of ski road 989061, fee usually payable.

(1) Muddy path contours SW round heathery hillside, beyond first burn strike S to prominent stony ridge, ascend to rounded summit, short descent to rocky arete, easy scrambling up to plateau edge, turning difficulties on RHS. Follow plateau lip R to summit cairn CAIRN LOCHAN (1215 m/3,983 ft).
4 kms/2½ miles

(2) Shallow ridge leads SW across plateau, gradual descent joins path near Lochan Buidhe, gradual ascent leads to large cairn trig pt BEN MACDUI (1309 m/4,296 ft★).
4 kms/2½ miles

(3) Descend easily ESE 1 km (½ mile), traces of path, to head 2nd shallow valley, burn. Avoid cliff edge ahead, descend NE by burn to outflow Loch Etchachan (925 m/3,035 ft). Levelish path leads NW, at lip LHS of burn descend steep rocky slopes diagonally L to boulder fields, Shelter Stone. (755 m/2,475 ft).
5 kms/3¼ miles

(4) Path descends E to Loch Avon S shore, continue to outflow. Cross to N bank, (**danger** – see text) path climbs diagonally W to Nethy Saddle whence strike up steep hillside ahead, avoid crag band, continue easier angle NW up dome to summit CAIRNGORM (1245 m/4,084 ft★).
6 kms/3¾ miles

(5) Descend W to flat area lip Coire Cas, cairn Pt 1141, descend narrow rounded ridge NW above ski area, continue direct to car-park.
3 kms/2 miles

Variations
(2.A+3.A) Direct route from Cairn Lochan to Shelter Stone: descend gradually E from summit into shallow basin Coire Domhain, follow LHS of stream steeply through gap in cliffs to head Loch Avon.
2 kms/1¼ miles

(2.B+3.B) To Shelter Stone via plateau: from Lochan Buidhe contour round NE flank – or cross – low hump Pt 1186 to shallow valley Garbh Uisge Beag whence scrambly descent RHS of burn over lip leads down R to Shelter Stone.
5 kms/3¼ miles

It needs little imagination to realise how difficult navigation can be and how fierce the weather on the wide arctic Cairngorm plateau in this April view towards Ben Macdhui from the slopes above Coire an t'Sneachda. Just right of centre appear the twin summits of Cairntoul and 'Angel's Peak' beyond the Lairig Ghru.

The Ladies of Kintail   Western Highlands

OS 1 50,000 Sheet 33

"We passed through Glen Shiel . . .
One mountain I called immense.
'No' said he 'but 'tis a considerable
protruberance.'"

James Boswell:
Journal of a Tour to the Hebrides, 1773.

Walled on either side by a high chain of
imposing peaks, its foot lapped by the salty
waters of beautiful Loch Duich and with a
modern highway running from end to end, Glen

We look back in this picture from Sgurr nan Saighead along
the narrow ridge towards Sgurr Fhuaran, highest of the Five
Sisters.

Shiel is the showpiece of the Western Highlands. The two great mountain crests to the north of the glen comprise the Kintail Forest, owned by the National Trust for Scotland thanks to the generosity of a past president of the Scottish Mountaineering Club, and there can be few mountain-lovers who are not familiar with the classic outline of the Five Sisters of Kintail, the stately crest forming the northern flank of the glen. Best seen from the Mam Ratagan, the steep pass to the west linking the head of Loch Duich to Glenelg, the ridge rises as five summits but the seaward top is shapeless and considerably lower than its sisters while a handsome sixth summit, with a distinct family resemblance, remains obscured at the far end of the crest. Should there not be Six Sisters of Kintail?

Notwithstanding their number, the traverse of these peaks is one of the very best ridge-walks in the entire Highlands. While hardly long and certainly not difficult, the expedition is sustained and fairly strenuous and the configuration of the six summits does not favour a circular trip. Public transport and taxis are virtually non-existent hereabouts but it is usually possible to arrange a lift or hitch-hike the eleven kilometres up the glen. An energetic party however could make a circular trip from Morvich using the good track up Gleann Lichd, whence an unrelenting climb of over 650 metres (2,200 feet) leads to the Bealach na Lapain: the penalty is an extra 11 kilometres (7 miles).

West Highland ridges are usually best traversed east to west, facing seaward for the views, and the Five Sisters is no exception. Bealach na Lapain is reached more easily than expected and from its grassy saddle views open up northwards to remote country at the head of Gleann Lichd. The first summit, Sgurr nan Spainteach – the 'Spaniards Peak' – is the outcast sister, rearing high above Telford's old bridge in the glen and the 1719 battlefield on the surrounding slopes. This engagement marked the end of a minor Jacobite uprising when a thousand MacKenzies aided by three hundred Spanish regulars were trounced by Government forces advancing from Inverness. The defeated clansmen and the Spaniards were driven up the hillside by bayonets and burning heather – and so the mountain got its name.

Sgurr na Ciste Duibhe – the 'Peak of the

The Five Sisters of Kintail

Length: from road 12 kms/7½ miles
Total ascent: from road 1500 m/5,000 ft
Difficulty: tough walking often over rough ground, some steep ascents, traces of path most of the way
Note: this route is **not** described as a circuit. Area belongs to NTS, Ranger's tel: Glenshiel 219

Start: good parking N side A87 at 009136 (180 m/590 ft).
Finish: limited parking on verge A87 beside Loch Shiel 9418, or by Shiel Bridge campsite 938187.

(1) At conspicuous gap in forest, narrow path, faint at first, climbs steep hillside to Bealach na Lapain (723 m/2,372 ft). On grassy crest go L following tenuous ridge path W rising to summit SGURR NAN SPAINTEACH (c 990 m/3,129 ft). Short rocky descent leads to twisting crest and complex saddle. Main path circles R round hollow before ascending L to rocky slopes and summit SGURR NA CISTE DUIBHE (1027 m/3,370 ft★).
4.5 kms/2¾ miles

(2) Descend wide grassy shoulder, ridge falls N to col (c 850 m/2,790 ft), stony ascent leads to SGURR NA CARNACH (1002 m/3,270 ft). Rocky path descends steeply to saddle (c 860 m/2,820 ft) before long steep climb continues to large cairn SGURR FHUARAN (1068 m/3,505 ft★).
2.5 kms/1½ miles

(3) Path descends W ridge c 200 m before descending traverse R across NW face joins N ridge above lowest point (810 m/2,660 ft). Narrow ridge bends NW, grassy path ascends to SGURR NAN SAIGHEAD (929 m/3,050 ft), descending thin ridge to NW summit. Descend NW ridge to wide boggy shoulder, contour R around knoll to lower saddle, drop L down steep slopes towards Loch Shiel, aiming for footbridge at 947181 to reach road.
5 kms/3 miles

Alternatives
(3.A) From NW ridge Sgurr nan Saighead follow L bank Allt a'Chruinn to path above rocky narrows, descend to Morvich road.

(3.B) From Sgurr nan Saighead cross narrow col NE to SGURR NA MORAICH (876 m/2,874 ft). Steep grassy NW shoulder descends to Morvich road.

In this classic view from the Mam Ratagan pass, the Five Sisters are seen above the head of Loch Duich and the mouth of Glen Shiel. The traverse virtually follows the skyline from right to left – the pointed peak at the centre is Sgurr Fhuaran.

Black Coffin' – is the next top and a line left of the curious hollow at its base gives a little scrambling *en route* to its chunky bosom-shaped summit. The huge slope dropping almost unbroken into Glen Shiel was described by a traveller in 1803 as 'an inclined wall, of such inaccessible height that no living creature would venture to scale it.' Obviously a tedious flank not recommended in descent? Now the ridge broadens, swinging northwards over the 'Pass of the Tree' – Bealach na Craoibhe – to Sgurr na Carnach and Bealach na Carnach beyond, the 'Stony Place': the grassy spur dropping west from this summit is a good escape route. Peculiar spines of rock decorate the slopes of Sgurr Fhuaran, the path avoids them but they offer some scrambling to enliven the long pull up the 'Peak of the Spring'; loftiest and most handsome

of the sisters and a superb viewpoint, its long northern flank hung with a succession of wild corries, sanctuary for both deer and wild goats. Here the route descends a little way west – down another potential escape line – to avoid a rocky step on the main north ridge, before crossing the shaly flank of the mountain to reach the 'Yellow Saddle', Bealach Bhuidhe. Spectacular cliffs of rotting schist fall eastward from the narrow crest of sharp Sgurr nan Saighead beyond: possibly these narrow slabs, bundled together like quivered arrows, account for its name 'Peak of the Arrows'. The way onwards depends on whether one considers this to be the fourth or fifth sister but the best is now over and the ugly sister, squat little Sgurr na Moraich – the 'Shellfish Peak' – can only be an anticlimax.

The Falls of Glomach and the Green Mountain Western Highlands

OS 1: 50,000 Sheet 33

'O Caledonia! . . .
Land of the Mountain and the Flood . . .'

Sir Walter Scott: 1805

Eas a'Ghlomaich – 'The Falls of the Chasm' – are the second highest in Britain and surely the most spectacular. They plunge nearly 150 metres (500 feet) into a rocky cauldron at the back of a deep ravine in a remote hillside in Kintail. Known sometimes as the Hidden Falls, they are accessible only to those prepared for a stiff walk and even then a view of their full majesty demands basic mountain skills and a steady head.

The shortest way to the falls lies up Glen Elchaig, reached by a narrow road from Ardelve on Loch Duich, but beyond Killilan the track is private and permission to drive to Loch na

The Falls of Glomach: this viewpoint on the lip of the chasm can be reached with great care by steep and exposed slopes below the path that descends the flanks of the gorge.

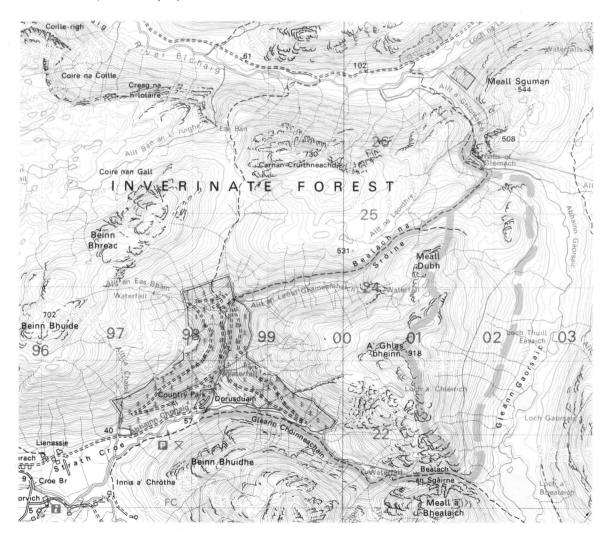

Leitreach must be sought from the Killilan Estate. From there a boggy ascent of some 250 metres (800 feet) leads to the head of the falls. But this is hardly a worthy expedition for a healthy hill-walker and the circular route described here, traversing a Munroe to boot, makes a more interesting approach. Gleann Choinneachain may mean the 'boggy Glen' but it is an especially fine one and a good path leads up it below the steep north-western slopes of the huge massif of Beinn Fhada – the 'Long Mountain' – to the narrow Bealach an Sgairne. Once known as St Duthac's Pass from the eleventh-century cleric who often used it, the present name translates as the 'Pass of Murmuring' and the eastern side falls to the barely perceptible Glen Affric watershed and a chain of shallow lochs which birth the Glomach, indeed the falls are sometimes reached this way, albeit rather boggily.

A'Ghlas-bheinn is a complex little mountain with a long and knobbly crest which rises directly from the Bealach, but the rocky outcrops are easily bypassed. It has been claimed that the mountain has sixteen tops, certainly the ridge is a fine grassy switchback with a succession of false summits and passes close above a charming little loch before climbing at last to the real summit. This aptly named 'Green Mountain' is a splendid viewpoint, both into the impressive corries of Beinn Fhada nearby and over Loch Duich to the pinnacled Cuillin, while Glen Elchaig points towards the Applecross hills rising beyond Loch Carron. Meall Dubh, the final shoulder on the descent northwards, presents a craggy flank to the Bealach na Sroine so the modest ridge dropping towards the falls should be carefully located.

There is little to be seen at the lip of the falls except boiling water disappearing into the void. But as the path leads across the steepening grassy flanks of the ravine the water can be seen plunging into a narrow chasm. There is nothing more to be seen as the main path winds down steep slopes to the ravine end, but with care a natural balcony of rock on the extreme edge of the chasm just beyond the base of the fall can be reached. The zigzag route to this vantage point is well used and will be fairly obvious but although not a scramble it is very steep and exposed and the ground is always slippery from the continual spray. From here the torrent is seen to drop in a single wild plunge before exploding on a great rock pillar and thundering down into an unseen cauldron of spray far below. It is possible for climbers to scramble to a still lower viewpoint from which the cauldron itself is visible, but this is hazardous ground. The waterfall has been climbed in winter when frozen hard.

The return route, following the excellent path over the wide Bealach na Sroine – the 'Pass of the Shoulder', no doubt that of Meal Dubh above – leads easily back to the dark pines and the car-park.

Over A'Ghlas-bheinn to the Falls of Glomach

Length: 16 kms/10 miles
Total ascent: 1400 m/4,600 ft
Difficulty: excellent path except on mountain itself, straightforward terrain, sometimes steep. **Caution** steep exposed wet grass surrounding Falls
Note: Dorusduain is Country Park, area surrounding Falls belongs to NTS, Ranger's tel: Glenshiel 219

Start: Forestry Commission car-park at Dorusduain 981224, reached from Morvich via Strath Croe (60 m/200 ft).

(1) Cross footbridge S over river, follow good path E up Gleann Choinneachain, zigzagging steeply to narrow Bealach an Sgairne. Now ascend N, avoiding rocky outcrops, to knobbly ridge NNE, continue to summit cairn A'GHLAS-BHEINN (918 m/3,006 ft★).
6 kms/3¾ miles

(2) Descend NW over narrow col, minor summit, to twin lochans. Avoiding cliffs to N and W, descend NE ridge to shallow boggy slopes, strike N to good path, continue NE to Falls of Glomach (c 340 m/1,100 ft). Steep exposed scrambly descent R below path gives best views from rock balcony. **Caution.**
4.5 kms/2¾ miles

(3) Reascend path crossing boggy ground to Bealach na Sroine (520 m/1,706 ft) before descending easily to Inverinate Forest. Forestry track S returns to start.
5.5 kms/3½ miles

The Great Corries of Applecross

Wester Ross, Northern Highlands

OS 1 : 50,000 Sheet 24

The Applecross road, a narrow ribbon of rough tarmac, flings round a tight corner and into a deep square-cut valley running far up into the mountain. On the left tiered red cliffs rear above the green receding valley floor, the road ahead, an incongruous ledge, climbs across a steep hillside of broken crags and grey scree. Then the cliffs close in to sweep round the blind headwall of the valley. A tight bend, a quick glimpse of blue sea framed by dark cliffs, a vicious hairpin – pink slabs curling back from the road – another hairpin and another and another. Suddenly the

top, desolate boulder-strewn slopes and a couple of tiny lochans. This is the Bealach na Bà, one of the highest road passes in Britain – the ancient Pass of the Cattle.

Until the early Seventies this tortuous highway, often blocked by winter snow and with grades of one in four, was the only road into the remote Applecross peninsula, jutting into the Minch between Lochs Torridon and Duich. But Applecross village, a line of white cottages along the shore of a shallow bay looking out towards Skye, is now linked to Shieldaig and Torridon by a new road joining the scattered crofts of the western and northern coasts. Applecross is a corruption of Aber Crossan, the mouth of the Crossan – the ancient name for Applecross River – but to the Gael the peninsula is known as A'Comaraich – 'the Sanctuary'. St Maelrubha, second only to Columba in the history of Scottish Christianity, established a sanctuary here in 673 as refuge for all manner of fugitives, and so it remained, under the aegis of the Church, until the Reformation, though Maelrubha's monastery was eventually destroyed by the Vikings.

The Applecross mountains, southernmost of the Torridonian sandstone summits, rise gently from the coast before hunching into an array of whale-back ridges which plunge eastward into a sequence of stupendous corries. Our suggested route, which traverses both the major peaks, starts unusually by descending the spectacular valley below the Beallach na Bà for over 500 metres (1,600 feet) to the mouth of Coire nan Arr – avoiding the road is no problem. A short walk to the sandy shore of Loch Coire nan Arr brings the great Cioch of Sgurr a'Chaorachain (pronounced 'Hoorichan') into view: Cioch means a breast or nipple, often used in Gaelic to describe such upstanding features, and its intimidating 300-metre (1,000-foot) skyline gives a famous rock climb of great character which many claim is the best anywhere on Torridonian sandstone.

Across the glen the ascent onto Beinn Bhan presents no difficulty and the lip of the first of the

From the South Top, sheer cliffs over 200 metres high – more than 700 feet – can be seen falling from the summit plateau of Beinn Bhan into the deep hollow of Coire na Feola.

five eastern corries is reached just before the South Top. From here the most exciting route lies closely and carefully along the abrupt cliff edge for the views down into the next two corries, Coire na Feola (Corrie of the Flesh) and Coire na Poite (Corrie of the Pot – pronounced 'potch') are sensational. In the latter, almost enclosed by two prodigeous castellated spurs, sheer walls drop over 300 metres (1,000 feet) from the mountain's summit to two tiny lochans. It is worth continuing north from the summit for the views into Coire nan Fhamair –

Corrie of the Giant – and the wider northern corrie which again cradles two small tarns.

The summit of Beinn Bhan – the 'White Mountain' – is a broad curving plateau with an extensive view stretching from Ben Nevis to Ardnamurchan, Rhum, South Uist and the Harris peak of Clisham. Set with craggy bands, areas of boiler–plate slabs and several tiny ponds, the west ridge leads to the Bealach nan Arr, a grassy saddle from which a rough path heads directly back to the road beyond the crest of the low ridge ahead. This is the long north ridge of

Sgurr a'Chaorachain, a rugged plateau of tangled rock, grassy dells and several small lochans. Its eastern lip should be followed past the North Top, with exceptional views over the Inner Sound towards the saw-toothed Cuillin, above the deep corrie below the jutting spur of the Cioch to the true summit on the massive southern spur. From here the controversial oil-rig construction yard on the Kishorn shore, dwarfed by the Cyclopean cliffs of Meall Gorm, is the only small scar on a wide vista of lochs and islands. The return to the Beallach na Bà is easy.

An Applecross Ambulation

Length: 19 kms/12 miles
Total ascent: 1100 m/3,600 ft
Difficulty: generally easy straightforward circuit, some going quite rough, paths poor or non-existent
Note: stalking information from Estate office, tel Applecross 209

Start: large parking area 775423, highest point Kishorn/Applecross road 1 km N Bealach na Bà. (626 m/2,053 ft).

(1) Descend SE from Bealach, cut corners R of hairpin bends, descend corrie floor, at rock-band go L of burn into lower corrie, make descending contour to road at Pt 250, continuing over hillside above and drop to road again at Russel Bridge (110 m/360 ft).
5 kms/3¼ miles

(2) Follow path R bank burn NW to Loch Coire nan Arr, cross burn, head ENE to long rounded S ridge Beinn Bhan, sustained ascent leads to SOUTH TOP (763 m/2,505 ft), continue easily round exposed corrie lips, short ascent to summit BEINN BHAN (896 m/2,938 ft).
5 kms/3¼ miles

(3) Faint cairned path descends long curling W ridge, wide stony slopes become rockier, glacis, rocky bands usually avoided near L edge. From obvious Bealach nan Arr (600 m/1,970 ft) path curves L to broad ridge crest, don't continue to car-park, but instead strike L over broken ground along corrie lip to radio mast NORTH TOP (776 m/2,545 ft) continuing S around lip, then E along narrow ridge to summit SGURR A'CHAORACHAIN (792 m/2,600 ft). Retrace steps to plateau, descend diagonally NW to join jeep track descending to road.
9 kms/5½ miles

The famous 'Nose' of Sgurr a'Chaorachain and the pinnacled arete connecting it to the main mountain look impressive from the west ridge of Beinn Bhan as it drops towards the Bealach nan Arr. The Cuillin of Rhum can be seen in the far distance behind the figure.

The Diadem of Torridon – Beinn Alligin

Wester Ross, Northern Highlands

OS 1: 50,000 Sheet 24

From the lower slopes of Tom na Gruagaich, Beinn Alligin appears more a range than a single mountain. The traverse follows the crest of the three Rathains of Alligin on the right before ascending the sharp main top, Sgurr Mor, distinguished by the great 'Gash' below the summit.

Beinn Alligin is best seen from the south, across the sheltered waters of Loch Torridon when the water is calm, the tide is low and golden sea-wrack rings the shore. It is a perfect marriage of two elements, the mountains and the sea, that so often embody the glory of Highland scenery. Most shapely of the three famous Torridon mountains, Beinn Alligin is also the smallest and most compact: an isolated tight-curving crescent of sandstone that rises to six distinct tops, a mountain range in microcosm. But do not underestimate Beinn Alligin, little it is not, it rises almost from the sea like a fortress, its slopes are steep and craggy, its crest narrow and its architecture imposing. Its name means jewel but when its arcing crest glistens with winter snow it

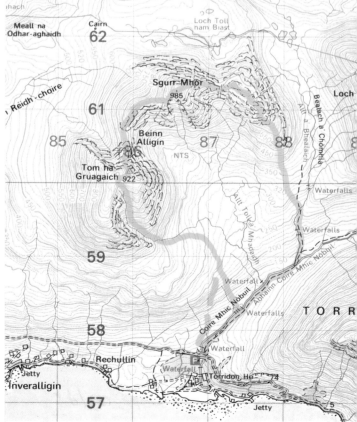

might well mean the Diadem of Torridon.

There is one obvious route on the mountain, a perfect – and popular – horseshoe above the elegant double-headed eastern corrie, which ideally should be followed north-east to south-west so that all difficulties are tackled in the ascent and the island-spread ocean is always ahead. The route starts on an excellent stalker's path leading at first through aromatic woods of Scots pine and rhododendron above the chasm of the Abhainn Coire Mhic Nobuil, the tumultuous river that drains the northern flank of Liathach. Passing a footbridge, deep pools and several waterfalls, the path eventually crosses the river at a second bridge and forks, the right fork continuing right round Liathach to the road – an easy walk through fine scenery if transport allows – the left climbing steadily northwards to a second fork where a climber's path leads up towards the looming eastern shoulder of Beinn Alligin. Ahead the badly eroded scar of the path can be seen winding up a succession of short scrambly rock steps and heathery ledges. This eastern peak of the mountain rises to three separate summits, the Rathains or 'Horns', and their entertaining and airy traverse involves some mild scrambling: the second Horn just appears the highest. Meanwhile Sgurr Mhor rears ahead resembling a snowless Monch with Tom na Gruagaich its neighbouring if emasculated Eiger.

A long but easy ascent, the only slog on the circuit, leads to the tall cairn of Sgurr Mhor, well named 'the Great Peak', a summit from which the free-standing isolation of Beinn Alligin is well appreciated. In perfect weather the view extends from Ardnamurchan to Cape Wrath, but especially impressive is the wild landscape immediately northward, a tangle of rugged outcrops, serpentine rivers and lochs of all shapes and sizes from which jut handsome blade-like peaks – the Shieldaig Forest. Some 200 metres beyond the summit the path descends past the yawning Great Cleft, a prodigious gully so deep and awesome and cutting back so far that it might have been slashed into the mountain's spine by a Cyclopean axeman. On close inspection its slabby floor is actually scramblable and a jumble of huge boulders in the corrie below – Toll a'Mhadaidh, the 'Fox's Hole' – appear to have fallen from it. Tom na Gruagaich is only

the second summit of Beinn Alligin though it bears the OS cairn, and as the 'Maiden's Hillock' it is appropriately graced with a gentle grassy top, though terraced crags fall from it and the ridge leading up to it is the rockiest on the mountain. Narrow but never really exposed, it provides some scrambling if the crest is tackled direct. Below the summit the path plunges from a sandy saddle into the narrow Coire an Laoigh – the 'Corrie of the Calves' – a V-shaped defile with one flank grass, the other scree, the floor steep and the path unrelenting. Eventually heathery moorland drops to the first footbridge over the river close above the beautiful waterfall of Eas Rob, and the stalker's path is rejoined.

The Beinn Alligin Traverse

Length: 11 kms/7 miles
Total ascent: 1400 m/4,600 ft
Difficulty: well-defined path throughout, direct route involves sections of mild but avoidable scrambling. Short expedition but a serious, steep and fairly committing mountain
Note: Rathains can be avoided by traverse path across steep S flank, escape possible S from col E Sgurr Mhor. Area owned by NTS, Ranger tel: Torridon 221

Start: NTS car-park 869576, river bridge 3 kms W Torridon village. (35 m/115 ft).

(1) Good path ascends E bank of river 2 kms to cross footbridge at stream junction. At fork go L, ascending N, cross burn, cairn LHS indicates rough path breaking L towards mountain, up steep spur on scrambly rocky steps, easier slopes lead to rocky tor, summit first Horn. Scrambly descents and ascents lead on over second Horn, highest of RATHAINS OF ALLIGIN

(c 866 m/2,840 ft) and over longer crest third Horn to col whence long grassy climb to SGURR MHOR (985 m/3,232 ft★).
5.5 kms/3½ miles

(2) Path descends ridge SSW past Great Cleft to flat grassy saddle. Cross grassy hummock to lower saddle, ascend steep narrow rocky ridge, scrambling avoidable, to OS cairn TOM NA GRUAGAICH (922 m/3,024 ft) on lip grassy plateau.
1.5 kms/1 mile

(3) Descend easily SW to wide saddle, drop L into narrow corrie, steep path descends grass, scree, to moorland, boggy slopes lead to footbridge, rejoins ascent path.
4 kms/2½ miles

The celebrated 'Gash' or 'Great Cleft' that splits the south ridge of Sgurr Mor close below the summit: it can provide a loose scrambly route up the mountain.

Liathach, the Grey One

Wester Ross, Northern Highlands

OS 1:50,000 Sheets 24, 25 (19)

'Liagush, rising sheer
From river-bed up to the sky,
Grey courses of masonry, tier in tier,
And pinnacles splintered on high.'

Cairngorm Club Journal: 1908

As the road from Kinlochewe down Glen Torridon swings west you will suddenly see Liathach. A bold blunt buttress rears skyward curving to a sharp peak, its tip glinting white in the sunshine, with another, sharper and whiter, beyond. Great slopes of tiered grey sandstone sweep down to dun-coloured hummocks on the glen floor. Liathach will be with you now for the next 12 kilometres (7 miles), steep, riven, and seemingly summitless, its massive flanks walling first the glen and then beyond to where the little white cottages of Fasag village find scant room to huddle between the mountain side and the tidal waters of Loch Torridon. Highest of the three mountains which line the northern side of glen and loch, and facing diffident lesser peaks to the south, arrogant Liathach dominates Torridon.

Liathach – pronounced 'Leeagach' – 'the Grey One': typically the upstanding mountains of terraced pinky-grey Torridonian sandstone that so characterize the North-West Highlands form complex massifs with several summits, and Liathach is no exception. Its 8-kilometre (5-mile) crest rises to no less than seven named tops,

Although a relatively straightforward scramble in summer, the winter traverse of the Fasarinen Pinnacles leading towards Mullach an Rathain, the second 'Munro' of Liathach visible beyond, can be quite formidable.

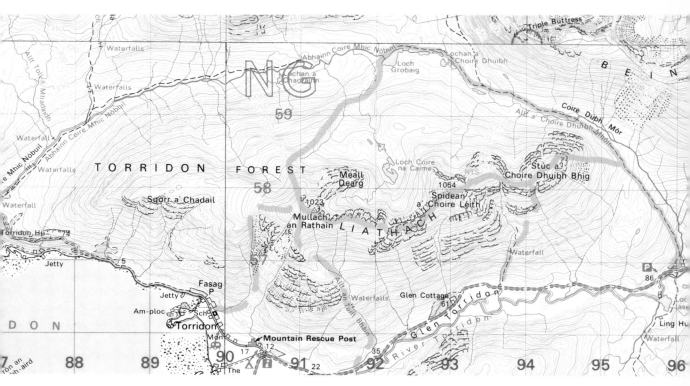

two of them classified as Munros, besides clusters of innominate pinnacles. Although it holds several high shallow corries, the southern flank appears as a continuously steep wall, leaving the northern side of the mountain, hung with five fine corries, to flaunt its real grandeur. Torridonian sandstone is an extraordinarily old rock, but older still – around 1,500 million years, the oldest known – is the Lewisian gneiss on which Liathach and its neighbours stand. Four of the Liathach summits are capped with white quartzite, rather younger than the sandstone, and from a distance they can seem plastered in glinting verglas, a bewildering sight in high summer, but the mountains themselves are of similar age to the Alps, were carved from an earlier plateau some thirty million years ago, and owe their detailed structure to more recent glaciation.

Some years ago a survey conducted by one of the mountain magazines indicated that Liathach

Stuc a'Choire Dhuibh Bhig is the most easterly summit of Liathach, and the ridge from it over Bidein Toll a'Mhuic beyond to the main summit – Spidean a'Choire Leith in the distance – appears deceptive for the going is not at all difficult.

was considered one of the two most magnificent mountains in Scotland (An Teallach was the other) and indeed its traverse is probably the best of the three great ridge scrambles on the Scottish mainland. It is usually taken east to west, a one-way outing necessitating pre-arranged transport or a road walk of at least four kilometres, but a far finer circuit can be made by using the excellent path that links the wide glens of Coire Mhic Nobuil and Coire Dubh Mor round the imposing northern side of the mountain: this is the route described here.

It starts logically, if a trifle fiercely, up the blunt easternmost shoulder which leads to the quartzite-crowned balcony of Stuc a'Choire Dhuibh Bhig, 'Peak of the little Black Corrie', no longer a Munro though it well deserves to be. The succession of tops ahead looks high and steep but proves to be neither, and the route, rough going over the quartzite blocks, leads over twin-topped Bidean Toll a'Mhuic – 'Pinnacle of the Pig's Hollow' – and easily on to Spidean a'Choire Leith with its monstrous southern shoulder jutting over Glen Torridon. The shallow 'Grey Coire' – Choire Leith – hangs below this shoulder providing an ascent route, though from above the steep terraced hillside appears as a green slope, the grassy ledges masking the rock walls, and a descent – as anywhere on sandstone – requires careful route-finding. The easier regular route reaches the crest further east. Spidean is the highest summit with an extensive view over much of the north and west, from Ben Hope via the Cuillin to Ben Nevis.

Now comes the *pièce de résistance*, the crossing of the Fasarinen Pinnacles. Fasarinen are 'teeth' or 'talons', and these sharp and shattered sandstone fangs drop sheer into magnificent Coire na Caime on the north side. Ringed by steep rock and another chain of fearsome *aiguilles* to the west and cradling a tiny lochan, this 'Crooked Corrie' is Liathach's most impressive feature. In summer the pinnacles are not difficult but their exciting crossing is real scrambling and requires a steady head, while in winter the easier bypass, a spectacular path that snakes across the gullies and grassy buttresses falling south from the pinnacles, may prove formidable indeed. The second Munro is Mullach an Rathain – the 'Summit above the Horns'. Which horns? A rocky north-

east ridge curls round Coire na Caime to Meall Dearg (c 960 metres/3,150 feet) a chunky cone of red rock, crossing *en route* the five so-called Northern Pinnacles. Steep and loose, these towers demand real though easy rock-climbing. Meall Dearg is among the few 'technical' summits in Britain and in good winter conditions its ascent from the north followed by the round of Coire na Caime over both sets of pinnacles is one of the best mountaineering expeditions of its kind.

Though it is now possible to descend gradually to the westernmost end of the Liathach crest, our route now takes the scree-draped northern ridge, dropping down its left flank until gentle slopes lead to the remote lochan-studded strath where the path is joined and from where the sculpted northern flanks of the mountain can really be appreciated. It is now downhill, through wild country but on a good path, back to the start.

A Circular Traverse of Liathach

Length: 15 kms/9¼ miles
Total ascent: 1340 m/4,400 ft
Difficulty: a rugged and strenuous traverse, sometimes a little exposed, serious scrambling can be avoided: a high mountain, potentially serious and committing in bad conditions, rope might be useful
Note: area owned by NTS, interesting Visitor Centre at 905557. Ranger phone: Torridon 221

Start: large NTS car-park at 958569 beside A896 bridge below Coire Dubh Mor.

(1) From bridge, path ascends W of burn towards Coire Dubh Mor. After c 1.5 kms large sapling-topped boulder marks toe of shallow glacis spur falling across moorland from E face Liathach. Ascend spur, occasional cairns, towards prominent Y gully in centre grassy face. Narrow path zigzags steeply up heather strip R of gully, trends R through craglets to nose NE buttress below narrow upper cliffs. Round corner R mild scramble 30 m leads to quartzite scree, blocks, summit STUC A'CHOIRE DHUIBH BHIG (913 m/2,997 ft).
3 kms/2 miles

(2) Gradual ridge then steep rocky ascent lead over vague cairned top to blocky summit, cairn, BIDEAN TOLL A'MHUIC

(975 m/3,200 ft). Descend R of crest, rocky, traverse L to col, stony ridge ascends to boulders, sharp rocky summit SPIDEAN A'CHOIRE LEITH (1054 m/3,456 ft★). Bouldery descent, grassy ridge lead to 5 Fasarinen Pinnacles: either traverse direct, sustained exposed scrambling, or take narrow traverse path across steep slopes S flank to narrow col past difficulties. Final highest pinnacle AM FASARINEN (c 930 m/3,050 ft) accessible from path, easy scramble up gully. Long undulating grassy crest leads to minor top, avoid outcrops on LHS, stony slopes lead to OS cairn MULLACH AN RATHAIN (1023 m/3,358 ft★).
3.5 kms/2 miles

(3) Descend W 200 m to sandy shoulder whence follow N ridge until easy diagonal scree descent L leads to steep grassy slopes. Lose c 300 m (1,000 ft) height, take grassy spur NE towards W end Loch Grobaig. Cross outfall easily, at good path N bank go E. Path crosses vague boggy saddle, descends SE down Coire Dubh Mor to start.
8.5 kms/5¼ miles

Alternatives

(1.A) Regular ascent/descent path leaves road 936566 2 kms climbing to col just W Stuc a'Choire Duibh Bhig.

(3.A) Descend S from Mullach an Rathain by S ridge whence steep grass tongues drop to SE corrie, follow burn to road at 918555, or by Fasag stone shoot 800 m W.

Loch Torridon runs deep into the sandstone mountains of Wester Ross. The shapely pyramid of Liathach rises over the head of the loch in this view eastwards from near the crofting village of Ardheslaig.

The Slioch Horseshoe

Wester Ross, Northern Highlands

OS 1:50,000 Sheet 19

Slioch they call it — the 'Spearhead' — from the array of great pointed buttresses that support its broad crown. One of the most distinctive peaks of the Northern Highlands, its shapely sandstone shoulders rise from a plinth of grey gneiss to dominate the southern end of Loch Maree while its summit looks out across the so-called Great Wilderness. Easily accessible, Slioch is of the Great Wilderness but not part of it.

Loch Maree is the finest of the large Scottish lochs, 19 kilometres (12 miles) in length, it lies in wilder and more rugged country than its rivals. It was known long ago as Loch Ewe — as is the sea-loch into which it drains — and our walk commences at the little village of Kinlochewe at its head. Initially the path leads down the broad strath of the Kinlochewe River, gentle green farmland which seems strangely out of place beneath the grey pyramids of Beinn Eighe rising beyond the dark alders of the river bank. This is

an ancient path, a right of way down the entire northern shore of the loch along which, before the advent of the Kyle of Lochalsh railway, all mail for the Outer Isles was carried by runner to the little port of Poolewe. It is still a very beautiful journey through the waterside woods of the Letterewe Estate, which can be reached only on foot or by boat.

At the head of the loch the path traverses the shores of a pretty bay which mirrors Slioch rising above the birches, before crossing a footbridge over the torrent which issues from a dark and narrow defile that splits the hillside: this is Gleann Bianasdail, also known as the Fasagh Glen, the major outlet of the sizable Lochan Fada beyond Slioch, and with an excellent stalker's path leading through it, an important cross-country link. Our route now ascends to the lip of Slioch's shallow eastern corrie, more of a wide hollow than a corrie, before gaining the shoulder of Sgurr Dubh – the 'Black Peak' – by steep and unrelenting slopes.

This is the southern extremity of the Slioch horseshoe and here the wary may well encounter wild goats among the boulders. The ridge continues easily past two curious green lochans before a short rocky step leads to slopes of wiry grass and the desolate and undulating plateau. Patches of gravel and grey sandstone lend a moonscape atmosphere to the scene while an incongruous sandy pool of the kind unlikely to survive the summer hides beneath the final summit hump. It is worthwhile visiting the far western spur of the plateau which gives a fine view back across the craggy south-western flanks towards Kinlochewe and the Torridon peaks, but it is essential to continue to the twin northern summit, an eyrie looking steeply down on the bright ribbon of Loch Maree. The loch takes its modern name from Isle Maree, one of more than a score of islands clustered in the distance, where once the druids worshipped among the oaks and where pagan rites and bull sacrifices are said to have continued until the seventeenth century. Certainly St Maelrubha

Sgurr an Tuill Bhàin is the eastern peak of Slioch and this is the view towards it from below the north summit: the final ridge is unexpectedly narrow. The southern end of the Lochan Fada is glimpsed on the left.

built a chapel there and the waters of his sacred well were believed, until quite recently, to cure mental disorders!

Be that as it may, the route onwards along the airy northern arm of the horseshoe provides a continuous prospect over the great trench of Lochan Fada and the serried ridges and summits of the alluring Great Wilderness beyond – some call it the Whitbread Wilderness after the redoubtable proprietor who while stalking the deer has eschewed the Land Rover and fought off variously the Hydro Board and mineral developers to continue the preservation of the finest area of unspoilt wild landscape in Britain. Long may it remain so. Beyond Lochan Fada rises A'Mhaighdean (pronounced Vyejun – 'The Maiden'), surprisingly similar to Slioch in outline and Scotland's most remote Munro, with the dark battlements of An Teallach (the 'Forge') easily identifiable in the distance.

A perfect narrow arete leads onto the final summit, pointed Sgurr an Tuill Bhàin ('Peak of the deep white corrie'), before easy slopes descend into the central corrie. The steep burn draining it gives the most interesting route down into the Fasagh Glen where the stalker's path is joined and followed back above the river with its delightful gorges, waterfalls, pools and bead-bonny rowans to the footbridge above the shore of Loch Maree.

A Circuit of Slioch

Length: 19 kms/12 miles
Total ascent: 1240 m/4,050 ft
Difficulty: excellent paths much of the way, otherwise good going though several very steep sections: the foot of this high mountain is a long way from habitation
Note: stalking information from Kinlochewe Estate, tel Kinlochewe 262.

Start: Incheril, limited parking at lane end 033624 before farm gate. (20 m/65 ft).

(1) Track passes R of farm, at junction go L to good path leading NW below hillside, by river, to shore Loch Maree and footbridge over river gorge.
4 kms/2½ miles

(2) Over bridge good path ascends NE towards narrow Gleann Bianasdail, levelling out after 500 m where rough trail forks L by twin cairns ascending hollow in hillside, steep, rocky in places, to Sgurr Dubh, after 200 m below rock buttress faint path contours L to very steep heathery slope, zigzag to summit Sgurr Dubh (c 730 m/2,400 ft). Follow broad ridge to saddle, round RH corrie edge to lochan, up steep narrow shoulder beyond to wide slopes leading to plateau and hump beyond, summit trig pt SLIOCH (980 m/3,217 ft★).
4.5 kms/3 miles

(3) Continue N 200 m to second hump, NORTH SUMMIT (980 m) whence strike E along plateau edge then narrowing ridge to thin saddle and summit SGURR AN TUILL BHÀIN (933 m/3,058 ft). Descend ridge SE to slight knoll where shallow spur falls R to corrie lip, small pond. Rough path descends steep ground L bank burn, cascades, waterfalls, to Gleann Bianasdail, take good path R to regain Loch Maree and retrace (1).
6.5 kms/4 miles to footbridge

Long-distance alternative
(3.A) In Gleann Bianasdail go L to outfall Lochan Fada, reach good path E side Gleann na Muice by stepping stones or crossing burn immediately N little Loch Gleann na Muice. Path becomes track to Heights of Kinlochewe and Incheril.
18.5 kms/11½ miles Slioch to Incheril

Dominating the southern end of Loch Maree, Slioch is seen here at dawn from the pine-scattered shore of Loch Maree near Grudie Bridge.

Ben More, the Roof of Mull Hebrides

OS 1 : 50,000 Sheet 48

'Mull . . . a hilly country, diversified
with heath and grass, and many rivulets.
Dr Johnson . . . said it was a dreary country,
much worse than Sky.' (sic)

James Boswell:
Journal of a Tour to the Hebrides 1773

'. . . in my journeys among the High Alps I
never found so much difficulty as here.'

Faujas de Saint Fond:
after attempting Ben More 1784

Perhaps the weather was terrible on both occasions and certainly Dr Johnson took little delight in mountains, but nevertheless both gentlemen were sadly mistaken for Mull is one of the most fascinating islands in the entire Hebrides while Ben More, its highest point and the only island Munro outside Skye, is a beautiful peak but in no sense a difficult one. The 'Big Hill', as its name translates, lords it over nine other summits rising above 2,000 feet (610 metres) besides a sizeable portion of the Inner Hebrides and is very much a

maritime mountain. Far gentler than the Cuillin of Skye yet still wild in a tempered way, these hills are geologically extremely complex and typically shapely: the conspicuous rocky cone of Ben Talaidh (761 m/2,496 ft) is surely one of Scotland's most imposing small peaks. Scots will need no prompting, but any southron mountain lover travelling the western Highlands should endeavour to climb Ben More and enjoy something of the intimate charm of Mull and its exceptional scenery: the road distance from Glencoe via the short ferries at Corran and Craignure is just 80 kilometres (50 miles).

Rising between two fine sea lochs and dominating them both, Ben More is a great pointed pile of basaltic lavas which radiates

Ben More is seen on a winter evening from the north-west across Loch Tuath. The traverse follows the skyline over A'Chioch on the left and then upwards to Ben More's summit before descending the left-trending ridge to the rounded shoulder of An Gearna.

several gracefully curving ridges and can be ascended easily from the roadside on either shore. However, the great grey scree slopes which the mountain throws down southwards are 'reminiscent of a slag heap' — to quote the SMC guide — and the discerning mountaineer will prefer to tackle the mountain from the lonely shores of the northern Loch na Keal — the 'Loch of the Cliffs'. The best route on the mountain, steep, spectacular but straightforward, traverses the worthy satellite of A'Chioch (the 'Breast') and leads directly to the summit.

The route starts almost on the beach where oyster-catchers dip and strut and seals watch quizzically from the lapping waves, at first ascending alongside the burn into Gleann na Beinne Fada. The torrent pours over a succession of waterfalls and several of its gorges must be 15 metres deep (50 feet) yet narrow enough to cross perhaps with an intrepid leap. Soon a steep ascent leads up to the crest of the Beinn Fhada ridge (the 'Long Mountain') where interesting going over grassy saddles and rocky outcrops leads towards the rugged wedge-shaped summit of Beinn Fhada rearing up like a miniature Gasherbrum IV.

From the broad saddle below Beinn Fhada the ridge up to A'Chioch appears deceptively easy but it soon steepens and narrows to become quite scrambly and the easiest route over the several little rock steps that bar the way is not always obvious. The bare domed summit of A'Chioch is surrounded by steep rocky slopes and from the small cairn the final arete rearing up to Ben More itself looks formidable — and in winter truly alpine. Taken direct in good conditions the ridge would be quite interesting for it is rocky and narrow and excitingly exposed on the northern side, but a series of grassy and mossy ledges on the steep enough southern flank turn all difficulties and the climbing ends suddenly on the plateau edge immediately below the summit cairn.

On a good day the astonishing view stretches from Ireland and the Outer Isles to the Cuillin and Ben Cruachan, but even in moody weather there is an inspiring panorama over the island-scattered western ocean. Awesome basalt cliffs close below frown down over Inch Kenneth and the mouth of Loch na Keal; Staffa rides the sea beyond like a viking galley while further still lies

Bac Mor — the bizarre Dutchman's Cap — the Treshnich Isles and perhaps a white sail or two. The ancient Cathedral of Holy Iona is clearly visible a full 25 kilometres (16 miles) distant. If you have ever sailed this sea you will find this the most exciting view in the Hebrides.

The descent to the flat-topped shoulder of An Gearna is straightforward with interesting glimpses back across the craggy face of Ben More toward A'Chioch and its arete, but there is still a sting in the mountain's tail. The northern flanks of An Gearna are girdled by a low band of vertical and dripping black cliffs and these must be bypassed to left or right before a safe descent can be made into Gleann na Beinne Fada and the burn followed down to the shore.

The Ben More Horseshoe

Length: 13 kms/8 miles
Total ascent: 1300 m/4,275 ft
Difficulty: rough ground most of the way, some quite steep sections and several easy but exposed scrambly moves: good tactical route-finding required
Note: Magnetic rock renders compass unreliable here. Stalking information from Estate office, tel: Aros 410

Start: easy roadside parking at bridge 507367 on B8035 (5 m/15 ft).

(1) Rough path ascends W bank of burn 1.5 kms to waterfalls (in spate ascend pathless E bank) cross and climb steepening hillsides to gain N end Beinn Fhada ridge, cairn Pt 563. Continue SE along undulating ridge crest then steep semi-scramble above tiny lochan leads to flat summit and cairn BEINN FHADA (702 m/2,303 ft).
4.5 kms/2¾ miles

(2) Return to final tarn, descend S to wide saddle (c 525 m/1,720 ft), ascend narrowing ridge ahead, scrambling over short rock bands exposed on LHS to summit A'CHIOCH (c 850 m/2,790 ft). Rocky crest descends W then SW to narrow col, ascend steep arete above, exposed on RHS. Traces of path avoid difficulties by ledge systems on LHS, leading steeply to summit cairn, BEN MORE (966 m/3,169 ft★).
3.5 kms/2 miles

(3) Cross summit plateau NW along crag edge, descend stony slopes to gradual ridge fork, keep R above corrie to An Gearna plateau. Descend WNW from small cairn Pt 563 to avoid steep cliff band around nose of ridge, descending diagonally R below difficulties to rejoin ascent route near waterfall. Alternatively 500 m before reaching Pt 563 descend steeply N to pass E of cliffs.
5 kms/3¼ miles

Ben More and its satellites look impressive from the north over the sheltered waters of Loch na Keal: Beinn Fhada is the apparently small wedge-shaped summit at centre almost hidden by its own shoulder while A'Chioch and Ben More rise to the right.

Skye: the Storr and the Trotternish Scarp

Hebrides

OS 1 50,000 Sheet 23

'A walk upon ploughed fields in England is a dance upon carpets compared to the toilsome drugery of wandering in Skie'

Dr Johnson:
A Journey to the Western Isles of Scotland: 1775

Skuyö the Vikings called it – 'Cloud Island' – its high ground so often wearing a cloud cap as the dragon ships swept down the Minch. It is an apt title for this magical island, largest of the Inner Hebrides and famed for its Cuillin mountains; but Skye rises to another lesser range, no less unique in its way, liable to better weather and well meriting the attention of any mountain lover. Its 25-kilometre (15-mile) escarpment forms the spine of Trotternish, the long northern

peninsula of the island, and while its summits rarely rise above 600 metres (2,000 feet) its crest falls eastward as a continuous line of frowning black precipices, often as high as 100 metres (300 feet), a landform as striking as any in Britain.

Apparently this incredible cliff is the largest landslip in these islands, dating to the retreat of the mainland ice-sheet in late Pleistocene times – the last ice-age – when the edges of a thick layer of basaltic lava, no longer supported by the ice, slipped away in a series of giant slivers like overlapping slices from a pre-cut bread loaf. Near the scarp's two extremities the rotting basalt has eroded into particularly spectacular architecture – the Storr and the Quiraing – both easily reached and well known. A long traverse

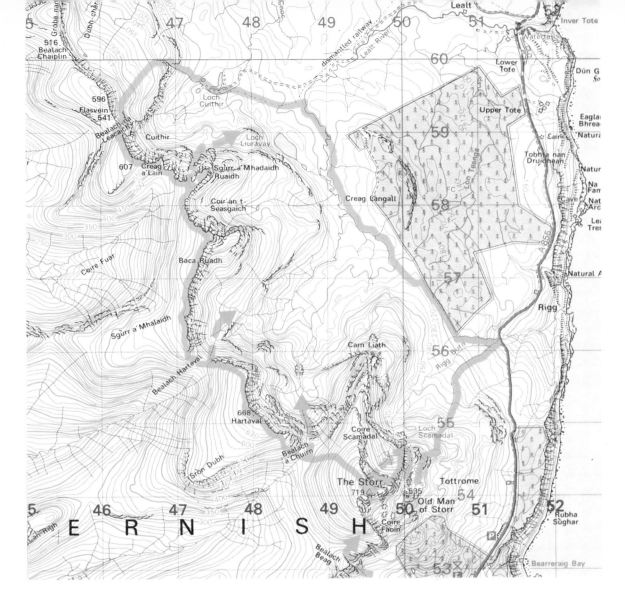

The extraordinary formations below the 180-metre (600-foot) cliffs of The Storr, the highest point in Trotternish, are usually reached from a Forestry Commission car-park at the northern end of Loch Leathan but for this route the Rigg Burn is followed, past banks of kingcups and through ferny dells to a delightful area of violet-

The Old Man of Storr and the weird basalt pinnacles surrounding it are seen to the south from the foot of the Storr cliffs: long island of Raasay lies in the sea beyond.

of the entire crest, a one-way trip of over 30 kilometres (20 miles) with no easy return, was first completed in 1901 but the route detailed traverses the best of the scarp in a reasonable day's circuit.

sprinkled lawn just north of a dozen or so rickety pinnacles. Chief among them is the Old Man of Storr, a slightly leaning cigar-shaped pillar some 60 metres high (200 feet) and a conspicuous Skye landmark first climbed in 1955 by Don Whillans and his mates: Storr is said to mean 'decayed tooth', an appropriate enough title. Ravens glide around the corrie while awesome dark cliffs split by black gullies tower above and exploration is suggested before climbing to the Storr summit, the best route taking a narrow path that works north round the base of the cliffs into hanging Coire Scamadal. Here the decaying basalt at the corrie lip provides a rich environment for arctic/alpine plants including the Icelandic pur-slane (*Koenigia islandica*) found only here and in

Mull. Sensational views unfold as the cliff edge is followed upwards until a final buttress forces a traverse across into a wide grassy gully which emerges suddenly onto the greensward of the wide summit plateau.

A fine viewpoint, the Storr summit is a gentle place but with the vertical always close at hand one moves around it with care. Convex slopes sweep down to the first bealach and the route onwards over Hartaval to Baca Ruadh is easy with firm turf underfoot. This wide and lonely country belongs to the soaring eagles: the tiny white crofts scattered along the sea-fringe and the black specks of ships out in the Sound are a different world, a world inaccessible beyond the omnipresent cliffs. Beyond the 'Red Ridge' of Baca Ruadh the cliff edge becomes convoluted and the rock architecture exceptional once more. An impressive prow guarded by a narrow scrambly rock band juts from the main crest, but Sgurr a'Mhadaidh Ruadh – the 'Peak of the Red Fox' – can be easily bypassed. Creag a'Lain is the next summit and before climbing steeply up the edge of its great cliffs the face of the bealach beyond should be closely studied for it must soon be descended. Creag a'Lain would appear to mean 'Enclosed Crag' and its imposing face rears out of a strange cliff-girt hollow named Cuithir – the 'Narrow Glen' – the next objective.

Despite its formidable appearance the descent from the Bealach na Leacaich – the 'Pass of the Stony slopes' – is safe and easy albeit requiring careful route-finding: a direct line down these craggy slopes would be horrific though a possible ascent route can be picked out from below. Lilies float on the impressively situated Cuithir lochans at the head of a rough road and the remains of an ancient tramway: diatomite was once excavated here, a siliceous deposit formed from the remains of microscopic algae which infilled small post-glacial tarns and which has industrial uses. The tramway gives a fairly dry descent to the river which provides an interesting route through otherwise extremely rough country most of the way back to the Rigg Burn.

The Storr and the Trotternish Scarp

Length: 20.5 kms/12¾ miles
Total ascent: 1130 m/3,700 ft
Difficulty: easy walking, usually excellent going, careful route-finding essential in places: serious high ground attracting fierce weather, cliff edges always close
Note: easy descent E possible Bealach a'Chuirn, Bealach Hartaval, green gully 471584 before Creag a'Lain, escape always possible W

Start: limited parking on verge A855 at Rigg Burn culvert, 512562. (120 m/395 ft).

(1) Follow S fork Rigg Burn then ascend grassy corridor to ridge, fence, crest path climbs easily to grassy plateau overlooking Storr pinnacles and lochan. Path traverses N below main cliffs to little col, stile, continuing round L up short rock step into Coire Scamadal. Either head directly up to obvious wide green gully or ascend rocky L edge above cliffs traversing R to gully on faint cairned path below final crags. Steep grass leads to summit plateau THE STORR (719 m/2,358 ft).
5 kms/3 miles

(2) Descend grassy slopes NW to Bealach a'Chuirn, ascend L of broken crags to join cliff edge leading to summit HARTAVAL (668 m/2,191 ft), continue along edge crossing Bealach Hartaval and shoulder Sgurr a'Mhalaich to BACA RUADH (637 m/2,091 ft). Descend round corrie lip to col, avoiding SGURR A'MHADAIDH RUADH (587 m/1,926 ft) as desired, ascend steeply near cliff edge to summit CREAG A'LAIN (607 m/1,993 ft).
7 kms/4½ miles

(3) Descend N to Bealach Leacaich. Immediately N of ancient wall drop R down soft scree c 100 m to horizontal sheep path traversing L above rocks to wide grass slopes. Descend steeply NE to twin ponds not on map. Follow stream descending from just N (469600) over lip to Loch Cuithir (180 m/590 ft).
2.5 kms/1½ miles

(4) E of lake old tramway route descends ESE, after 1 km break R to River Lealt, follow river bank S, then E fork, to gain forestry edge, follow fence SE along to Rigg Burn, descend to start.
6 kms/3¾ miles

Eastward-facing basalt cliffs line the edge of the Trotternish escarpment as it stretches northward over Hartaval towards Baca Ruadh.

Blaven and the White Strand   Hebrides

OS 1: 50,000 Sheet 32

'Oh Blaven! rocky Blaven!
How I long to be with thee again,
to see lashed gulf and gully
smoke white in the windy rain . . .'

probably Alexander Smith: 1865

Immortalized in song, celebrated by repute and beloved of climbers, the spectacular Cuillin mountains of Skye are unique in Britain. Despite the romantic claim of the Gaellic bards that this savage range was named from Cuchullin, the legendary 'Hound of Ulster' who ran a war-school on Skye, a more likely derivation is Norse and means 'Keel-like Ridges', a name applied also to similar mountains in Scandinavia.

There is the Black Cuillin and the Red, the former a serrated horseshoe ridge of thirty-one formidable gabbro summits surrounding remote Loch Coruisk, the Red an array of smaller more genteel conical peaks cloaked in granite screes and ranged to the north-east. There is but a single anomaly, one major peak, very definitely a Black Cuillin, which rises beyond the main range in what is really Red Cuillin territory at the head of the lovely Strathaird peninsula. To the Gaels it was Bla Bheinn, variously the 'Blue Mountain' or the 'Hill of Bloom', Sherrif Alexander Nicolson – he who first climbed Sgurr Alasdair, Skye's highest summit in 1873 – claimed it was the island's finest mountain, more recently Hamish Brown described it as the 'portentous postcript to the gabbro galaxy', but

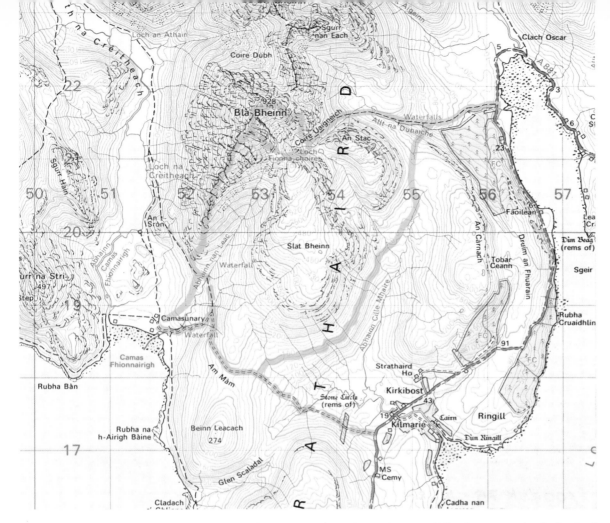

to most mountain lovers it is just plain Blaven.

But Blaven is hardly plain, a superb peak, maritime and aloof, its great tousled head throws down rent and rugged walls east and west to form the spine of narrow Strathaird between the great sea lochs of Slapin and Scavaig. A long southern ridge sweeps down to the sea while four black satellites guard the summit from the encroaching Red Cuillin clustered to the north. Notwithstanding the explorations of Sherrif Nicolson and others, serious mountaineering started in Scotland with the discovery by members of the Alpine Club in the 1870s that there were still virgin peaks to climb in the Cuillin, though Blaven had been climbed in 1857 by a Professor Nicol and the young poet Charles Swinburne, which suggests that by its easiest

way Blaven is not a difficult mountain. Indeed of all the major Black Cuillin summits, Blaven is the most suitable objective for the experienced hill-walker with no rock-climbing pretensions.

The regular route takes the line of least resistance up the great steep and craggy shoulder that falls east from the summit into Coire Uaigneich (the 'Lonely Corrie'?) from where it is fairly unobtrusively marked and cairned: in poor conditions route-finding skills will be useful. Carefully avoiding precipitous ground, the ascent requires only a little mild scrambling, avoidable by diligent reconnaissance and never exposed. There are tantalizing glimpses northward over a narrow rocky bealach to Clach Glas, the jagged tower-like peak that appears as Blaven's smaller twin in most distant views: the traverse of the pair is a classic expedition first made in 1888 by the famous alpinist Charles Pilkington but it involves difficult and exposed scrambling.

As it zigzags up the eastern shoulder of Blaven, the route passes the head of this great gully which frames the summit of Clach Glas, the next peak to the north.

It has been claimed that Blaven is the best viewpoint in the Cuillin. Certainly the view includes a superb – and rare – panorama of the entire Main Ridge, all three Red Cuillin groups, the distant Storr, the Rough Bounds of Knoydart, the western seaboard and the Inner Isles. The summit ridge leads over the head of the great gully that splits the eastern flank to a second top before commencing a long descent to Camasunary where the white beach beckons towards the shimmering sea and pinnacled Rhum.

Continuously interesting, the ridge is sustained at harder-than-walking standard until grassy slopes – crossed by a good traverse path which can be used to avoid Camasunary itself – lead to the deserted lodge in the green meadows behind a lonely and beautiful bay. The name of this place appears to mean 'Bay of the White Strand' and it would be worth the climb just to walk the shore and swim maybe in the deep river channel at its western extremity. Unfortunately this tranquillity was interrupted in 1968 by an ill-

conceived military attempt to blast an unwanted jeep-road over the Strathaird peninsula and through to Coruisk, luckily it was prevented but not before the old track linking Camasunary to Kilmarie on the Elgol road had been ruined. This is the final section of the route: stalwart walkers will cut the corner through the 'corrie of St Mary's kirk' – Cille Mhaire – which is rough terrain though shorter than the road which is easy going and unfrequented.

Over Blaven to Camasunary

Length: 15 kms/9½ miles
Total ascent: 1300 m/4,250 ft
Difficulty: rugged strenuous itinerary involving tactical route-finding, a little mild scrambling on a large craggy mountain deserving respect: a committing route once started on S ridge descent

Start: good roadside parking at 561215, old bridge beside A881 Elgol road. (10 m/30 ft).

(1) From bridge 200 m N, path ascends N bank Dunaiche burn to lower corrie thence climbing steeply L into Coire Uaigneich beneath craggy E face Blaven to fade by stream fan below dark cleft. Turn N, ascend grassy rake diagonally up great shoulder, traces of path, cairns, take easiest line through rocky outcrops to crest of shoulder. Path zigzags up steep stony slope past tops two big gullies, across scree, boulders, to gap in summit rock band, short scramble to OS cairn and summit BLA BHEINN
(928 m/3,044 ft★).
4 kms/2½ miles

(2) Descend S 20 m to little col, easy airy rock ramp leads L across broken crag to shallow gully and cairn SOUTH SUMMIT
(924 m/3,033 ft) continue S avoiding blind ridge on R, best route cairned. Descend stony crest, dropping L round each difficulty, occasional mild scrambly moves, to steep grass slopes, flat green spur crossed by traverse path, and Camasunary shore.
3.5 kms/2¼ miles

(3) Rough track climbs E to wide saddle Am Mam
(185 m/610 ft). Descend c 400 m, strike L across hillsides following c 150 m contour into Cille Mhaire corrie. Short ascent leads to wide boggy saddle (210 m/690 ft), descend steep grassy slopes N to Dunaiche burn near waterfall, rejoin ascent path far bank.
7.5 kms/4¾ miles

Alternatives
(2.A) From c 200 m S of S summit descend steep stony slopes SE, hints of path, avoiding cliffs on L, to narrow col above head Coire Uaigneich, broken slopes drop into corrie, rejoin ascent route.
4.5 kms/2¾ miles

(3.A) From Camasunary cross Am Mam but continue track to Kilmarie (4 kms/2½ miles), follow road N to start.
9.5 kms/6 miles

The south ridge of Blaven rises from the lonely meadows of Camasunary where a deserted cottage stands behind the white strand.

South Uist, the Mountains and the Sea

Hebrides

OS 1: 50,000 Sheet 22

'Of these islands it must be confessed
that they have not many allurements, but
to the lover of naked nature.'

Dr Johnson:
A Journey to the Western Isles of Scotland, 1775.

Tir nan Og, the Isles beyond the Sunset, the
Islands of the Blessed. The mysterious islands of
the far west have ever beckoned, even if the
Long Island – the chain of the Outer Hebrides
that stretches from Barra and the Uists to Harris
and Lewis – is not quite as remote as that
mythical archipelago of Gaellic legend. Few folk
associate South Uist with mountains, nevertheless the mountaineer on Skye or the yachtsman
beating across the Little Minch will be well
aware of two prominent peaks rising high above
the horizon-hugging shape of the island. Actually three worthy summits spring from a rugged
mass of gneiss which rears steeply from the
indented eastern shore to drop more gently
westward towards the loch-scattered moorland,
green machair and white sands of the Atlantic
coast.

This walk traverses all three of these summits
and must be one of the finest expeditions of its
kind in Britain, long but always interesting and
atmospheric with that heady mixture of mountains and sea. A strong party could obviously
complete the circuit in a long day, but would be
unable to linger to enjoy the stupendous views
of ocean and islands or to explore the intricacies

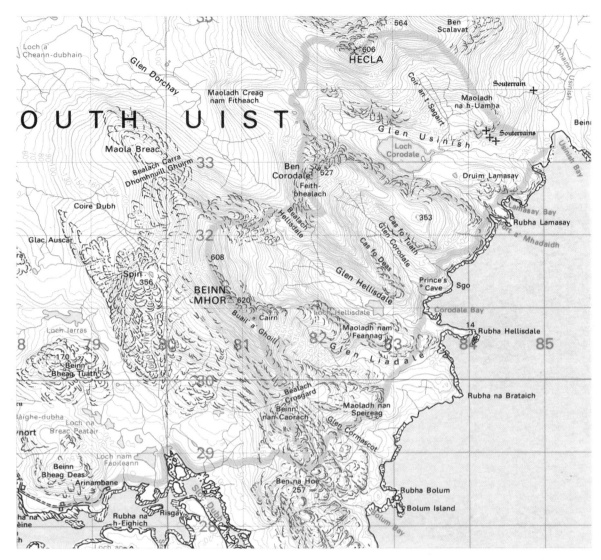

of summit and glen. Starting along the attractive serpentine shores of Loch Eynort, the route climbs through the heather towards the obvious low point in the long southern ridge of Beinn Mhor and the hidden ravine which crosses it. At first the ridge is wide and peaty but soon becomes scabbed with patches of glacis and sprinkled with erratic blocks. Beyond a second strange defile the mountain asserts itself and the

Though only of modest height, Beinn Mhor and its surrounding peaks are visible throughout the Uists. Here Hecla (left) and Beinn Mhor are seen to the south from Benbecula across the shallow Bagh nam Faoilean – the South Channel.

narrowing ridge sweeps up towards Beinn Mhor: a glance back down this elegant grassy crest reveals sea on three sides, a spectacular situation. From the summit however the view is truly unique, to the north-west the great stacks of St Kilda rise from the wide ocean a full 100 kilometres (62 miles) distant, while the jagged crest of Rhum floats south-eastward on the Hebridean Sea.

A narrow arete leads onwards from the summit, a row of small spiky tops and miniature gendarmes liberally hung with moss and grass from which a succession of bold buttresses and deep gullies fall some 250 metres (850 feet) into Glen Hellisdale: several climbs have been made here. A short way past the northern summit the regular route takes the north-west ridge but our traverse descends round the head of Glen Hellisdale to a wide saddle at the base of a craggy wall guarding the route upwards onto Ben Corodale. This barrier can be turned on the left by a steep uncompromising slog but a more entertaining route follows the obvious grass rake, more spectacular but quite straightforward, which ascends rightwards across the cliff to gain the pleasant seaward ridge of the mountain. Best described as a rocky turret, the top of Ben Corodale is reached easily or by a good scramble, but the way ahead is not obvious for sizable crags almost encircle the summit. However a grassy rake descends to easy slopes below and a wide shoulder littered with rock outcrops and little ponds. Ben Corodale is also known as Feaveallach, probably a corruption of Feith Bealach – the Boggy Pass – which succinctly describes the next saddle, a waste of peat hags below the slopes rising to Hecla. Obviously once a Viking sailing mark – for there is a prominent Icelandic volcano of the same name – Hecla is another fine mountain, and the west ridge leads first to a craggy top which provides a satisfying – but avoidable – scramble, whence a short rocky arete leads to the bouldery main summit. Now the view northwards opens up, over a myriad lochs, lochans, islands and islets towards the distant blue line of the Harris peaks, while Skye looms across the Little Minch behind the massive bulwark of Waterstein head. From the northern summit, Point 564, the grassy ridge falling seawards gives an excellent descent to Usinish Bay.

Set into the grassy hillside, the stone-walled Usinish bothy is not difficult to find. Kept shipshape by shepherds and military parties from the Benbecula rocket range, thus snug refuge is pretty basic but storm-proof and furnished with rough bunks and a few old pots and pans: visitors should bring sleeping bags and a light stove. In good weather the return route southward down the coast is delightful with steep hillsides above the sea linking a series of narrow and secluded glens. Particularly evocative is Glen Corodale where now ruined cottages gave sanctuary to the Young Pretender for three weeks in 1746 while the authorities combed the island for him. In rain and mist however the return route can be hard going and the navigation tricky, especially through the complex terrain leading to the Bealach na Hoe.

A Traverse of Beinn Mhor and Hecla

Length: 28 kms/17 miles, plus
Total ascent: outward – 1250 m/4,100 ft; return – 650 m/2,150 ft
Difficulty: virtually trackless, rocky in places but usually excellent going on high ground, several slightly exposed sections but direct route avoids any scrambling. Return route more rugged, some bog and deep heather: tactical navigational skills useful. This is wild, remote and unfrequented country and should be respected accordingly
Notes: compass unreliable on Hecla. For NW ridge regular route leave road at 768341. B&B at Arinambane roadhead, phone Bornish 379. Hut-type accommodation nearby at Gatcliff Trust's Howmore hostel, warden at cottage 759564

Start: roadhead beyond North Locheynort, ask permission to park at Arinambane croft 789283.

(1) Good path leads E above shore but disappears at 805287 Sloc Dubh inlet E side, ascend by burn E into shallow corrie, more steeply into defile Bealach na Hoe (c 200 m/660 ft). Follow broad ridge N over Beinn nan Caorach, cross deep gash Bealach Crosgard, climb narrowing ridge N to flat shoulder, continue NW along cliff edge to trig point BEINN MHOR (620 m/2,034 ft).
7 kms/4½ miles

(2) Continue NW along narrow arete, almost scrambly, exposed on R, to minor top, then wider ridge leads to shelter circle NORTH SUMMIT (608 m/1,994 ft). Descend N to flat shoulder then E down broad slabby ridge – best route is wide grassy gully R of crest – to broken ground of wide Bealach Hellisdale (c 290 m/955 ft). Grassy rake, wide, easy, ascends diagonally R across cliffs above Bealach to join broad ridge above leading NW to rocky turret BEN CORODALE (527 m/1,729 ft).
3.5 kms/2¼ miles

(3) Grassy rake immediately below summit drops W through cliffs to easy slopes, then broad broken ridge descends to wide peaty saddle Bealach Corodale (c 300 m/985 ft). Ascend steep slope above to wide ridge leading E to rocky summit HECLA (606 m/1,988 ft).
3 kms/2 miles

(4) Descend NE to col, contour R to gain grassy ridge dropping SE towards Usinish Bay. Bothy at 850333 below grassy hillside (c 10 m/30 ft).
3.5 kms/2¼ miles

(5) Return route follows rugged coast S, steep slopes above sea link glens, rough going. From boggy Glen Liadale strike SW, pass lochan Glen Cormascot whence broken rocky hillsides lead to Bealach na Hoe. Retrace route to start.
10.5 kms/6½ miles, plus

The pride of Beinn Mhor is its narrow summit crest, this is the view south-eastward from the north summit towards the main top: Loch Eynort is seen below on the right and the Cuillin of Rhum in the far distance, left.

Safety and Access

'The recollections of past pleasures cannot be effaced ... There have been joys too great to be described in words ... Climb if you will, but remember that courage and strength are nought without prudence, and that a momentary negligence may destroy the happiness of a lifetime. Do nothing in haste; look well to each step; and from the beginning think what may be the end.'

With those memorable lines Edward Whymper closed his book *Scrambles Amongst The Alps* which despite being first published in 1871, is still probably the greatest mountain narrative in the English language. He had good cause to voice such sentiments for after reaching the coveted summit of the Matterhorn for the very first time, he was involved in a tragedy which clouded his entire life thereafter; an avoidable accident which killed four of his companions and which today we would blame on negligence. Admittedly he was writing of the Alps and we have no Matterhorns here in Britain, not quite, but the same strictures apply, for although the hazards among our subtle British hills are far less apparent, they are no less real.

Our hills are all things to all men – there is a challenge somewhere among our wonderful uplands for everyone and the routes described in this book cover a fantastic variety of terrain. Certainly first-time ramblers ambling along the Malvern crest on a balmy summer's day may be forgiven for wondering how they could possibly be courting danger. As a teenager my own introduction to the Malverns was a double traverse with the hills plastered in hard frozen snow, and we made the return leg after nightfall by moonlight. It was an unforgettable experience that Whymper would surely have understood, but in such conditions even a sprained ankle could have resulted in an exposure case and an epic midnight rescue by gum-booted policemen, themselves at risk slithering around steep and icy slopes. However, we were aware of the hazards – such as they were – and we were suitably equipped for traversing the small mountain chain that the Malverns had temporarily become.

How much fiercer the Cairngorms, where I have skied on the very route described on page 160, in April, stripped to the waist. Yet in the same place, in the same month but another year, I have fought for survival, hardly able even to crawl, in a blizzard of Antarctic ferocity. Such are the British hills. The lesson to be learnt from these experiences is that even our most apparently benign terrain can swiftly become hostile in the extremes of weather and conditions which can afflict our hills at all seasons. Every walker should be aware of this fact and it must temper his activities and his equipment.

But this is a where-to-do-it rather than a how-to-do-it book, and safety – indeed sanity – on the hills are subjects well covered elsewhere: Whymper so succinctly summed up the ethos. Suffice to stress three obvious yet salutary points. First is that the selection of equipment and clothing is 'horses for courses': if spare clothing and emergency gear is not in your rucksack you can't use it, but too much of it –

unneeded – will ensure a miserable overloaded trip at the best, and at the worst engender a situation where it eventually becomes necessary. Second is that skill with map and compass is a prerequisite for virtually all upland walks, indeed useful on any country outing. In poor visibility or confusing country such skills are obviously essential, but even following a right of way across the fields in perfect visibility they can save time, effort and frustration. Thirdly, and perhaps most important of all, we must each know our personal limitations. We can know them only by experience, to which there is no short cut. Thus never be afraid to retreat: it is better to have tried and turned back than never to have started. Nothing ventured, nothing gained. But pushing beyond one's limits is a foolish game.

In modern Britain, with continual and growing pressure on our countryside and wild places, access is often a contentious issue. All land belongs to someone or somebody, and in England and Wales, even in our ten National Parks, the only access, as of right, is along properly registered Rights of Way or to certain, but by no means all, classes of common land. However, many landowners, such as the National Trust or Forestry Commission, will allow virtually unrestricted access to their wild country holdings, other private landowners kindly permit access to certain areas of open country while the National Parks or local authorities have negotiated access agreements to yet others. Sometimes it is just a matter of habitual use and 'nobody minds' but often somebody does mind, and with good reason, for instance when lambing is taking place, or grouse shooting, or – like it or not – the military are exercising.

In short, if you enter land you do not own, against the wishes of the landowner, you are committing the civil offence of Trespass.

In Scotland the situation is slightly different. The act of 'trespass' is not defined in statute and, unlike south of the Border, is not in itself an offence and although a landowner has the right to insist that 'trespassers' leave his land, he must prove nuisance or damage in court before an individual can be excluded or sued. Over the years responsible behaviour and mutual respect between landowner and hill-walker have resulted in a situation whereby, in practice, there is reasonable freedom to roam open country at will.

Having said that, in Scotland's upland regions – besides the sheep and grouse that (as in England and Wales) provide employment and income for local people, there is also deer stalking on which the local economy may well depend. Typically from mid-August to mid-October indiscriminate hill-walkers in many parts of the Highlands may do damage by disrupting stalking and at this season the walker should contact the local gillie or estate office before venturing onto the hills to ensure that the legitimate interests of both parties do not clash. Thankfully in Scotland there is still a tradition of courtesy, consideration and amicable co-operation between landowner and walker.★

Unfortunately this is not always true in the different

★An extremely useful Access booklet is published by the Scottish Landowners Federation in conjunction with the Mountaineering Council of Scotland and may be obtained from the Federation at:
18 Abercromby Place, Edinburgh EH3 6TY

situation that exists south of the Border. Thoughtless behaviour and vandalism by a minority have sometimes soured the goodwill of many landowners towards the majority. But – certainly in my own experience – if approached politely most landowners will be pleased to allow him or her reasonable access across their open land. In one notable instance we were threatened by an irate gentleman, complete with shotgun, who – it transpired – had been rudely abused by trespassing yobbos on his own land the previous weekend and wanted no more with walkers. But we were all country lovers and man-to-man diplomacy resulted in an invitation to a tea at the castle!

'I know there is no right of way, but do you mind if we go up the hillside here onto the ridge?' I once asked a farmer in a remote Welsh valley. 'Oh naw, go up wheneffer you like, as long as it's only you and your mate – but I don't want big partiss up there, see.' Was the very reasonable reply.

To sum up, always be courteous in the countryside, if in doubt ask permission, scrupulously observe the Country Code, leaving the hills as you would wish to find them, with no sign of your passing.

The Country Code
- Guard against all risks of fire
- Fasten all gates after you
- Keep dogs under proper control
- Keep to footpaths across farm land
- Use gates and stiles – avoiding damage to walls, fences and hedges
- Leave no litter – take it with you
- Be careful not to pollute water supplies
- Protect all flora and fauna – do not pick flowers
- Respect the countryside and the way of life of the folk who live there

Bibliography, Useful Addresses, Acknowledgements

Interesting Background Reading for Hill-Walkers

A Cambrian Way – Richard Sale (Constable 1983)

The Country Life Guide to Weather Forecasting – S Dunlop and F Wilson (*Country Life* 1982)

Fitness on Foot – Peter Gillman (World's Work/*Sunday Times* 1978)

Geology and Scenery in England and Wales – A E Trueman (Penguin 1971)

Geology and Scenery in Scotland – J B Whittow (Penguin 1977)

Guide to Britain's Nature Reserves – Jeremy Hywel-Davies and Valerie Thom (Macmillan 1984)

Hamish's Mountain Walk – Hamish Brown (Gollancz 1978, Paladin 1984)

The High Mountains of Britain & Ireland – Irvine Butterfield (Diadem 1986)

The Hill Walker's Manual – Bill Birkett (Oxford Illustrated Press 1988)

Mountain Navigation – Peter Cliff (published by Peter Cliff 1987, distributed by Cordee, Leicester)

Mountaincraft and Leadership – Eric Langmuir (Scottish Sports Council/MLTB 1984)

The Mountains of England & Wales (so called 'Bridge's Tables') – George Bridge (Gastons/Westcol 1973)

Mountaineering – John Cleare (Blandford 1980)

Munro's Tables – (Scottish Mountaineering Trust)

Countryside Commission Official Guides to the National Parks (ten volumes) – various authors (Webb & Bower 1987)

Ordnance Survey Outdoor Handbook – Michael Allaby (Macmillan 1987)

Speak to the Hills, an anthology of twentieth-century British and Irish mountain poetry – ed Hamish Brown and Martyn Berry (Aberdeen University Press 1985)

Walking, Hiking & Backpacking – Tony Greenbank (Constable 1985)

Relevant Magazines

Climber – Holmes McDougall Ltd, 7th floor, Plaza Tower, East Kilbride, Glasgow G74 1LW

Great Outdoors – also Holmes McDougall Ltd

High – Springfield House, The Parade, Oadby, Leicester LE2 5BF

Country Walking – Emap Pursuit Publications Ltd, Bretton Court, Bretton, Peterborough PE3 8DZ

Odyssey – Chatsworth Publicity Services, High Bridge House, High Bridge Street, Newcastle upon Tyne NE1 1EW

Some Useful Addresses

British Mountaineering Council – Crawford House, Precinct Centre, Booth Street East, Manchester M13 9RZ

Long Distance Walker's Association – 29 Appledown Road, Alresford, Hants SO24 9ND

Mountaineering Council of Scotland – Hon Sec, Rahoy Lodge, Gallanach Road, Oban PA34 4PD

National Trust – 36 Queen Anne's Gate, London SW1H 9AS

National Trust for Scotland – Suntrap, 43 Gogarbank, Edinburgh EH12 9BY

Nature Conservancy Council – Northminster House, Peterborough PE1 1UA

Rambler's Association – 1/5 Wandsworth Road, London SW8 2LJ

Youth Hostel's Association – Trevelyan House, 8 St Stephen's Hill, St Albans, Herts AL1 2DY

Useful Weather Forecast Telephone Numbers

Special Mountain Forecasts:

Peak District	– 0742–8091
Lake District National Park	– 099–662–5151
North Yorkshire	– 0632–8091
Snowdonia National Park	– 0286–870–120
Scottish Highlands	– 041–248–5757

Weathercall Regional Forecasts:

Devon and Cornwall	– 0898–500–404
Malverns and Shropshire	– ,, ,,–410
Central Wales	– ,, ,,–414
Yorkshire Dales	– ,, ,,–417
Southern Uplands	– ,, ,,–422
Galloway	– ,, ,,–420
NW Highlands/Hebrides	– ,, ,,–425

Acknowledgements

It would be impossible to research, write and photograph a book like this without the assistance, co-operation and hospitality of many people all over the country: I wish I could list them all. I am especially grateful for the crucial assistance of: David Archer (Snowdonia National Park), Mike Couling (Welsh Water Authority), Brian Griffiths (Clwyd County Council), Andrew Jenkinson (Scenesetter Publications), Mr Lane (Dyfed County Council), Graham Little (Mountaineering Council of Scotland), Rae Lonsdale (Yorkshire Dales National Park), Beverley Penney (Ramblers Association), Mrs Richards (Powys County Council), Roland Smith (Peak National Park), Lesley Smithson (British Mountaineering Council), Roger Stevens (Brecon Beacons National Park), Tom Wall (NCC Stiperstones Ranger), and indeed that of many other officers and officials of National Parks, County Councils, the Nature Conservancy Council, and the National Trust besides so many private individuals. Thank you all.

My thanks are due also to the contemporary poets from whose work I've received inspiration and from which I've taken the liberty of quoting a few apposite words: Ivor Brown, Huw Jones and Roger Redfern.

Index

Page numbers in *italics* refer to illustrations.
Where appropriate summits and tops ascended
during walks have been indicated in **bold** type.